Ice humanities

MANCHESTER
1824

Manchester University Press

Ice humanities

Living, working, and thinking in a melting world

Klaus Dodds and Sverker Sörlin

Manchester University Press

Published by Manchester University Press
Oxford Road, Manchester M13 9PL
www.manchesteruniversitypress.co.uk

British Library Cataloguing-in-Publication Data
A catalogue record for this book is available from the British Library

ISBN 978 1 5261 5777 5 hardback

First published 2022

Typeset
by New Best-set Typesetters Ltd

Contents

List of figures

List of tables

List of contributors

Christine Bichsel is Professor of Human Geography at the Department of Geosciences, University of Fribourg. Her research interests and expertise are in political geography, environmental history, and critical water studies. She is the author of *Conflict Transformation in Central Asia: Irrigation Disputes in the Ferghana Valley* (Routledge, 2009). Her recent research focuses on the history of Soviet glaciology in Central Asia. She is the principal investigator of the project 'Timescapes of ice: Soviet glacier science in Central Asia, 1950s–1980s', funded by the Swiss National Science Foundation (2021–25).

Jordan Barton is an undergraduate student at the University of Oregon majoring in biology and minoring in legal studies, expected to graduate in June 2021. She joined the Glacier Lab in 2019 researching topics including iceberg tourism in Greenland and how icebergs impact the oil and gas industry. She and other Glacier Lab members have co-authored the article 'Justicia glaciar en los Andes y más allá' (Glacier Justice in the Andes and Beyond), published in the Peruvian journal *Ambiente, Comportamiento y Sociedad*. She is currently completing a thesis on environmental law and youth activism against climate change.

Mark Carey is Professor of Environmental Studies and Geography at the University of Oregon. His research, funded by the National Science Foundation, addresses environmental justice issues related to climate change, glaciers, and water in the Andes and Arctic. He has co-authored several Intergovernmental Panel on Climate Change (IPCC) chapters and won the Elinor Melville Book Award, the Leopold-Hidy article prize, and the King Albert Mountain Award for lifetime contributions to mountain conservation and peoples. He currently runs the Glacier Lab for the Study of Ice and Society.

Klaus Dodds is Professor of Geopolitics at Royal Holloway, University of London, and a Fellow of the Academy of Social Sciences in the UK. He has written several books including *Ice: Nature and History* (Reaktion, 2019),

The Scramble for the Poles (Polity, 2016 with Mark Nuttall) and, most recently, *The Arctic: A Very Short Introduction* (Oxford University Press, 2021 with Jamie Woodward). In 2016–20 he held a Major Research Fellowship on Arctic geopolitics and governance, which was generously funded by the Leverhulme Trust.

Bruce Erickson is an Associate Professor in the Department of Environment and Geography at the University of Manitoba and the author of *Canoe Nation: Race, Nature, and the Making of a National Icon* (University of British Columbia Press, 2013). His work addresses the cultural politics of outdoor recreation, tourism, and contemporary environmentalism, especially as they relate to colonialism, race, and neoliberalism in Canada.

Sam Flanzer is a third-year Environmental Studies major and Women's, Gender and Sexuality Studies minor at the University of Oregon's Clark Honors College. She is a member of the University of Oregon's Glacier Lab, a collaborative research group interested in ice and society. Sam will be applying to law school in the fall of 2021 to pursue a career in environmental law. She is particularly interested in how climate change disproportionately impacts women across the globe.

Jasmin Höglund Hellgren is a doctoral student at the Division of History of Science, Technology and Environment at KTH Royal Institute of Technology, Stockholm, Sweden. She is part of the European Research Council-funded research project SPHERE – Study of the Human-Environment Relationship – that addresses humanity's relation to planetary conditions and how it has become understood as a governance issue. Her research focuses on environmental governance, particularly the making of governable environmental objects, explored in the context of debates around climate change and resource extraction in Latin America.

Dani K. Inkpen is Assistant Professor of History at Cape Breton University, Unama'ki (Cape Breton), Nova Scotia. Trained in the history of science, she works on the history of glaciology, exploration, and recreation in mountain environments. She is presently working on a history of repeat glacier photographs as icons of global warming. From 1 July 2022, Dani will be starting as an Assistant Professor of History at Mount Allison University.

Erik Isberg is a Doctoral Candidate at the Division of History of Science, Technology and Environment at KTH Royal Institute of Technology in Stockholm, Sweden. With a background in history of ideas and science, Erik's current research concerns the history of paleoclimatology and the rise of planetary-scale environmental knowledge.

Liam Kennedy-Slaney is a PhD student at Simon Fraser University (territory of the Musqueam, Squamish, Tsleil-Waututh, and Kwikwetlem Nations).

Having worked as a wildlife biologist for several years, he has turned his focus to critical geographies of conservation in Canada. He is currently working on a political ecology of wildlife translocation.

Alexei Kraikovski graduated from St Petersburg University and the European University at St Petersburg and received his PhD degree from the St Petersburg Institute of History of the Russian Academy of Sciences. He worked for twenty years in research projects directed towards the research of environmental history of the Russian North and Northwest. His research interests revolve around the interrelations between Russian society of the early modern and modern period and the water environment, be that marine harvesting, hydraulic technologies, or shipping. Since 2017 he works at the Laboratory for Environmental and Technological History at the National Research University Higher School of Economics, and in 2020 also received a position of the MSCA Seal of Excellence Research fellow at the University of Padua.

Julia Lajus is a Visiting Scholar at Max Planck Institute for the History of Science in Berlin. Before spring 2022 she was a Head of Laboratory for Environmental and Technological History and Associate Professor at the Department of History, National Research University Higher School of Economics (HSE), St Petersburg, Russia. In 2014–2019 she was also an Academic Head of the International MA Programme in Applied and Interdisciplinary History "Usable Pasts" at HSE. In 2011–2015 she served as vice-president of the European Society of Environmental History, and in 2009 was a Marie Curie Research Fellow at the University of Birmingham (UK). Her research focuses on the history of field sciences such as fisheries science, oceanography, and climatology, as well as environmental history of biological resources, especially in marine and polar areas. Her publications include chapters in *Competing Arctic Futures: Historical* and *Contemporary Perspectives and Eurasian Environments: Nature and Ecology in Imperial Russian and Soviet History*.

Elizabeth Leane is Professor of English and Associate Dean (Research Performance) in the College of Arts, Law and Education, University of Tasmania. She is Arts and Culture editor of *The Polar Journal* and a past recipient of an Australian Antarctic Arts Fellowship. Her publications include three monographs, including *Antarctica in Fiction* (Cambridge, 2012) and *South Pole: Nature and Culture* (Reaktion, 2016), four co-edited collections, most recently *Anthropocene Antarctica* (2019) and *Performing Ice* (2020), and articles in a wide range of journals, such as *Polar Record*, *Studies in Travel Writing* and *ISLE: Interdisciplinary Studies in Literature and Environment*.

Ruth Maclennan is an artist and researcher. She exhibits internationally in exhibitions and film festivals. Her films, *Cloudberries*, *Call of North*, and *Hero City* shot in the Russian Arctic premiered at the London Film

Festival. She studied modern languages and art, and has a PhD from the Royal College of Art. She teaches at Central Saint Martins and is Institute Associate at Scott Polar Research Institute, University of Cambridge. She recently completed a film, *Treeline*, using collectively sourced footage of forests, for COP26 (FVU/Forestry England), and contributes to the collective art project, www.crownproject.art. LUX Artists' Moving Image distributes her films.

Astrid Ogilvie's research focuses on climate history and current Arctic issues. She enjoys building bridges between the arts, humanities, and the natural sciences, in order to foster interdisciplinary cross-fertilization. Current projects include: Arctic Climate Predictions: Pathways to Resilient Sustainable Societies (ARCPATH) www.svs.is/en/projects/arcpath; and Reflections of Change: The Natural World in Literary and Historical Sources from Iceland *c.*AD 800 to 1800 (ICECHANGE) www.svs.is/en/projects/icechange. She is the author of around a hundred scientific papers and two edited books. Her primary affiliations are: Senior Scientist at the Stefansson Arctic Institute, Akureyri, Iceland; Fellow of the Institute of Arctic and Alpine Research (INSTAAR) at the University of Colorado; and Visiting Professor at the University of the Highlands and Islands, Orkney, Scotland.

Jessica O'Reilly, Associate Professor of International Studies at Indiana University Bloomington, is an environmental anthropologist who studies Antarctic and climate scientists and policymakers. She is the author of *The Technocratic Antarctic: An Ethnography of Scientific Expertise and Environmental Governance* (Cornell University Press, 2017) and co-author of *Discerning Experts: The practices of Scientific Assessment for Environmental Policy* (Chicago University Press, 2019). O'Reilly is currently undertaking ethnographic research on expert decision-making within the Intergovernmental Panel on Climate Change.

Jonathan Rees is Professor of History at Colorado State University, Pueblo. His many books include three titles that deal with ice and refrigeration: *Refrigeration Nation: A History of Ice Appliances and Enterprise in America* (Johns Hopkins, 2013), *Refrigerator* (Bloomsbury, 2015), and *Before the Refrigerator: How We Used to Get Ice* (Johns Hopkins, 2018).

Thomas Simpson is Research Associate on the 'Making Climate History' project at the University of Cambridge, Research Fellow at Gonville and Caius College, and author of the monograph *The Frontier in British India: Space, Science, and Power in the Nineteenth Century* (Cambridge University Press, 2021). He is currently working on a history of climate cartography during the nineteenth and twentieth centuries and a study of climate sciences in South and Central Asia and the Indian Ocean World from the high imperial to the postcolonial era.

Sverker Sörlin is Professor of Environmental History at the KTH Environmental Humanities Laboratory at the KTH Royal School of Technology in Stockholm. He is also a non-fiction author, and contributor to the Stockholm daily *Dagens Nyheter*. Sverker's interest in ice spans a long career with a particular focus on the history of glaciology and other Arctic environmental field sciences. He launched concepts such as 'cryo-history' and the 'Arctic humanities', and has been engaged in advice and organization of polar research serving as President of the Swedish Committee for the Internal Polar Year 2007–09. His current research includes the historical and contemporary science politics of climate change, the cryosphere, and the Anthropocene. Forthcoming books include *Resource Extraction and Arctic Communities: The New Extractivist Paradigm* (2022), and *The Human Environment: Stockholm and the Rise of Global Environmental Governance* (under contract, co-authored with Eric Paglia), both with Cambridge University Press.

James Wilt is a PhD student in the Department of Environment and Geography at the University of Manitoba. His master's thesis examined the role of ice science by two fossil fuel consortiums in the Canadian Arctic during the 1970s and 1980s, while his current research focuses on the historical and contemporary Arctic oil spill science.

Acknowledgements

As editors, we are deeply grateful to the multiple social and intellectual contexts of which we have been part in the work on this book. Sverker acknowledges the members of the European Research Council Advanced Grant project SPHERE – Study of the Planetary Human-Environment Relationship: The Rise of Global Environmental Governance (grant number ERC-ADG 787516), several of whom read and commented on a draft version of the book's introduction. Klaus Dodds acknowledges the support of a Major Research Fellowship from the Leverhulme Trust (2016–20) and the conversations with polar-focused colleagues and friends in multiple workshops and networks. We have also benefitted immensely from our ice humanities (avant la lettre…) colleagues over the years, some of whom have contributed to the present volume. Among many names we could have mentioned (and who probably know themselves!), here are a few who stand out, for their assistance in reading and commenting in the final stages of preparing the introductory chapter: Mark Carey, Julia Lajus, and Jessica O'Reilly. We thank all the editorial staff at Manchester University Press for their goodwill and support, including our editor Robert Byron, our production editor David Appleyard, and assistant editor Lucy Burns. Finally our thanks to Rudy Leon for their help in preparing the index for this book.

Ice humanities: living, working, and thinking in a melting world

Sverker Sörlin and Klaus Dodds

> Many years later, as he faced the firing squad, Colonel Aureliano Buendía was to remember that distant afternoon when his father took him to discover ice.
>
> Gabriel García Márquez *One Hundred Years of Solitude* (1967)

> 'The thing is, it fizzed,' he said. 'It was melting with the warmth of my palm, and the air was under such pressure that it exploded out of its pockets. It fizzed,' he repeated, 'then it melted, and I just wiped it on my shirt.'
>
> A scientist holding a piece of Vostok ice core as quoted in Nancy Campbell's *Library of Ice* (2018)

Ice humanities is an emerging interdisciplinary field of inquiry that bestrides many academic disciplines ranging from environmental history, geography, maritime and polar studies, glaciology, anthropology, and permafrost science, as well as science and technology studies exploring low temperatures and 'life on and with ice'. It draws on what is already a sizeable body of work, most of it produced before the concept itself very recently (see Dodds, 2018, 2021) came into circulation (for example, Spufford, 1996; Hains, 2002; Gosnell, 2005; Orlove *et al.*, 2008; Carey, 2010; Radin, 2017; Radin and Kowal, 2017; White *et al.*, 2018; Jackson, 2019; Ruiz *et al.*, 2019; Chu, 2020; Wadham, 2021). Taking inspiration from other emerging scholarly fields, such as blue humanities (Steinberg, 2001; Gillis, 2012), critical ocean studies (Armitage *et al.*, 2017; DeLoughrey, 2019) and plant humanities (Batsaki and Humphreys, 2018), it takes as its subject matter something that the Nobel laureate Gabriel García Márquez recognized in his first novel, *One Hundred Years of Solitude*. The famous opening sentence of the book has a childhood recollection of 'discovering' ice – which begs the immediate question: what on earth does discovering ice involve? Where would one need to be for such a discovery to be surprising rather than banal? Why would a childhood memory involving ice be invoked when one was about to face imminent death? What source of ice did his father have in mind?

Thinking about ice like that, as a formative episode in a single person's life, evokes questions about the transformative powers of ice on societal and global levels. How do people and societies invent, create, and narrate ice (including snow) so it becomes not just physical but embedded in our minds and identities? This reveals the myriad relations human communities have had and continue to have with ice, relations that warrant the rise of 'ice humanities' to engage reflexively and critically with this emphatically cultural element. Ice humanities expands physical ice into the societal and cultural realms where it combines earlier 'integrative humanities' (Sörlin and Wynn, 2016; Sörlin, 2018) such as environmental and climate humanities into a new domain of inquiry. The physical object is still there as an essential 'element', a more-than-physics entity. What is added are sensibilities such as seeing, smelling, touching, understanding, and remembering – the multiple ways ice enters human and more-than-human life, and life in all its diversity encounters ice. Ice physics has meant spectacular progress in making sense of our human predicament under climate change. In order to deal with ice as a global challenge we should become reenchanted with it, partly to make

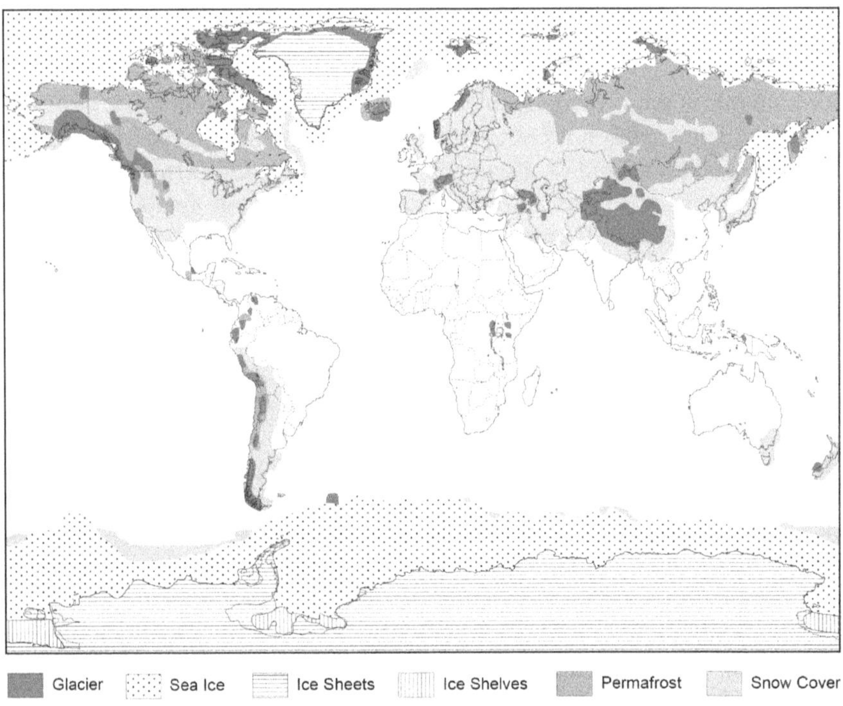

Figure 0.1 Global location of ice.

it come alive as a 'partner species' to us humans and to formulate a politics of ice to engage and sustain the life forms that have co-evolved with it. Ice is an element turned environmental object, and now very much part of the future equation with humanity, alongside many knowns and unknowns.

Discovering ice

'Discovering ice', either as a child or as an adult, is not a straightforward affair. Let us presume for a minute that we are not referring to artificial ice such as an ice cube or ice rink. The physical ice that is found on a mountain, across a frozen lake, or trapped in a glacial mass is extraordinarily diverse in shape, texture, density, size, and durability. How we have chosen to live with ice in large part reflects that material diversity. Some ice lends itself well to travelling over and some ice is better suited as ice storage. Some ice is safe to ski and skate on and some is dissolving, useless, even hazardous. Living with ice has gone hand in hand with working with ice – from hosting ice markets on the one hand to dealing with the messy consequences of spring ice melt and the blockage of infrastructure due to avalanches.

But perhaps it is thinking with ice that has revealed the promissory potential of ice humanities. Ice is evidentiary of past and present, human and natural, and presence and absence. We don't assume ice to be 'natural' per se because what is 'natural' about say Greenlandic ice that is punctuated by traces of PCBs (polychlorinated biphenyls), kerosine, trace metals, and even evidence of ancient human life? In that sense we might argue that ice is chiding humanity for its past excesses. Since the early 1990s, the term 'ice patch archaeology' was coined as sites of archaeological interest emerged in the wake of glacial retreat and ice disappearance. Past ice has left an indelible mark on contemporaneous landscapes and seascapes. It might be invisible to the naked eye, but our landscapes and seascapes bear witness to what was once here and there. But none of this is immediately obvious to the readers of *A Hundred Years of Solitude*. There is in the novel a confusing encounter between ice and diamonds, 'icy hands' make a fleeting appearance, and ice serves as a metaphor for a permanently frozen future, where anything is possible even if the ending is likely to involve a solitary death (Faris, 1985).

Trading with such ambiguities became a defining character of Márquez's literary canon. It captures well the spirit and purpose of ice humanities – where the ordinary and extraordinary qualities of this element provoke new ways of thinking about ice, living with ice, constructing (with) ice, and ascribing meaning, significance, and value to ice. Ice is much more than simply frozen water, and the geophysical and geocultural binaries (so often inflamed by declinist/ruinous narratives) that are so popular and prevalent in Euro-Western media and journalism need unpacking (Jackson, 2015; Carey *et al.*, 2016). Still, there is no denying that ice is part of the Earth's freshwater household and an essential ingredient to reproducible life. But if we just said that and started mapping where we would have drought and where we would have flooded if it all melted, we would not have moved ice humanities much further forward.

Thinking with ice starts with recognizing that changes involving ice may appear on any scale, from the individual to the continental and global. If ice preserves past life, it is also complicit with our artificially chilled lifestyles in the here and now. If ice is recognized as a geological force, then let us also acknowledge the extraordinary ecological service it performs to the planet by moderating the power of the sun. If ice is a 'prop' for illustrating the urgency of the climate emergency, then it is also a recorder of our pre-industrial pasts. The fate of ice is intimately linked to ongoing climate change and the spectre of rapid, catastrophic, and irreversible ice loss is not beyond the realm of short-term possibility. Ice is the interlocutor par excellence in that respect. Change might not be as rapid as depicted in *The Day After Tomorrow* (2004), where the Gulf Stream turns decisively. New York City freezes, and in no time fills up with ice to the top of the Empire State Building (with interesting parallels to the end of *Planet of the Apes* (1968)) when exhausted human survivors witness the ruination of the Statue of Liberty poking out of the shoreline. Nonetheless, the polar ice sheets could change at a rate that far outpaces previous disintegrations, and inundate a global network of coastal settlements.

Climate change fiction (cli-fi) has proven itself a rich medium to frame and represent ice as both in crisis and as an earthly 'tripwire' that is more than capable of anticipating and articulating what lies ahead for humanity and planet Earth (Evans, 2017; Leane and McGee, 2019). Living, working, and thinking with ice goes hand in hand with speculating about what might yet come to pass. For the American residents in the fictional future Earth in *The Day After Tomorrow*, dramatic climate change means dispossession and southerly retreat towards Mexico as environmental refugees. All of this sits uneasily with the lived realities of native Alaskan villages who are already bearing the brunt of a warming Arctic. Small coastal settlements such as Kivalina (located to the north of the small town of Kotzebue in western Alaska, population around 380) and Quinhagak (located in southwest Alaska, population around 720) are facing threats on all sides. Permafrost thawing is leading to subsidence and critical infrastructure damage. Diminishing sea ice is making the coastline more vulnerable to violent winter storms. Indigenous representatives such as the Bering Sea Elders Group are warning of ongoing disruption to access to traditional food supplies such as seals and walruses. Sea ice, like the mangroves, are crucial buffer zones, which help to regulate their respective coastal environments. The bill for re-village location will run into hundreds of millions and over thirty settlements have been identified as in imminent danger (Ristroph, 2017). Yet for most people worldwide who live close to it, ice is not just an element of alarm, but also a daily interaction integral for cultures that support food security, transportation, livelihoods, values, traditions, and kinship. We must see ice as more than just a substance in recess and crisis that acts in response to the climate change wrought largely by the wealthy strata of this world, but fundamentally – and more enduringly over the millennia – as a crucial part of life: a part that comes and goes and changes alongside societies inhabiting icy regions (Krupnik *et al.*, 2010; Gearheard *et al.*, 2013; Hastrup, 2013).

Alaskan communities are involved in ongoing struggles to insist that they participate actively in shaping those relocation plans. However, Indigenous groups have warned that these narratives of disruption should not reduce communities as simply victims of an element rather than societies and individuals with their own agency, and indeed long-standing history of change and adaptation. More egregiously, the resilience of Indigenous peoples in the Arctic has also been framed as a 'resource' for non-Indigenous settler colonies to harness in the face of ongoing climate change (Lewis and Reid, 2014). Zoe Todd has cautioned against appeals to environment as a 'hyper-object' and the more-than-human as marginalizing Indigenous knowledge and perspectives that have long held that there are limits to human mastery over things (Todd, 2016 and noting Morton, 2010). There are plenty of ways of living, working, and thinking with ice, much of which continues to bear the brunt of settler colonialism and the marginalization of Indigenous knowledges, experiences, and perspectives in favour of settler control of land, water, and ice (see Chapter 5 this volume).

Ice also acts as a witness for the Earth's past dynamics and is credited as an anticipatory medium par excellence (Scranton, 2015). As a historical medium, ice cores from Antarctica and Greenland's interior are proving memorable, acting as a 'kind of library' as the American scientist Roger Revelle noted to the US House of Representatives during the 1957–58 International Geophysical Year (Revelle, quoted in Berkman, 2002: 55; Campbell, 2018; Isberg, this volume). Ice remembers and records a dazzling array of past geological transformations and more recent human interventions. Trapped in ice bubbles, sediments and chemicals offer up the tantalizing prospect of reconstructing past climate and 'cryo-histories', including the composition of air thousands of years ago (Revelle and Suess, 1957; Sörlin, 2015).

Careful analysis of the ice bubbles can and does reveal how much methane, carbon, and other gases were present when the ice formed. Within carefully extracted ice cores, we find evidence of past human activity. Through ice we can better understand deep time as well as human time. Greenland's ice carries with it traces of metal smelting in the Roman era, raising intriguing issues about what kind of human activity makes itself more legible than others. Antarctica's ice records dramatic changes in the Earth's planetary history including ice ages, volcanic eruptions, and periods of warming. Ice, as Robert Macfarlane reminds us, is a recording medium and a storage medium par excellence (Macfarlane, 2019). It stores earthly intelligence for millennia and offers up a rich and inviting archival source. Still, as we shall see, it is more than a library or an archive that awaits Euro-Western scientific interrogation. It is a prehistoric warning system, which gave us a better sense of critical leaps and planetary drama, ominously flagging the worst, which is yet to come, as part of a growing 'politics of anticipation' (Jónsdóttir, 2013). The entire idea of planetary boundaries (Rockström *et al.*, 2009; Steffen *et al.*, 2015) rests on the short-term scale jumps that were perhaps the most revealing finding in the early ice cores. The entire Earth can change drastically, and the idea was born that maybe humans, too, could bring about that change. This frightening more-than-archival insight came from a tall, slender column of ice. But the ice cores are more than just physical objects and specimens. The way people discuss, represent, and otherwise circulate scientific conclusions from ice cores is rearranging conceptualizations of Earth's temporalities, creating new narratives of social change over time, and reframing not only the human place in the planet's past but also the historical impact of climate on societies around the world (Antonello and Carey, 2017; Salazar, 2018).

The elemental exchanges between ice and water are providing visceral evidence of a substance in apparent crisis. Language on and of ice reveals deeper value patterns, and the use of metaphor and visual representation has a rich history in Euro-Western cultures – from the early modern fear and frozen 'mountain gloom' that Marjorie Hope Nicolson observed in her 1958 classic *Mountain Gloom and Mountain Glory*, to the pilgrimages to the melting ice edge in Greenland's Ilulissat in the present century (Nicholson, 1958; Salim *et al.*, 2021). It is already more than thirty years since Robert P. Sharp could talk of ice as dynamic and 'alive', in *Living Ice: Understanding Glaciers and Glaciation.* Organic and vital metaphor reach much further back, reflecting a need among scholars of ice to translate the distant and alien element to processes and conditions known to their audiences in tempered climates (Sharp, 1988). Glaciers can be in 'health' or be sick or downright dying (Carey, 2007). To represent their disappearing and their shimmering vulnerability has become almost a visual industry (Balog and Williams Tempest, 2012). To render glaciers and ice sheets as 'icons' has become common, but no longer as terrifying geographies of 'the farthest ends of the earth' – rather as icons of a certain *Welt von Gestern*. The trope carries the existential reflection of a self-imposed deep loss, linking ice to other elements that are sacrificed as our juggernaut capitalist-inspired civilization continues to bulldoze the spheres of life (Glasberg, 2012; Boyer and Howe, 2018).

If the Anthropocene is making itself felt and known through elemental state-change, ice would be a good place to collect the damning evidence; but it is not the only form of evidence that deserves further consideration (Pollack, 2010; Lewis and Maslin, 2018). Ice can of course be measured and represented as significant through expressions of disappearance and loss, thawing and retreat. The fate of ice and humanity are tied up with one another but perhaps not in the highly deterministic and alarmist manner that is commonly assumed. For one thing the 'we' in all of this needs urgent clarification as many scholars working on the intersection of climate change and social justice have correctly insisted on feminist and Indigenous scholarship (for example, Farhana 2021). Put bluntly, the more ice that disappears the more the sea level rises, albeit with very uneven geographical and social-cultural consequences. Low-lying and coastal areas that lack resilient infrastructure or political power will be particularly vulnerable to the rise of seawater and the decline of ice. Elsewhere in the world, glacial retreat and instability carries with it risks of devastating meltwater release, which can in turn flood and overwhelm downstream communities, where vulnerabilities will be closely aligned with indices of inequality (Carey, 2010). Elemental divisions involving land, sea, and ice become scrambled. Islands disappear and coastlines inundate, cold oceans lose their ice cover, and retreating glaciers expose rocks and soil. International legal frameworks predicated on a reliable distinction between land, sea, and ice become confused and confusing, with potential new kinds of 'border wars' (Dodds, 2021). Populations, human and nonhuman, have had to adapt, flee, or find new ways of co-living in a world remade by 'friction' (Tsing, 2005).

This demand for new ways of living, thinking, and working with ice could not be more apposite given a near-constant diet of stories about sea ice shrinking, permafrost thawing, and glacial retreat, with permafrost thawing arguably being the least charismatic example of 'loss' (Bouffard *et al.*, 2021). Snow barely gets a mention and river and lake ice that has been so integral to human cultures rarely merits more than a footnote. What is left behind in this documentary auditing of shrinkage and loss are the symbolic properties of ice and snow – the 'cryosphere cultures', as we term it, that exist in so many parts of the world. In the chapters that follow, our authors embrace the cultural richness of ice and snow and investigate how, why, and where frozen water matters – from histories of refrigeration, to the back and forth of ice markets in northwest Russia, and everyday life in and around glaciated regions.

Ice as a 'crisis concept'

Turning to ice is not just an opportunity to engage with it; there has, after all, been no shortage of ice-based studies ranging from scientific fields such as glaciology, permafrost science, and cold weather engineering to humanities scholarship on popular and literary cultures of ice. They keep appearing in new fields, such as cinema studies (MacKenzie and Westerstahl Stenport, 2015; Kaganovsky *et al.*, 2019). In the history of visual representation, the relatively few studies that exist (e.g. Matilsky, 2013; Heuer, 2019) suggest that the representation of ice confounded artistic conventions and challenged European ideas of 'pictorial composition, space, selfhood'. For example, in the very first known attempt – by the sailor Thomas Ellis travelling with Martin Frobisher to what is now Baffin Island in the 1570s – to render an illustration of an iceberg, a 'great and monstrous peece of yce' that, thanks to its perplexing irregularities, required no less than four representations, all poor (Heuer, 2019: ch 1, 10).

The ice humanities turn is also a means of staging an encounter with the temporality, materiality, and spatiality of ice. Some of this must involve a fundamental recognition of Indigenous and settler communities and their experience and knowledge of ice and snow, whether that be in the Arctic or in other cold and mountainous communities around the world. Some take into account the way living and working with ice is generative of more global conversations and encounters that don't fetish the cultural encounter with ice and snow as simply involving those settler communities who live and work in cold and high places.

This is particularly relevant when recognizing the extraordinary importance of Indigenous experience, memory, and knowledge of ice, which runs into millennia. Indigenous communities have been working with the conceptual and embodied terrain long before glaciologists or ice humanities scholarship encountered it (Krupnik *et al.*, 2010; Hastrup and Olwig, 2012). Conventional ice metrics as such cannot do justice to the lively and animated encounters with ice, as Julie Cruikshank revealed in her work with Athapaskan and Tlingit oral traditions in the Pacific Northwest. Glaciers are sentient. Ice is animate. Glaciers are responsive to encounters with humans. As Cruikshank noted, they 'i.e. the glaciers listen, pay attention, and respond to human behaviour' (Cruikshank, 2001: 378; 2005).

Indigenous and local communities living with ice have long favoured a view of it as capricious and capable of being 'calmed' with respectful action.

In northwest India, for example, local communities recognize that they must care for glaciers and take care of them while nurturing their own spiritual interactions with ice and each other (Gagné, 2019). As humanities and social science scholars have recognized with some vigour, making sense of climate change is too important to be left to the environmental and physical sciences alone. They can detect the melting, the species loss, and the microplastics in places where we thought they should not be, such as the Central Arctic Ocean. Our edited collection, in contrast as well as collaboration, works with the materiality of ice, the ontological claims made on ice, and the representative ecologies that have enabled and sustained it (Bravo and Rees, 2006). Indigenous communities are so far bearing the existential brunt of ice loss, permafrost thawing, and what one notable Indigenous Canadian activist Sheila Watt-Cloutier (2018) described as the 'right to be cold'.

Living with ice does not just involve human communities. From microbial life in and around glaciers (Sharp *et al.*, 1999; Wadham, 2021) to animals and plants that depend on ice and snow, taking ice seriously will in turn demand that we reflect on our (in the broadest sense) co-dependencies and kinship. Without ice our planet would be radically different. Its absence is as significant as its presence. Ice is integral to global climate systems and the global exchange of ice, ocean, and air. The surface albedo of Earth is critically dependent on ice; ice reflects rather than absorbs solar radiation. Around the world, hundreds of millions of people depend on the spring melt of glaciers for essential water supplies, reaching from Andean mountain villages to farming families on the plains and deltas of the Indus, Ganges, and Brahmaputra. In his water history of South Asia, Sunil Amrith writes about the enormous Himalayan 'mountain core' of the region: 'So massive are the mountains, so heavy is their concentration of snow, ice, heat, and melting water that they shape Earth's climate. — Over centuries the rivers have carried silt and sediment from the mountains; they have deposited them along Asia's valleys and floodplains to sustain large human populations' (Amrith, 2018: ch 1). What Karl Wittfogel (1957) once called 'hydraulic societies' were to a large degree societies whose critical supply of water was based on ice – for example, in China where geopolitical interest in Tibetan and other 'Third Pole' glaciers remains and is growing (Ternes, 2020). The transition of ice to water is integral to well-being, but also hazardous to human communities if glacial bursts and flash flooding follows. How we tell stories about ice reveals our relationship to it, but also our capacity and willingness to listen and learn from other storytelling communities (Morehouse and Cigliano, 2021).

Ice is neither ahistorical nor, as we are discovering to our growing cost, is it impervious to intensifying anthropogenic climate change. Ice needs to be understood as a 'crisis concept' rather than reflective of a 'state of crisis', which simply invites us to witness its disappearance and fragility – the latter being particularly prone to selectively focusing on that 'disappearing optic' around ice that is easier to grasp, such as sea ice and glacial ice, rather than ice underground such as permafrost. This optic at the same time privileges Euro-American perspectives on ice science and ice metrics. As a 'crisis concept', we argue that a focus on ice, rather than say 'environment', heralds a shift in the scale and intensity of human–Earth historiography.

A focus on elemental substances such as ice, rock, and water signals two major fractures. First, the move to 'decolonize' the Earth and physical sciences that underpin dominant Euro-Western epistemological frameworks and heroic explorer-scientist figures and European timescales and framings such as 'ice age' (Robinson, 2006; Lewis-Jones, 2017; Knight, 2019). Earth sciences, such as geology and specialist fields such as glaciology, have been critiqued for their complicities with settler colonialism and scientific racism (Carey *et al.*, 2016). One of the leading figures of modern glaciology, the Swiss scientist Louis Agassiz, combined an interest with ice ages in earthly histories alongside an advocacy of racist polygenism (which posits that the human species has multiple origins) and anti-Darwinian creationism. At about the same time Agassiz was writing his pioneering book about glaciers (*Études Sur Les Glaciers*, 1840) the use of coal-fired steam engines exploded. Sadi Carnot in France (in *Reflections on the Motive Power of Fire*, 1824) and James Joule in England (in his heat–friction experiments 1843) laid the groundwork of what became, respectively, the second and first laws of thermodynamics. Marvelling at ice coincided with the invention of energy concepts and technologies that were and are complicit with its accelerated disappearance, and ideologies that advocated Western supremacy. The European Geophysical Union renamed its 'Louis Agassiz Award' in 2019 after initially naming it so in 2005. Decolonization is ongoing and the forces that extract resource value from land, ice, and sea are still very much part of a terra-altering system otherwise known as capitalism, geopolitics, and settler colonialism (Grove, 2019).

Second, ice has now been recognized as a 'matter of concern' and even a 'concerning matter' rather than an object of curiosity or inconvenience. It is an 'environmental object' that is now a matter of and for environmental and legal governance, in the form of legislation, to protect it (Taillant, 2015; Jackson, 2019; Höglund Hellgren this volume), or treatment, as if it were a patient (see Chapter 4 this volume), which in a sense it is also, sliding ever deeper into the crisis. The malady of ice is approaching that classical turning point that Hippocrates described, and which Reinhart Koselleck took as the starting point in his *Begriffsgeschichte* of 'crisis' as a concept (Koselleck, 2006).

To become a 'crisis concept' ice first has had to go through a phase of epistemic concern – become an object of Eurocentric knowledge. This crisis-laden history is longer, but accelerated in the middle of the nineteenth century by the European discovery of the 'Ice Ages'. A Euro-Western fascination with glaciers and past evidence of ice-free landscapes in the 1830s onwards elicited concern that ice was not only changing, but perhaps even changing as an effect of human forcing rather than natural variations in the Earth's climate. How solid was it? It cracked and was heavily pressured, argued John Tyndall (1860). Others thought, more accurately as it turned out, that it was a viscose substance (Hevly, 1996). What no one thought of doing was asking Indigenous and local communities who lived, worked, and thought about ice what they thought. It was nineteenth-century debates in Europe about deforestation and ditching affecting climate that spurred interest in climate change indicators, hence in ice and snow, not Indigenous communities in Peru or Tibet, despite them being much closer to the ice. Projections of ice diminishing as an effect of anthropogenic emission of greenhouse gases followed immediately on Svante Arrhenius's seminal 1896 paper establishing the connection between atmospheric CO_2 and global temperature (Arrhenius, 1896). His interest was, however, not deglaciation. On the contrary, he wanted to understand the causes of ice ages, at the time a prime concern in explaining natural conditions in resource-rich Sweden, a country born out of 'ice and snow' (De Geer, 1896: 150; Nordlund, 1999, 2001; Sörlin, 2002). Besides, Arrhenius flatly denied the possibility that humans could burn enough carbon to seriously impact on climate.

This Euro-Western idea returned in spurts of concern in the following decades (Fleming, 1998; Weart, 2008; Oppenheimer *et al.*, 2019). In the 1950s, the first projections came of an ice-free Arctic Ocean and large-scale melting of the ice sheets of Greenland and Antarctica (Siple, 1953). It coincided with the early rise of 'the environment' as a crisis concept, identifying an ever-growing number of features of the Earth as subject to anthropogenic change (Warde *et al.*, 2018). More than natural 'rates of change' signalled the anthropogenic addition. Humans were, to cite the title of an important conference in Princeton in June 1955, 'changing the face of the earth' (Thomas Jr, 1956). Ice, with its ten per cent of the total global terrestrial area, was gradually acknowledged as one of the parts that changed most. It was a theme in the International Geophysical Year (1957–58), the largest ever scientific enterprise akin to a 'scientific Olympics' with sixty-seven countries participating in research into the Earth and heavens and twelve countries dedicating considerable resources to investigating the subterranean worlds of the Antarctic ice sheet (Collis and Dodds, 2008). In the following decades, when computer-based climate models became available (Edwards, 2010) and ice cores demonstrated drastic changes in global temperatures in the deep past (Dansgaard *et al.*, 1969; Martin-Nielsen, 2013; Elzinga 2017) ice was drawn into the growing data-modelling complex of climate change science (Bolin, 2007).

It was in the aftermath of these insights at a time when global interest had focused on the environment through the 1972 UN conference in Stockholm – when climate was again on the agenda – that ice, or the cryosphere, became an environmental object, subject to concern and consideration, just like other environmental objects such as atmosphere, oceans, deserts, or biodiversity. We might perhaps say that it happened at the same moment as Wally Broecker coined the concept 'global warming' and used the Dansgaard ice core to make his warning (Broecker, 1975; see also Isberg this volume). Changes in ice became an anticipated feature of each of the Intergovernmental Panel on Climate Change's assessment reports, beginning in 1990 (Oppenheimer *et al.*, 2019). Their chronologies and trajectories as environmental objects differ, though they share their dependence of the overarching understanding that there is 'the environment' out there that growing portions of humankind dangerously engage, extract, and exploit. But for many decades still, how glacial change affected local and Indigenous communities in Greenland, the Andes, or the melting Siberian tundra north, was a concern kept in the margins.

In the case of ice, it also required an epistemic shift. Ice as an object of knowledge had been dominated by a 'horizontal' and incremental thinking. Ice was primarily a flat element that shrank or grew in area, and it did so through natural variation that ranged from short-term oscillation to multimillennial glaciations. With the verticality of ice cores, the deeper timescales, and the augmented human agency in the age of 'the environment' a profound mind shift came and turned the narrative of ice around: a volumetric and deep time appreciation of ice (Antonello and Carey, 2017; Salazar, 2018; Achermann, 2020). The shift was reinforced by other temporalities also emerging, for other elements, environmental objects, and world regions introducing a 'chaos of times' around the middle of the twentieth century (Sörlin and Isberg, 2021). An outcome of this was that ice transformed, from a slow, oscillating variable in the geophysical panorama coming out as one of the most rapidly responding bellwethers of the anthropogenic forcing on the global scale. It was also relatively easy to monitor, especially with the arrival of orbiting satellites from 1972 (Gärdebo, 2019). Satellites, buoys that were increasingly used, and comprehensive measurement of terrestrial ice using a range of documentary techniques served as 'environing technologies' that quantified and 'scientized' the epistemic transformation of ice into an environmental object (Sörlin and Wormbs, 2018). Perhaps it was, in some magically realistic fashion, precisely this emerging turn of

Figure 0.2 '1765'. An ice core with air bubbles filled with air dated to around 1765, prior to the modern industrial era.

events that García Márquez had in mind when he put that last little word into his opening sentence – 'ice': a magical element that opened the eyes of the children of Macondo and made it possible for them to imagine worlds beyond the spatial and temporal horizon that they did not know existed. The element that showed us humans to ourselves. We are, in the very same moment, both the poor Aureliano Buendía and his executioners, and it behoves us to always remember the day when we in earnest discovered ice, before and after the industrial era.

Privileged communities in the Global North have shifted from thinking of ice as simply an archive or a library to something that challenges our routinized application of frameworks and understanding. Ice has been and continues to be complicit with colonialism and extractive projects, what Michael Bravo and Gareth Rees have called 'cryo-politics' (Bravo and Rees,

2006; Bravo, 2017). We might even speak of a *settler cryosphere* and understand 'icy geopolitics' to be an operating system that has probed ice, removed ice, and assaulted the cold and darkness and regard that operating system as part of a continuum of activities that cherishes access, mobility, and profitability. In Jarius Grove's terms, ice and Indigenous peoples have for centuries been framed as part of 'savage ecologies' that need taming and improving and/or simply ignoring (Thrush, 2014; Grove, 2019). The nineteenth-century ice trade for example saw ice from the US end up in British India – and we can think about how the trade in natural ice acted as accomplice to colonial/tropical medicine and racialized tropes about the health of colonial authority. Imported ice had a distorting effect on local ice markets. One has the ultimate irony: ice seen as obstacle to settler colonial progress and ice being extracted and exported around the world as part of a thriving economy to bring ice from the Americas to the British colonizers in India.

Part of the urgent task of ice humanities is to reveal to further scrutiny those stories/experiences/legacies that don't simply reduce ice to data points or objects passively awaiting extraction/disappearance/auditing by Euro-Western scientific bodies.

A centennial 'Ice Age'

The term cryosphere – from Greek *krúos*: ice, cold – was coined in 1923 by the Polish geophysicist and meteorologist Antoni Bolesław Dobrowolski (1872–1954) in his book *The Natural History of Ice* (*Historja naturalna lodu*). He noted that 'the ice comprises in truth a *global envelope*, extremely uniform in composition, extremely diverse in appearance. This envelope of ice – the *cryosphere* – enters into a close, definite and peculiar relationship with the hydrosphere, lithosphere and atmosphere' (translated from the original Polish by the authors). This appeal to an 'envelope' is significant in the manner it conveys both a sense of coverage of terrestrial and maritime surfaces, but also to 'envelope' conjures up a sense of ontological security. As Dobrowolski and others recognized, to 'envelope' does not mean that nothing could penetrate the cryosphere; rather, its relationship with the hydrosphere, lithosphere, and atmosphere was integral to the functioning of the global envelope of ice. At the point of the term's coinage, the cumulative effects of steam-engine power and modern industrial development were not commonly linked to anthropogenic climate change, although industrial fossil engines had been living alongside with the knowledge of ice ages for almost a century. Global warming and its deleterious effect on ice was, as we have noted, to be recognized far later. The 2019 IPCC special report on the

Ocean and Cryosphere in a Changing Climate was decidedly less assured about the composition and resilience of this global envelope of ice:

> The cryosphere refers to frozen components of the Earth system. Around 10% of Earth's land area is covered by glaciers or ice sheets. The ocean and cryosphere support unique habitats, and are interconnected with other components of the climate system through global exchange of water, energy and carbon. The projected responses of the ocean and cryosphere to past and current human-induced greenhouse gas emissions and ongoing global warming include climate feedbacks, changes over decades to millennia that cannot be avoided, thresholds of abrupt change, and irreversibility (IPCC, 2019).

In the intervening period between the cryosphere being coined and the IPCC report warning about climate feedback loops and abrupt even perilous change, the 'career' of ice has been varied. In the 1920s and 1930s, the ice superpower, the Soviet Union, was embarking on an unprecedented assault on northern territories. Stalin's march to the north was ambitious, brutal, and transformative. Ice, frozen ground, and cold were problems to be managed. Recognizing that permafrost and sea ice were cunning, even treacherous, foes, Soviet planners and scientists were tasked with mobilizing the engineering and military might of the Soviet Union. In his magisterial *Red Arctic: Polar Exploration and the Myth of the North in the Soviet Union*, John McCannon writes about the extraordinary efforts the Soviet Union made to exploit, develop, and even conquer the 'frozen North' (McCannon, 1997). The will of the Soviet people could not be allowed to be blocked by a recalcitrant nature.

The environmental historian Pey-Yi Chu's *The Life of Permafrost: A History of Frozen Earth in Russian and Soviet Science* (2020) offers a detailed reading of how Soviet scientists conceptualized permafrost. She argues that there was a patriotic Russian and Soviet framing of permafrost/frozen ground. If frozen earth was a barrier to development, then someone or something had to be held responsible. Were there subversive elements in the Soviet North secretly undermining attempts to develop and exploit Soviet resources? The problem posed by permafrost was not one, as Soviet researchers later noted, that could easily be 'defeated'. In his *The Conquest of the North (in the Region of Permafrost)*, the scientist Sumgin Demchinskii wrote in 1938 that frozen earth was framed as a highly dynamic and challenging opponent. Permafrost was dangerous because of its capability to manipulate the intersection of ice, water, soil, land. Could it be removed? Thawed? How could the Soviet Union overcome it? It might be framed as a 'cunning adversary' by Communist Party officials, but what emerges is a more complicated story involving adaption and concession to this ice envelope. Soviet scientists and planners began to move away from 'conquest' to a series of pragmatic accommodations including de-icing roads, elevating buildings, and avoiding accidental thawing by an overconcentration of infrastructure (Chu, 2020).

As Soviet expertise and experience of ice and permafrost grew, so interest in their scientific and technical findings mounted. In the past, ice was traded between ice suppliers and markets around the world. By the 1940s, knowledge about ice was being translated and exchanged. A Russian-born scientist based in the US, Siemon William Muller, was instrumental in coining the term permafrost. Reading relevant Russian language reports and investigations in too frozen ground, Muller produced a field guide and technical report on the subject matter (Muller, 1945). About the same time, Russian literature on sea ice was also being translated into English and distributed to relevant agencies such as US Geological Survey and the United States Army Corps of Engineers (e.g. Kolchak, 1928; Smith, 1932; Sörlin and Lajus, 2013). Ice and snow were becoming matters of national strategic importance. World War II revealed the geostrategic value of Alaska and other northerly territories such as Svalbard. Japanese and German forces were seeking to invade, disrupt, and control resources and establish dominance in and around the Bering Strait and North Atlantic.

The 'Cold War' was aptly named. Ice, snow, permafrost, and ice cores were integral to the military-industrial-academic complex (Senator William Fulbright's modification of President Eisenhower's warnings about changes affecting the United States in 1961) on both geopolitical sides of the Arctic Ocean. To the Americans, the northern latitudes were hugely important for hemispheric defence purposes. Soviet bombers might fly over the ice cap and enter North American air space, as the Bering Strait divided the USSR and Alaska by less than a hundred kilometres. Understanding polar meteorology was one element to this northern recasting, but so was ensuring that the United States had suitable infrastructure in place to cope with the ice, cold, and darkness of the High North (Farish, 2010). Later there were fears that the Soviet Union might send its nuclear-powered submarines under Arctic ice and emerge in Hudson Bay to launch ballistic missiles towards the United States. Greenland was a notable 'coldspot' for military and scientific investment (Doel *et al.*, 2016). Danish, Swiss, and US glaciologists used the massive engineering infrastructures in the Camp Century site in northwest Greenland to produce the tallest ice core to date: almost 1,400 metres. While scientists were extracting ice cores from the inland interior and sending the cores to the US Army's Snow, Ice, and Permafrost Research Establishment (later the Cold Regions Research and Engineering Laboratory, CRREL), the US military were hatching plans at the same site as Camp Century to hide missiles under the ice via Project Ice Worm. Knowledge and understanding of ice mechanics and physics was integral to scientific and military agendas, and as far as funders such as the Office of Naval Research were concerned the academic progress of cold weather engineering and ice studies in general aligned well with one another (Herzberg *et al.*, 2018). Both Project Ice Worm and Camp Century were later abandoned because the subglacial base camp was being crushed by glacial force and plasticity (Nielsen and Martin-Nielsen, 2014; Nielsen and Nielsen, 2021).

If scientists were learning how to extract ever-longer ice cores from Greenland in the 1950s and 1960s, satellite technology was enabling extraction of data from a great height rather than depth. The 'satellite era', dating from the 1970s onwards, was a game changer for how sea ice could be charted and even understood in terms of distribution, area, and behaviour. Aerial survey was not replaced by satellite as the more recent NASA's Operation Ice Bridge missions over Greenland and Antarctica demonstrate. However, satellites did allow for increasing precision in capturing sea-ice minima and maxima. Underwater data collection was not redundant either because satellite coverage struggled to collect information about volume and thickness. Submarines and later drones were integral to that volumetric work. The role of infrastructures and resources in enabling knowledge production of ice is integral to the era, raising important questions about who gets to generate that remote sensing and in turn providing opportunities for other polar powers such as China and the European Union to contribute to public and scientific debates about the state of ice. Earlier techniques, such as repeat photography of glaciers, paid little to no attention to local communities and their knowledges and experiences of that remote-sensed ice (Garrard and Carey, 2017).

The growth and development of remote sensing lent itself well to not only the rise of systems modelling but also the public visualization of ice. In the 1980s and 1990s, the Arctic and Antarctic were popularized and visualized through charts, maps, imagery. While the Brazilian-hosted Earth Summit (1992) had the Amazon rainforest as its immediate material and visual backdrop, it was the trio of jungles, poles, and mountains that were increasingly recognized as proxies for global environmental change, with the ice core, in particular, acting as a Rosetta Stone of the Anthropocene. Looking backwards, the '1980s' was a pivotal decade of change for the cryosphere. Until the 1980s, for example, Greenland rarely experienced any summer season melt at all. What started at the lower latitudes became evident in the highest areas of the interior ice. While the Americans were celebrating their ice hockey team's 'miracle on the ice' victory over the Soviet Union at the 1980 Winter Olympics, the state of 'natural ice' was faring less well. Strikingly, environmental campaigning was becoming more vociferous around protecting the polar regions from mining and waste management rather than loss per se and even disappearance. Notably, however, the first

IPCC report (Houghton *et al.*, 1990) remarked upon ice loss and the likely impact on ecosystems and communities in the light of 'global warming'. The report made clear the prevailing scientific understanding of the Earth's ice sheets at the time: that they were stable for the foreseeable future. Because of this, more pressing matters, like the rapid deforestation of the Amazon, gained more popular and political attention than the cryosphere.

In the last two decades, ice has been widely recognized to be in crisis. In 2002, the splintering and then collapse of the Larsen B Ice Shelf turned 'slow thinking' upside down, providing powerfully charismatic imagery of rapid collapse, even as experts cautioned that the disintegration could not be causally linked to anthropogenic climate change (O'Reilly, 2017). No longer thought of as either a protective envelope, impenetrable barrier, or a source of frustration to human design and mobility, the focus is now resolutely on shrinkage, disappearance, retreat, and thawing. A new generation of satellites and drones are monitoring the state of the world's ice sheets and glaciers and newer actors such as China are now substantive 'cryosphere powers' (*South China Morning Post*, 2020). Ice is and will remain hyperpoliticized and culturalized – we should not forget that there is no shortage of commentaries on the unsettled legal status of ice and accompanying speculative plans to tow icebergs from the polar regions in response to the needs of water-stressed cities and environments. Beyond that extractive logic, ice is not only generative of anthropogenic thinking but also caught up in a wholesale production of narratives, tracing its impact on communities living with it and those who will be affected by either disappearance or replacement by other ecologies. Interlinking the local and the global was in part achieved by the idea of 'Planetary Boundaries', published as an article in *Nature* in 2009 with many follow-up studies since (Rockström *et al.*, 2009). It argues that planetary stability can be subsumed under nine critical dimensions – such as rate of species loss, CO_2 levels in the atmosphere, ocean salinity, and six others – for which quantified boundaries are presented. Each of the boundaries is defined against a background of previous change and conclusions drawn from their past 'performance.' Their timescales of change differ wildly, and they did so in pre-Anthropocene, or even pre-human, times of (mostly) natural variability as well. The now-emerging Anthropocene *Weltanschauung* builds significantly on environmental times identified and investigated in the past. Such organizational devices are abundant in Anthropocene research where they serve the purpose of demonstrating, often in sharp, sometimes baffling, detail how environmental times are predicated on human agency.

The 2007–08 International Polar Year, following on from the 1957–58 International Geophysical Year, emphasized the importance of ice knowledge to both scientists and Indigenous communities. Community participation, especially in the Arctic, was integral to the programme of work as local communities express concern that the disappearance of ice invites two reactions from Arctic states and non-Arctic states alike: first, a desire to extend ever greater control over open land and water; second, to overdetermine what activities should take place on land and underwater. There is a residual danger that a new sort of 'settler cryosphere' takes hold where the powerful and privileged materially and epistemologically dispossess those with traditional Indigenous knowledge and experience – the loss of ice acting as a proxy to justify more external intervention at a time when Indigenous peoples are demanding and securing their rights to consultation, land, resource development, and self-determination.

Danish anthropologist Kirsten Hastrup, working for a long time in the community of Qaanaaq in northwestern Greenland, observed that the integration of the world is just as tangible and existentially central to these seven hundred Inuit as it is to the denizens of Copenhagen. They live in a frontier zone of global transformative forces (Hastrup, 2020). Their local ice worlds have become intertwined with elemental worlds elsewhere, connected by 'teleconnections' (the concept coined by meteorologist Anders Ångström in 1935) or long-distance geophysical influences on the Earth system scale so typical of the Anthropocene (Paglia, 2020). In the same vein, ice has gained centrality in Copenhagen, as well as in Cape Town, Canberra, Shanghai, and Caracas, and in so doing further decentres the dominant narratives and symbolism about ice that reflects and inflect Euro-Western experiences with high and cold places (Cosgrove and Della Dora, 2008).

While the polar regions and mountainous areas of the world are elemental to discussions of climate change and the Anthropocene, temporality and spatiality of earthly change is recognized as transcending traditional disciplinary and geographical boundaries. No longer resilient nor long-lasting, the melting and thawing of ice have provoked new interactions between humanities and science scholars interested in how creative and literary modalities of expression might make sense of the fate of the world's ice (Brady, 2017). Artists are increasingly drawn to ice as a topic of inquiry into human fate, folly, and enigma. Danish-Icelandic artist Olafur Eliasson had an installation, 'Ice Watch', consisting of thirty blocks of glacial ice from the waters surrounding Greenland. They were placed to melt outside Tate Modern (twenty four) and outside the Bloomberg Headquarters in the City of London (an additional six), as a visual reminder of climate change on the environment and as an opportunity for people to interact with a substance usually so distant from urban lives. A second installation took place in Paris during the negotiations over the Paris Agreement.

Ice loss is not only causing discombobulation for those who live with ice and snow on a semi-permanent basis, but distant and low-lying others will suffer from ice cap and glacial melting. Beyond that, we are learning that the large-scale release of meltwater, the disappearance of sea ice, and the retreat of glacial ice has complex ramifications for human and nonhuman communities and ecologies. As the Arctic Ocean becomes progressively devoid of sea ice, new agreements on scientific co-operation (2017) and commercial fisheries (2018) are being negotiated to help manage this emerging volume of open water. The Arctic Ocean is no longer a reliably frozen ocean (Wang and Overland, 2009) and speculation continues as to how the region's geopolitics will shift in the likely event of ever more interest in trading, shipping, and extractive futures. Some of this articulation and speculation will bring further potential for ice humanities scholarship to emerge from the Global South, where a diversity of experiences, knowledges, and perspectives will make themselves felt.

Following the ice

In this volume, our contributors follow ice and snow. They tell stories about its relationship to ecology, geopolitics, history, Indigenous and communal politics, and socio-cultural relationships to ice, frozen ground, and high places. Dani Inkpen's chapter, for instance, offers up a deft examination of how the mountain cryosphere is richly suggestive for ice humanities scholarship – our relations with it, intimate (Raffles, 2002) and immediate as well heterogenous and diverse, telling much more than simply stories of collapse and retreat. Our authors more generally find ice in strange and not so strange places, and they find other things in ice such as microplastics, plant matter, and soot. Our collection addresses the capacity of ice to be generative of subjectivities, ways of knowing, trade and geopolitical intrigue, marker of time, object of and for governance, and exemplar of the state of Earth and humanity. Ice is fecund, fertile, and not without life. It is an assemblage of nature, culture, and moral economy. The responses thus far have tended to revolve around saving ice (e.g. artificial engineering schemes), making artificial ice (e.g. technical fixes), and living with a world where there is less ice, or even no ice (Wadhams, 2017).

While most contributions are preoccupied with the period when ice turned into an environmental object – the Cold War and the Anthropocene present (a smaller 'ice age' of sorts) – there are several who extend further back in time. Astrid Ogilvie describes how learned men in Iceland reported on sea ice in the sixteenth and seventeenth centuries, sometimes including data they could gather from previous centuries. Alexei Kraikovski demonstrates the multiple functions and uses of ice in early modern Russia, where frozen rivers and lakes were important for mobility and vernacular fishing, and where ice was used for cooling and food maintenance in the cities and in the warmer periods of the year. Ice was also a seasonal stage, an arena for society life, showing off, and for skating and other sports. Christine Bichsel shows how Pamir glaciers were consciously turned into assets in the Russian nineteenth-century imperial conquest of Central Asia. Glaciers became what she terms 'scientific objects', mapped first by military topographers, and studied by mining engineers carrying out geological surveys in Central Asia, contributing the prestige of epistemic and geotechnological competencies to the imperial project.

Ice became a key element to British imperialism as well, encountered in the high Arctic, in Canada, and subsequently in Antarctica. In his chapter on colonial British India in the nineteenth century, Tom Simpson identifies multiple configurations of ice in the consumption of food and drink, in Himalayan mountaineering and exploration, but also in applying ice more generally as a marker of racial and gendered distinction. The point is corroborated by Inkpen in her chapter of the mountain cryosphere, 'Ever higher': 'British mountaineers ranked Himalayan peoples according to altitude and usefulness to British expeditions.' The higher – in both altitude and latitude – the more ice, more elevated properties, and nobler people. The British Empire, in all its presumptuous and perversely hierarchical self-understanding, built and visualized itself in stark contrast to its colonial subjects, the lowland people without ice, to paraphrase Eric Wolf's 'people without history' (Wolf, 1982).

The 'nineteenth century ice' that ended up in the glasses of the imperial higher strata and their families on Bombay and Burmese veranda bars did not, however, come from nearby glaciers. Instead, it arrived from a growing overseas commercial trade in the cooling element. Given the settler colonial origins of the American ice trade, we might even consider it an example of the Kleptocene – stealing not just land and materials but also ice for the pursuit of profit. Ice had been harvested and stored in both ancient Rome and ancient China (David, 1994; Heidbrink, 2020), but, as Jonathan Rees describes, the first modern ice industry started in 1806, when the American merchant Frederic Tudor sent a shipment cut from Fresh Pond, near Cambridge, Massachusetts to the island of Martinique in the Caribbean. The ice trade grew and proliferated up until modern ice-making and cold chain technologies were introduced in the twentieth century (Hamilton, 2003). Rees demonstrates that market demands and multiple uses led to careful examination of the various properties of ice and how these corresponded with segments of the market. Ice was always complex and diverse and speaking to multiple needs. So, when George Orwell asks, in *Burmese Days*

(1931) 'How much ice have we got left?', the question can be read literally, or as a metaphor for the temporalities of empire or of the planetary stability of the Holocene (Roberts, 2014) – granted that this geochronological neologism, proposed by the French palaeontologist and entomologist Paul Gervais in 1850, was probably unknown to Orwell.

It stands out as one of the interesting findings of this volume – and one that we did not anticipate when we set out to compose it – how much of an imperial concept ice also is. Ice represents, in many contexts, power and control, and as its flip side, the lack and loss of power and control among those who are adversely affected by its melting and exploitation. Further evidence can be found in the chapter by Julia Lajus and Ruth McLennan who explore Soviet era museum representations of ice conquest. A half century after the imperial glaciology in Central Asia, the Soviet state set up grand research institutes and launched ice breakers and airborne research teams into the high Arctic. This was part of a strategic and economic agenda to establish the Northern Sea Route, but it was also part of a propaganda industry to build confidence in the Soviet model of human and resource development. The museum exhibitions played on these notions.

With the gradual turn towards ice assuming the role of an environmental object, the representational logic transforms. When icebergs originating in Antarctica travel the world oceans, they are rather kin with the tragic stranded whales the media observe, and can cause widespread emotion. As Elizabeth Leane puts it in her chapter on icebergs as planetary travellers, icebergs in the Anthropocene have moved to the centre of global public consciousness. These objects, possible to track and observe through satellites and GPS technologies, signal the waning stability of the polar regions, and hence question the long-term comfort and security of human existence. The huge tabular bergs, always a fascination for Antarctic travellers (Pyne, 1998) have turned into political events, framed by the media worldwide, she argues, 'not simply as spectacle but also as a source of communal guilt, fear and anger'. Part of a growing 'eventization' (Christensen *et al.*, 2013) of ice phenomena, some of these feelings harken back to the sinking of the *Titanic* in April 1912. But if fear of icebergs then was located to the instant encounter between human technology and the sheer force of ice mass, and empathy to the victims and their loved ones, the feelings now have become hugely different. Local disaster has turned into global drama and feelings of empathy and anxiety are fuelled by apocalyptic imaginaries of what these peripatetic nomadic icebergs may tell us of who we are and what we do, a recurring theme of a global environmental object such as ice. A 'Lonesome George' turtle lacking a partner could not be more lonesome than an iceberg doomed to disappearance in the ocean, which will be rising and threatening low-lying islands like Vanuatu, just as melting glaciers are threatening small, often original, populations in the glacier regions.

As Jessica O'Reilly observes about the yearlong journey of A-68, an iceberg rifting from the Larsen C Ice Shelf in Antarctica in 2020 and breaking up near South Georgia in February 2021, this follows a pattern of notable events featuring elemental manifestations of climate change. The flipside of this story is of course that icebergs can be towed, and thus change location – a history dating back to the 1850s and 1860s when small icebergs and glacial ice were towed from Laguna San Rafael in Chilean Patagonia to ports such as Valparaíso in Chile. Provisions of fresh water to the Earth's arid zones, for example oil rich desert states, was a 1970s techno-utopian dream, addressed by both the RAND Corporation and the *Bulletin of Concerned Scientists*, the latter with the appropriately sceptical question 'Manna or Madness?' Anil Agarwal asked, in the *New Scientist*, 'Will Saudi Arabia Drink Icebergs?' (7 July 1977). The global freshwater towing idea brought new unknowns into the already complex equations of uneven geopolitical exchange, so far not weighed up by the fact that the number of available icebergs will only grow in the Anthropocene. They are no longer the 'perplexitie' that Thomas Ellis could not make sense of in the 1570s, but have risen to potential charismatic goods in the global household (Birkhold, 2020).

Similar instances/processes of high attention directed to the environmental object undergoing its fateful, inevitable change have been taking shape over several decades, and with glaciers the focus has been on protection rather than use. Early interest in glacier preservation started in the European Alps in the 1960s, although most of the experimentation has taken place after 2000, when the idea of ice as a threatened species has gained ground. Mark Carey, Jordan Barton, and Sam Flanzer cover several such efforts in many continents, including ice blankets in Europe, legislation in Argentina, and the creation of ice-Buddhas or stupas – artificial glacial mounds designed to facilitate water storage – in the Himalayas. These are interesting stories in their own right, but they are also read as technological and legal interventions into a melting and thawing world. These ambitions, however effective or futile they prove to be, tend to draw attention away from the anthropogenic causes that concern bigger geopolitical and structural issues of how modern societies, and their economies, are organized. Salvation campaigns prioritize techno-scientific interventions, commodify environments, and exacerbate social inequalities.

Jasmin Höglund Hellgren has looked further into the background stories to the Argentinian legislation and found that the decision itself, while significant, importantly mobilized a major knowledge effort to map Argentinian glaciers and build a comprehensive body of knowledge of ice in order to assess their cultural and legal status and turn them into 'governable objects'.

This process may turn out to be the most crucial outcome of the Argentinian case, illustrating how actors negotiate new multidimensional and multiscaled conceptualizations of ice at the intersection of scientific investigation, politics, and cultures living on the fringe of existence, thanks to ice. At the same time, Höglund Hellgren's analysis demonstrates the diabolical dilemmas that have entered national environmental governance as the encounters of global environmental objects, such as the atmosphere and the cryosphere, take place in domestic parliaments and courtrooms. The anthropogenic forces of change are everywhere; national jurisdiction is somewhere and can only do so much.

Building national atlases and archives of glaciers, like they did in Argentina, could however be repeated in multiple countries, and ultimately serve as collective, common-pool evidence to underpin a global governance of solidarity that can put pressure on those resisting more progressive forms of climate change politics. How this plays out in each single case requires scrutiny. In the Canadian North, Bruce Erickson, Liam Kennedy-Slaney, and James Wilt find the concern about the future of Arctic ice has encouraged further assertions in neocolonial jurisdiction and provided an appealing rationale for increasing state jurisdiction and the continuation of the settler colonial state, even at the cost of Indigenous ways of life. Environmental degradation in the form of changing ice can serve as an excuse to pursue policies, in this case the Arctic Pilot Project (an initiative from the 1970s designed to transport natural gas from the Canadian Arctic to southern consumer markets) and the rise of polar bear deterrence programmes that would not otherwise be possible, somewhat like what governments in many countries did to limit freedom and democratic rights to their citizens during the 2020–21 COVID-19 pandemic. Diminishing ice provokes colonial acceleration and further deferment of climate justice for Indigenous peoples (Whyte, 2020).

Dobrowolski's cryosphere concept was a prerequisite for the concerted concerned conversation about ice that we can now conduct. It has become a century old. Will it survive in the Anthropocene? This is a different question than the one about the future of ice. It seems clear that ice will keep waning for a long time still, making the ice humanities an ever more necessary – even urgent – enterprise. Humans are now the major melting and thawing force, in some bizarre ways stronger than even the sun. How humans have engaged with ice in the past is already a fantastically rich topic to explore, as we hope this book can demonstrate. With the waning of the ice the ice humanities will, in a way sadly, only grow in volume and acuteness. The current rise in interest in this shrinking element is, we believe, only the beginning of a tidal wave of crisis and care of and for ice (Ruiz *et al.*, 2019).

What about the *concept* of the cryosphere in the Anthropocene? In his chapter on the implications of ice core drilling, Erik Isberg notes that ice cores in the 1970s started to travel away from the ice sheets they originated from and into another, even bigger, sphere of concerns than the cryosphere. His chapter, 'Frozen archives on the go', charts the move from the ice itself to the new, synthetic Earth system science approach to studying human and natural systems in tandem that started to emerge in the 1970s and 1980s. The product was, however, not the ice itself but its temporalities, or perhaps even its 'verticality', its capacity to bring stories of deeper timescales and what they could tell us about major, sometimes very abrupt, changes in the planetary system. Glacier ice, Isberg argues, 'became a scientific object beyond glaciology'. He goes on: 'The role of the ice itself shifted from being geographically limited to the cryosphere into a proxy for increasingly volatile planetary pasts and futures.'

That is of course not to say that the cryosphere will not remain a centrepiece of the discourse. It does suggest though that with the territorial expansion of temporalities the conversation itself is jumping scales. For the better part of the last century, ice humanities have been about humans and ice in their local and regional settings and cultures. The shift to the Earth system means that these variations, which remain, may be placed in a planetary logic where the ice has taken its full and justified place as part of the human domain. Ice is no longer just a 'science' object.

A book, a field – and a future

We have organized this book around three words that position ice firmly in humanities categories: living, working, and thinking. Like probably any categorization, it is slightly arbitrary; words like these simply must overlap. They are not a taxonomy; rather they serve as ideal types in Max Weber's sense. Together they constitute a big chunk of what we mean by 'humanities'.

The section called *Living* gathers chapters that show how ice encounters life and life enters ice. Much of it is the experience of people who lead their lives close to the ice and have adapted to it. Ogilvie and Kraikovski talk about how life in Iceland and northwestern Russia used to be shaped and even determined by ice. Inkpen says the same about ice in high elevations. She also notes how that life changes, and in so doing they provide timely reminders of how fluidity, dynamism, and the material realities of frozen water intersect with physical and imaginative features of verticality. Carey and Erickson, with their respective colleagues, outline how a life with ice increasingly means no longer living in the same way, which brings an elegiac tone to the ice humanities that is not only empirically fitting but also morally and politically commending.

Working with ice certainly includes living with it. What we seek to explore in this section is therefore something more than the work that is part of everyday life. We identify practices in the form of interventions that could be periodic, sporadic, yet essential for the understanding of the element. Bichsel explores glacier mapping as a field practice, which is at the same time a work to define and enhance imperial geography in Russia. Rees talks very concretely about the counter-cyclical work of ice, cooling things in warm seasons and places, which in turn requires work in harvest, trade, and transport. Lajus and Maclennan note the curatorial and communicative work performed in a museum where memorializing is the central practice by the visitors. O'Reilly has spotted a practice on behalf of the global commons: searching for the rifts and ruptures in the no longer so continuous Antarctic shelf ice. It is a field practice too, but one that requires mediation and environing work with satellites and computers.

Ultimately, all practices require *Thinking*. In this last section, we have assembled pieces that emphasize intellectually innovative dimensions, whereby ice is part of cognitive revolutions in time and space. Simpson grapples with a cryosphere version of the British Empire, which above all requires a leap of imagination to include what was considered impediment. Höglund Hellgren examines a quantum leap in governance thinking, which required a redefinition of the ontological and legal status of what used to be 'just ice'. In Elizabeth Leane's chapter on travelling tabular ice it is the ice itself which is at loss. A 'fright' reversal. In the early years of Euro-Western exploration, it was the humans who feared being taken over by the infinite cold and deadly white. If an iceberg could think 'like a mountain', as environmentalist Aldo Leopold suggested in *A Sand County Almanac* (1949), what would it think? Quite naturally, that it is now the ice that should beware of the humans (Cruikshank, 2005). Erik Isberg wrings his chapter around perhaps the largest of all these scale jumps of *human* thought with ice as their common element: the sideward move of the temporalities of the ice core to the oceans and onward to other times of the Anthropocene.

We hope that these chapters will make the reader as confident as we are that the ice humanities is a viable sub- or cross-field of the integrative humanities. We also hope the book invokes the sense of excitement we have enjoyed in working with its authors to craft a volume that gives a sense of the diversity of a humanities and social approach to the cryosphere in its many manifestations. Yet, at the same time, we understand these chapters can only serve as a sample of the ice humanities that could be imagined. Despite a certain sense of frustration this is probably a good thing, since it points forward to the potential that future ice humanities can hold. This is not a programmatic volume, so will not offer a plan for how to build the field; it will be best shaped by continuous collaboration and critique among its own practitioners with diverse roots. But we would like to say that there is every reason to believe that there is a meaningful intellectual space for ice humanities. It is already happening. The revival and renewal of climate history (e.g. White, 2017; Lane *et al.*, 2018) and the tremendous rise of the environmental humanities have led the way.

The cryosphere as a critical zone of global environmental change offers lasting relevance. Just as the planet, with its atmospheric ailments and its precarious biodiversity, the worlds of ice and snow are sliding down into an abyss of climate emergency. If we are right in assuming that ice, like once 'the environment', is a crisis concept, it will be a 'crisis humanities' that will take on the work. Such an added element of acuteness and concern will only raise the stakes for the field's practitioners. The ice humanities will not primarily be about ornament and beauty, real or lost. Nor are the ice humanities just about the cultures of ice, or the knowledge of ice, or the practices of ice from the drinking glass to the icebreaker. It is about ice as it is becoming, and as we knew it: whether we as societies can put our act together so that we will still have ice, or if ice humanities will turn into a purely historical field of knowledge. We hope not. Part of the mission of this book, and a true source of energy and motivation as we have been working on it, is to help provide the friction.

References

Achermann, D. (2020) 'Vertical Glaciology: The Second Discovery of the Third Dimension in Climate Research', *Centarus* 62(4): 720–743.

Amrith, S. (2018) *Unruly Waters: How Rains, Rivers, Coasts, and Seas Have Shaped Asia's History*. New York: Basic Books.

Antonello, A. and M. Carey (2017) 'Ice Cores and the Temporalities of the Global Environment', *Environmental Humanities* 9(2): 181–203.

Armitage, D., A. Bashford, and S. Sivasundaram, eds (2017) *Oceanic Histories*. Cambridge: Cambridge University Press.

Arrhenius, S. (1896) 'On the Influence of Carbonic Acid in the Air upon the Temperature of the Ground', *Philosophical Magazine and Journal of Science, Series 5* 41(251): 237–276.

Balog, J. and T. Williams Tempest (2012) *Ice: Portraits of Vanishing Glaciers*. New York: Rizzori.

Batsaki, Y. and A. Humphries (2018) 'How Have Plants Shaped Human Societies?', *Scientific American* 13 November. Available at: https://blogs.scientificamerican.com/observations/how-have-plants-shaped-human-societies/ (accessed 21 March 2022).

Berkman, P. (2002) *Science into Policy: Lessons from Antarctica*. London: Elsevier.

Birkhold, M. A. (2020) 'Towing Icebergs in the Anthropocene', *Geschichte und Gesellschaft* 46(4): 662–681.

Bolin, B. (2007) *A History of the Science and Politics of Climate Change: The Role of the Intergovernmental Panel on Climate Change*. Cambridge: Cambridge University Press.

Bouffard, T. J., E. Uryupova, K. Dodds *et al.* (2021) 'Scientific Cooperation: Supporting Circumpolar Permafrost Monitoring and Data Sharing', *Land* 10(6): 590. Available at: www.mdpi.com/2073-445X/10/6/590 (accessed 21 March 2022).

Boyer, D. and C. Howe (2018) *Not ok* [film]. Available at: www.notokmovie.com/ (accessed 21 March 2022).

Brady, A. (2017) 'Why Climate Fiction Matters: An Interview with Malcolm Sen', *Chicago Review of Books*, 8 May. Available at: https://chireviewofbooks.com/2017/05/08/why-climate-fiction-matters-an-interview-with-malcolm-sen/ (accessed 21 March 2022).

Bravo, M. (2017) 'A Cryopolitics to Reclaim Our Frozen Material States', in J. Radin and E. Kowal, eds, *Cryopolitics: Frozen Life in a Melting World*. Cambridge, MA: MIT Press, 27–58.

Bravo, M. and W. Rees (2006) 'Cryopolitics: Environmental Security and the Future of Arctic Navigation', *Brown Journal of World Affairs* 13(1): 205–215.

Broecker, W. (1975) 'Are We on the Brink of a Pronounced Global Warming?', *Science* 189(4201): 460–463.

Campbell, N. (2018) *Library of Ice*. London: Scribner.

Carey, M. (2007) 'The History of Ice: How Glaciers Became an Endangered Species', *Environmental History* 12(3): 497–527.

Carey, M. (2010) *In the Shadow of Melting Glaciers: Climate Change and Andean Society*. Oxford: Oxford University Press.

Carey, M., M. Jackson, A. Antonello, and J. Rushing (2016) 'Glaciers, Gender, and Science: A Feminist Glaciology Framework for Global Environmental Change Research', *Progress in Human Geography* 40(6): 770–793.

Christensen, M., A. Nilsson, and N. Wormbs (2013) 'Globalization, Climate Change and the Media: An Introduction', in M. Christensen, A. Nilsson, and N. Wormbs, eds, *Media and Arctic Climate Politics: Breaking the Ice*. New York: Palgrave Macmillan, 1–25.

Chu, P.-Y. (2020) *The Life of Permafrost: A History of Frozen Earth in Russian and Soviet Science*. Toronto: University of Toronto Press.

Collis, C. and K. Dodds (2008) 'Assault on the Unknown: The Historical and Political Geographies of the International Geophysical Year (1957–8)', *Journal of Historical Geography* 34(4): 555–573.

Cosgrove, D. and V. Della Dora, eds (2008) *High Places: Cultural Geographies of Mountains, Ice and Science*. London: IB Tauris.

Cruikshank, J. (2001) 'Glaciers and Climate Change: Perspectives from the Oral Tradition', *Arctic* 54(4): 377–393.

Cruikshank, J. (2005) *Do Glaciers Listen?: Local Knowledge, Colonial Encounters, and Social Imagination*. Vancouver: University of British Columbia Press.

Dansgaard, W., S. J. Johnsen, J. Møller, et al. (1969) 'One Thousand Centuries of Climatic Record from Camp Century on the Greenland Ice Sheet', *Science* 166(3903): 377–380.
David, E. (1994) *Harvest of the Cold Months: The Social History of Ice and Ices*. London: Michael Joseph.
De Geer, G. (1896) *Om Skandinaviens geografiska utveckling efter istiden* [On the Geographic Development of Scandinavia after the Ice Age]. Stockholm: Norstedts.
DeLoughrey, E. (2019) 'Toward a Critical Ocean Studies for the Anthropocene', *English Language Notes* 57(1): 21–36.
Demchinskii, S. (1938) *The Conquest of the North*. Leningrad: Publishing House of the USSR Academy of Sciences.
Dobrowolski, A. B. (1923) *The Natural History of Ice* (Historia naturalna lodu). Warszawa: Kasa Pomocy im.
Dodds, K. (2018) *Ice: Nature and Culture*. London: Reaktion Books.
Dodds, K. (2021) 'Geopolitics and Ice Humanities: Elemental, Metaphorical and Volumetric Reverberations', *Geopolitics* 26(4): 1121–1149.
Doel, R. E., K. C. Harper, and M. Heymann, eds (2016) *Exploring Greenland: Cold War Science and Technology on Ice*. New York: Palgrave Macmillan.
Edwards, P. N. (2010) *A Vast Machine: Computer Models, Climate Data, and the Politics of Global Warming*. Cambridge, MA: MIT Press.
Elzinga, A. (2017) 'Polar Ice Cores: Climate Change Messengers', in B. B. Vincent, S. Loeve, A. Nordmann, and A. Schwarz, eds, *Research Objects in their Technological Setting*. London and New York: Routledge, 215–231.
Evans, R. (2017) 'Fantastic Futures? Cli-fi, Climate Justice, and Queer Futurity', *Resilience: A Journal of the Environmental Humanities* 4(2–3): 94–110.
Farhana, S. (2021) 'Climate Change, COVID-19, and the Co-production of Injustices: A Feminist Reading of Overlapping Crises', *Social and Cultural Geography* 22(4): 447–460.
Faris, W. (1985) 'Icy Solitude: Magic and Violence in Macondo and San Lorenzo', *Latin American Literary Review* 13(25): 44–54.
Farish, M. (2010) *Contours of the Cold War*. Minneapolis, MN: University of Minnesota Press.
Fleming, J. (1998) *Historical Perspectives on Climate Change*. Oxford: Oxford University Press.
Gagné, K. (2019) *Caring for Glaciers: Land, Animals, and Humanity in the Himalayas*. Seattle, WA: University of Washington Press.
Gärdebo, J. (2019) 'Environing Technology: Swedish Satellite Remote Sensing in the Making of Environment, 1969–2001'. PhD dissertation, Division of History, KTH Royal Institute of Technology, Stockholm.
Garrard, R. and M. Carey (2017) 'Beyond Images of Melting Ice: Hidden Stories of People, Place, and Time in Repeat Photography of Glaciers', in J. Bear and K. Palmer Albers, eds, *Before-and-After Photography: Histories and Contexts*. New York: Bloomsbury Academic, 101–122.
Gearheard, S. F., L. K. Holm, H. Huntington *et al.* 2013. *The Meaning of Ice: People and Sea Ice in Three Arctic Communities*. Hanover, NH: International Polar Institute Press.

Gillis, J. (2012) *The Human Shore: Seacoasts in History*. Chicago, IL: University of Chicago Press.

Glasberg, E. (2012) *Antarctica as Cultural Critique: The Gendered Politics of Scientific Exploration and Climate Change*. London: Palgrave.

Gosnell, M. (2005) *Ice: The Nature, the History, and the Uses of an Astonishing Substance*. New York: Alfred Knopf.

Grove, J. (2019) *Savage Ecologies*. Durham, NC: Duke University Press.

Hains, B. (2002) *The Ice and the Inland: Mawson, Flynn, and the Myth of the Frontier*. Carlton, Victoria: Melbourne University Press.

Hamilton, S. (2003) 'The Economies and Conveniences of Modern-Day Living: Frozen Foods and Mass Marketing, 1945–1965', *Business History Review* 77(1): 33–60.

Hastrup, K. (2013) 'The Ice as Argument: Topographical Mementos in the High Arctic', *Cambridge Anthropology* 31(1): 51–67.

Hastrup, K. (2020) 'Thule as Frontier: Commons, Contested Resources, and Contact Zones in the High Arctic', *Anthropological Journal of European Cultures* 29(1): 1–19.

Hastrup, K. and K. F. Olwig, eds (2012) *Climate Change and Human Mobility: Global Challenges to the Social Sciences*. Cambridge: Cambridge University Press.

Heidbrink, I. (2020) 'The Natural Ice Factory at Røsneshamn, Norway: How to Compete with Big-Tech by Using Nature' Environment & Society Portal, *Arcadia* 43(Autumn 2020). Rachel Carson Center for Environment and Society. Available at: https://doi.org/10.5282/rcc/9155 (accessed 21 March 2022).

Herzberg, J., C. Kehrt, and F. Torma, eds. (2018) *Ice and Snow in the Cold War: Histories of Extreme Climatic Environments*. New York, Oxford: Berghahn Books.

Heuer, C. P. (2019) *Into the White: The Renaissance Arctic and the End of the Image*. New York: Zone Books.

Hevly, B. (1996) 'The Heroic Science of *Glacier* Motion', *Osiris* 11: 66–86.

Houghton, J. T., Jenkins, G. J., and Ephraums, J. J., eds (1990) *Report Prepared for Intergovernmental Panel on Climate Change by Working Group I*. Cambridge: Cambridge University Press.

IPCC (2019) 'Summary for Policymakers', in H.-O. Pörtner, D. C. Roberts, V. Masson-Delmotte *et al.*, eds, *IPCC Special Report on the Ocean and Cryosphere in a Changing Climate*. Cambridge, New York: Cambridge University Press, 3–35. Available at: www.ipcc.ch/srocc/chapter/summary-for-policymakers/ (accessed 21 March 2022).

Jackson, M. (2015) 'Glaciers and Climate Change: Narratives of Ruined Futures', *WIREs Climate Change* 6(5): 479–492.

Jackson, M. (2019) *The Secret Lives of Glaciers*. Brattleboro, VT: Green Writers Press.

Jónsdóttir, A. (2013) 'Scaling Climate: The Politics of Anticipation', in K. Hastrup and M. Skrydstrup, eds, *The Social Life of Climate Change Models: Anticipating Nature*. New York: Routledge, 128–143.

Kaganovsky, L., S. MacKenzie, and A. Westerstahl Stenport (2019) *Arctic Cinemas and the Documentary Ethos*. Bloomington, IN: Indiana University Press.

Knight, P. (2019) *Glacier*. London: Reaktion Books.

Kolchak, A. (1928) 'The Arctic Pack and the Polynya', in W. L. G. Joerg, ed., *Problems of Polar Research*. Special Publications vol. 7. New York: American Geographical Society, 125–141.

Koselleck, R. (2006) 'Crisis', trans. from German by Maria Richter [orig. 1972–1997], *Journal of the History of Ideas* 67(2): 357–400.

Krupnik, I., C. Aporta, S. F. Gearheard, G. J. Laidler, and L. K. Holm, eds (2010) *SIKU: Knowing Our Ice: Documenting Inuit Sea-Ice Knowledge and Use*. New York: Springer.

Lane, M., S. Sörlin, R. H. Socolow, and J. R. McNeill, eds (2018) *Responding to Climate Change: Studies in Intellectual, Political, and Lived History*. Special issue of *Climatic Change* 151(1): 1–78.

Leane, E. and J. McGee (2019) *Anthropocene Antarctica*. London: Routledge.

Leopold, A. (1949) 'Thinking Like a Mountain', in *A Sand County Almanac: And Sketches Here and There*. New York: Oxford University Press, 129–133.

Lewis, B. and J. Reid (2014) *Resilient Life: The Art of Living Dangerously*. Oxford: Wiley-Blackwell.

Lewis, S. and M. Maslin (2018) *The Human Planet: How We Created the Anthropocene*. Harmondsworth: Penguin.

Lewis-Jones, H. (2017) *Imagining the Arctic: Heroism, Spectacle, and Polar Exploration*. London: IB Tauris.

Macfarlane, R. (2019) 'Landscape and the Memory of Ice', *Times Literary Supplement*, 15 May.

MacKenzie, S. and A. Westerstahl Stenport, eds (2015) *Films on Ice: Cinemas of the Arctic*. Edinburgh: Edinburgh University Press.

Márquez, G. (1967) *One Hundred Years of Solitude*. London: Jonathan Cape.

Martin-Nielsen, J. (2013) '"The Deepest and Most Rewarding Hole Ever Drilled": Ice Cores and the Cold War in Greenland', *Annals of Science* 70(1): 47–70.

Matilsky, B. C. (2013) *Vanishing Ice: Alpine and Polar Landscapes in Art, 1775–2012*. Seattle, WA: University of Washington Press.

McCannon, J. (1997) *Red Arctic: Polar Exploration and the Myth of the North in the Soviet Union*. Oxford: Oxford University Press.

Morehouse, H. and M. Cigliano (2021) 'Cultures and Concepts of Ice: Listening to the Other Narratives in the Anthropocene', *Annals of the American Association of Geographers*, 111(3): 913–920. doi: 10.1080/24694452.2020.1792266.

Morton, T. (2010) *Hyperobjects: Philosophy and Ecology after the End of the World*. Minneapolis, MN: University of Minnesota Press.

Muller, S. (1945) *Permafrost or Permanently Frozen Ground and Related Engineering Problems*. Strategic engineering study, Special Report #62. Ann Arbor: US Army Corp of Engineers.

Nicholson, M. (1958, reprinted 1997) *Mountain Gloom and Mountain Glory: The Development of the Aesthetics of the Infinite*. Seattle, WA: University of Washington Press.

Nielsen, H. and J. Martin-Nielsen (2014) 'City under the Ice: The Closed World of Camp Century in Cold War Science', *Science as Culture* 23(4): 443–464.

Nielsen, K. H. and H. Nielsen (2021) *Camp Century: The Untold Story of America's Secret Arctic Military Base Under the Greenland Ice*. New York: Columbia University Press.
Nordlund, C. (1999) 'On Going Up in the World: Nation, Region and the Land Elevation Debate in Sweden', *Annals of Science* 58(1): 17–50.
Nordlund, C. (2001) 'Det upphöjda landet: Vetenskapen, landhöjningsfrågan och kartläggningen av Sveriges förflutna, 1860–1930' [The elevated land: Science, land elevation and the formulation of a Swedish past, 1860–1930]. PhD dissertation, Department of Historical Studies, Umeå University.
Oppenheimer, M., N. Oreskes, D. Jamieson *et al.* (2019) *Discerning Experts: The Practices of Scientific Assessment for Environmental Policy*. Chicago, IL: University of Chicago Press.
O'Reilly, J. (2017) *The Technocratic Antarctic: An Ethnography of Scientific Expertise and Environmental Governance*. Ithaca, NY: Cornell University Press.
Orlove, B., E. Wiegandt, and B. Luckman, eds (2008) *Darkening Peaks: Glacier Retreat, Science, and Society*. Berkeley, CA: University of California Press.
Orwell, G. (1931) *Burmese Days*. London: Victor Gollancz.
Paglia, E. (2020) 'A Higher Level of Civilisation? The Transformation of Ny-Ålesund from Arctic Coalmining Settlement in Svalbard to Global Environmental Knowledge Center at 79° North', *Polar Record* 56: e15. https://doi.org/10.1017/S0032247419000603.
Pollack, H. (2010) *A World without Ice*. New York: Avery.
Pyne, S. (1998) *The Ice: A Journey to Antarctica*. Seattle, WA: University of Washington Press.
Radin, J. (2017) *Life on Ice*. Chicago, IL: University of Chicago Press.
Radin, J. and E. Kowal, eds (2017) *Cryopolitics: Frozen Life in a Melting World*. Cambridge, MA: MIT Press.
Raffles, H. (2002) 'Intimate Knowledge', *International Social Science Journal* 54(173): 315–335.
Revelle, R. and H. Suess (1957) 'Carbon Dioxide Exchange Between Atmosphere and Ocean and the Question of an Increase of Atmospheric CO_2 during the Past Decades', *Tellus* 9(1): 18–27.
Ristroph, E. (2017) 'When Climate Takes a Village: Legal Pathways toward the Relocation of Alaska Native Villages', *Climate Law* 7(4): 259–289.
Roberts, N. (2014) *The Holocene: An Environmental History*, 3rd ed. Malden, MA: Wiley-Blackwell.
Robinson, M. (2006) *The Coldest Crucible: Arctic Exploration and American Culture*. Chicago, IL: University of Chicago Press.
Rockström, J., W. Steffen, K. Noone *et al.* (2009) 'A Safe Operating Space for Humanity', *Nature* 461(September): 472–475.
Ruiz, R., P. Schonach, and R. Shields (2019) 'Beyond Melt: Indigenous Lifeways in a Fading Cryosphere', *Journal of Northern Studies* 13(2): 7–15.
Salazar, J. F. (2018) 'Ice Cores as Temporal Probes', *Journal of Contemporary Archaeology* 5(1): 32–43.

Salim, E., L. Ravanel, P. Deline, and C. Gauchon (2021) 'A Review of Melting Ice Adaptation Strategies in the Glacier Tourism Context', *Scandinavian Journal of Hospitality and Tourism* 21(2): 229–246. doi: 10.1080/15022250.2021.1879670.
Scranton, R. (2015) *Learning to Die in the Anthropocene: Reflections on the End of a Civilization*. New York: City Light Books.
Sharp, M., J. Parkes, B. Cragg, I. Fairchild, H. Lamb, and M. Tranter (1999) 'Widespread Bacterial Populations at Glacier Beds and their Relationship to Rock Weathering and Carbon Cycling', *Geology* 27(2): 107–110. https://doi.org/10.1130/0091-7613(1999)027<0107:WBPAGB>2.3.CO;2.
Sharp, R. (1988) *Living Ice: Understanding Glaciers and Glaciation*. Cambridge: Cambridge University Press.
Siple, P. A. (1953) *Proposal for Consideration by the US National Committee (UGY): 1 May 1953, C1, USNC-IGY*. Washington, DC: National Academy of Sciences.
Smith, E. H. (1932) 'Ice in the Sea', in *Physics of the Earth V: Oceanography*. Washington: The National Research Council, 384–408.
Sörlin, S. (2002) 'Rituals and Resources of Natural History: The North and the Arctic in Swedish Scientific Nationalism', in M. T. Bravo and S. Sörlin, eds, *Narrating the Arctic: A Cultural History of Nordic Scientific Practices*, 73–122. Canton, MA: Science History Publications.
Sörlin, S. (2015) 'Cryo-history: Narratives of Ice and the Emerging Arctic Humanities', in B. Evengård, J. Nymand Larsen, and Ø. Paasche, eds, *The New Arctic*. Berlin: Springer, 327–339.
Sörlin, S. (2018) 'Humanities of Transformation: From Crisis and Critique towards the Emerging Integrative Humanities', *Research Evaluation* 27(4): 287–297.
Sörlin, S. and E. Isberg (2021) 'Synchronizing Earthly Timescales: Ice, Pollen, and the Making of Proto-Anthropocene Knowledge in the North Atlantic Region', *Annals of the American Association of Geographers* 111(3): 717–728.
Sörlin, S. and J. Lajus (2013) 'An Ice Free Arctic Sea?: The Science of Sea Ice and Its Interests', in M. Christensen, A. Nilsson, and N. Wormbs, eds, *Media and Arctic Climate Politics: Breaking the Ice*. New York: Palgrave Macmillan, 70–92.
Sörlin, S. and N. Wormbs (2018) 'Environing Technologies: A Theory of Making Environment', *History & Technology* 34(2): 101–125.
Sörlin, S. and G. Wynn (2016) 'Fire and Ice in the Academy: The Rise of the Integrative Humanities', *Literary Review of Canada* 24(6): 14–15.
South China Morning Post (2020) 'China Planning to Launch Satellite to Monitor Arctic Shipping Routes', *South China Morning Post*, 10 December. Available at: www.scmp.com/news/china/diplomacy/article/3113376/china-planning-launch-satellite-monitor-arctic-shipping-routes (accessed 21 March 2022).
Spufford, F. (1996) *I May Be Some Time: Ice and the English Imagination*. London: Faber and Faber.
Steffen, W., K. Richardson, J. Rockström *et al.* (2015) 'Planetary Boundaries: Guiding Human Development on a Changing Planet', *Science* 347(6223): 736–746.
Steinberg, P. (2001) *The Social Construction of Oceans*. Cambridge: Cambridge University Press.
Taillant, J. D. (2015) *Glaciers: The Politics of Ice*. New York: Oxford University Press.

Ternes, B. (2020) 'Exploiting Shangri-La: Assessing the Tibetan Plateau's Natural Resources and the Work of Karl Wittfogel', *Review of European Studies* 12(1): 1–11.

Thomas, W. L. Jr., with the collaboration of Carl O. Sauer, Marston Bates, and Lewis Mumford, eds (1956) *Man's Role in Changing the Face of the Earth*. Chicago, IL: University of Chicago Press.

Thrush, C. (2014) 'The Iceberg and the Cathedral: Encounter, Entanglement, and Isuma in Inuit London', *Journal of British Studies* 53(1): 59–79.

Todd, Z. (2016) 'An Indigenous Feminist's Take on the Ontological Turn: 'Ontology' is Just Another Word for Colonialism', *Journal of Historical Sociology* 29(1): 4–22.

Tsing, A. (2005) *Friction: An Ethnography of Global Connection*. Princeton, NJ: Princeton University Press.

Tyndall, J. (1860) *The Science of Glaciers*. London: Murray.

Wadham, J. (2021) *Ice Rivers*. London: Allen Lane.

Wadhams, P. (2017) *A Farewell to Ice: A Report from the Arctic*. Oxford: Oxford University Press.

Wang, M. and J. Overland (2009) 'A Sea Ice Free Summer Arctic Within 30 years?', *Geophysical Research Letters*, 3 April. Available at: https://agupubs.onlinelibrary.wiley.com/doi/full/10.1029/2009GL037820 (accessed 21 March 2022).

Warde, P., L. Robin, and S. Sörlin (2018) *The Environment: A History of the Idea*. Baltimore, MD: Johns Hopkins University Press.

Watt-Cloutier, S. (2018) *The Right to Be Cold*. Minneapolis, MN: University of Minnesota Press.

Weart, S. (2008) *The Discovery of Global Warming*. Cambridge, MA: Harvard University Press.

White, S. (2017) *A Cold Welcome: The Little Ice Age and Europe's Encounter with North America*. Cambridge, MA: Harvard University Press.

White, S., C. Pfister, and F. Mauelshagen, eds (2018) *The Palgrave Handbook of Climate History*. London: Palgrave Macmillan.

Whyte, K. (2020) 'Too Late for Indigenous Climate Justice: Ecological and Relational Tipping Point', *WIREs Climate Change* 11: e603.

Wittfogel, K. (1957) *Oriental Despotism: A Comparative Study of Total Power*. New Haven, CT: Yale University Press.

Wolf, E. (1982) *Europe and the People without History*. Berkeley, CA: University of California Press.

Part I

Living with ice

1

Writing on sea ice: early modern Icelandic scholars

Astrid E. J. Ogilvie

Iceland, sea ice, and scholars

> The sea ice ... has been the most important causal factor in the dearth-years, price-rises and famines, and has done more harm to the Icelandic population than all the volcanic eruptions and earthquakes.[1] (Thoroddsen, 1914: 205)

This chapter is concerned with the sea ice that reached the coasts of Iceland in times past and the remarkable Icelandic scholars who 'wrote on sea ice' during the late sixteenth and early seventeenth centuries: an early form of 'ice humanities'. Although Iceland also has its fair share of terrestrial ice, there are good reasons to focus on the marine type. Sea ice impacts landscapes and life on land, and it holds a central place in the Icelandic psyche, even today. This account follows the writings of four scholars: Arngrímur Jónsson (1568–1648); Oddur Einarsson (1559–1630); and Oddur's son, Gísli Oddsson (1593–1638), who all wrote in Latin, and Björn Jónsson (1574–1655) who wrote in the vernacular Icelandic.[2]

Sea ice is far more than just a fascinating phenomenon. It is one of the most important and variable components of the planetary surface and is the key to understanding many basic questions concerning the energy balance of the globe (Ogilvie, 2017; Miles *et al.*, 2020). In order to put climatic change into context it is necessary to use long climate proxy records, including records drawn from historical, documentary evidence. This can provide detailed and precise information on climatic variations for the past several centuries (Ogilvie and Jónsson, 2001; Ogilvie, 2005, 2010, 2020; Nicholls, 2010; Hartman *et al.*, 2017; Pfister *et al.*, 2018). Iceland is unusually rich in such evidence and has a unique geographical location. Created by geological activity some 18 million years ago, and still experiencing the ongoing effects of its volcanic birth, it is appropriately known as a land of fire. It is also a land of ice, both with regard to its glaciers (now melting fast) but also because of the sea ice that reached its coasts at frequent intervals on currents from

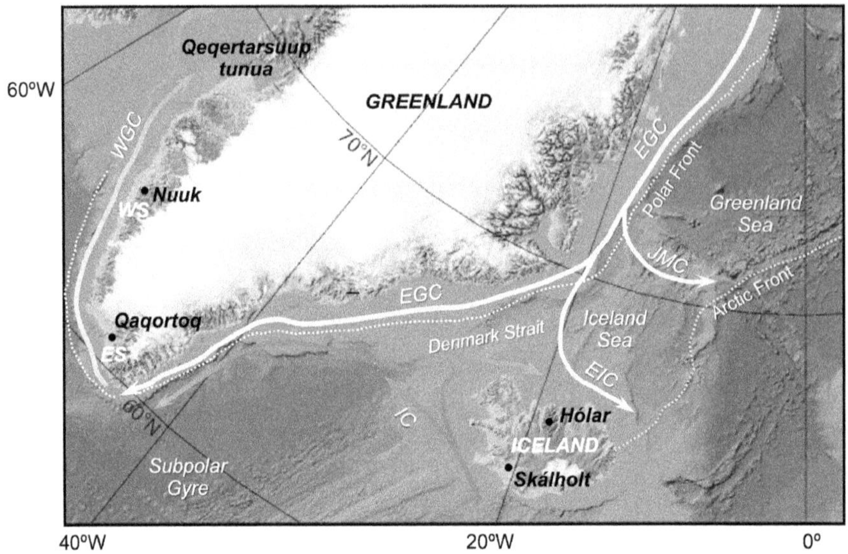

Figure 1.1 Iceland, Greenland, and the North Atlantic Ocean (bathymetric and topographic map of region based on Jakobsson *et al.*, 2012). Major surface currents are indicated: East Greenland Current (EGC), Jan Mayen Current (JMC), East Icelandic Current (EIC), Irminger Current (IC), and West Greenland Current (WGC). The polar front indicates the modern mean limit of polar waters and sea ice of Arctic Ocean origin. The map of Greenland shows modern locations that are close to where the Norse Greenland settlements were located. These are: Qaqortoq (formerly Julianehåb) in the region of the Eastern Settlement (ES); the capital Nuuk (formerly Godthåb) close to the area of the Western Settlement (WS); and Qeqertarsuup tunua (Disko Bay) in the region of the northern hunting grounds of the Norse.

Greenland until the very recent past (Thórarinsson, 1956). Iceland's very name is a powerful symbol of the country's special relationship with ice.[3]

It has been very rare that sea ice has reached the shores of Iceland in recent times. However, at certain times in the past, sea ice was a frequent visitor. It could bring benefits in the form of driftwood and also whales and other marine mammals that were trapped close to shore by the ice and that augmented peoples' food supply.[4] However, for the most part, it brought extreme hardship in its wake. This was associated with three main factors. One was that the presence of the ice would prevent people from fishing – fish being a staple food source – both for home consumption and for export. Another was that trading vessels were prevented from landing, and thus another food supply was cut off. The third and most serious effect was that the presence of the ice had the effect of lowering temperatures on land. This adversely affected the all-important grass crop. As Iceland was marginal for agriculture, for the most part the only crop that flourished was composed of native grasses. This provided summer grazing for the livestock and winter fodder in the form of hay. The cold temperatures that accompanied the sea ice could damage the grass, especially when the ice came during the growing season in the spring and early summer.

Sea ice most frequently visited the north, the northwest and the east coasts. The south of Iceland was only affected in extreme ice years. The ice generally did not reach the west of Iceland due to the milder ocean currents in that region. Sea ice visited most frequently in the winter and spring. Although economic and political factors played their part, famines and dearths that accompanied sea-ice years occurred many times throughout Iceland's history (Ogilvie, 1982 *et seq.*) Such events were described in a wide variety of historical writings, and also in literary form, including poetry. Many poems were written about sea ice, the most famous one in Iceland being *Hafísinn* or 'The Sea Ice' where the ice is described as *landsins forni fjandi* 'the country's ancient enemy'. This was written in 1888 by the Icelandic poet and playwright Matthías Jochumsson (1835–1920) after a series of severe years with much sea ice (Jochumsson, 1988; Ogilvie, 2015; Ogilvie *et al.*, 2021). Several of the scholars discussed here also wrote poems regarding the weather and ice.[5] However, it is their prose descriptions of sea ice that we are concerned with here.

Initially, a brief discussion will offer further context for the writings of these sea-ice scholars. Iceland was first settled in the late ninth century as part of the Viking Age expansion across the North Atlantic, with the early population coming primarily from Norway and the British Isles.[6] Christianity was accepted in the year 1000, and learning and scholarship became closely associated with the Church which brought with it writing skills. Thus, for example, most of the magnificent Icelandic sagas that were composed in the thirteenth and fourteenth centuries were written down in monasteries and related scriptoria. Schools and centres of learning were located in particular at the two bishoprics: Skálholt in the south (established in 1056) and Hólar in the north (established in 1106). These were first intended for the education of the clergy, but they soon began to educate secular leaders as well (Júlíusson, 2002). What the chapter reveals through these explorations of scholarly writings is a recognition that sea ice was framed as both an elemental blessing in the form of access to marine mammals (food) but also a harbinger of extreme weather and hazards. Living well in Iceland required fortitude and resilience.

The sea-ice writing catalysts: Icelandic bishops and Danish scholars

Although this was doubtless not his primary intention, a key catalyst in the production of early descriptions of sea ice was Guðbrandur Þorláksson (1542–1627). He became Bishop of the northern see of Hólar in 1571 and held this office until his death. Guðbrandur was also one of Iceland's earliest cartographers. Some time before 1575, he constructed a celestial globe accommodated to the latitude of Iceland. Although Guðbrandur's name does not appear on the maps there is little doubt that the maps of Iceland published in Ortelius's Atlas in 1590 and Mercator's Atlas in 1595 were taken from originals, possibly just one, drawn by him (Hermannsson, 1926; Sigurðsson, 1971, 1978). He played a critical part both in Iceland's intellectual development and in the life and writings of Arngrímur Jónsson. Guðbrandur was also well acquainted with Oddur Einarsson.

The Ortelius map (Figure 1.2) is embellished with a variety of sea monsters and sea ice is shown on the east coast of Iceland. These additions were almost certainly added by Ortelius, who never visited Iceland, and not by

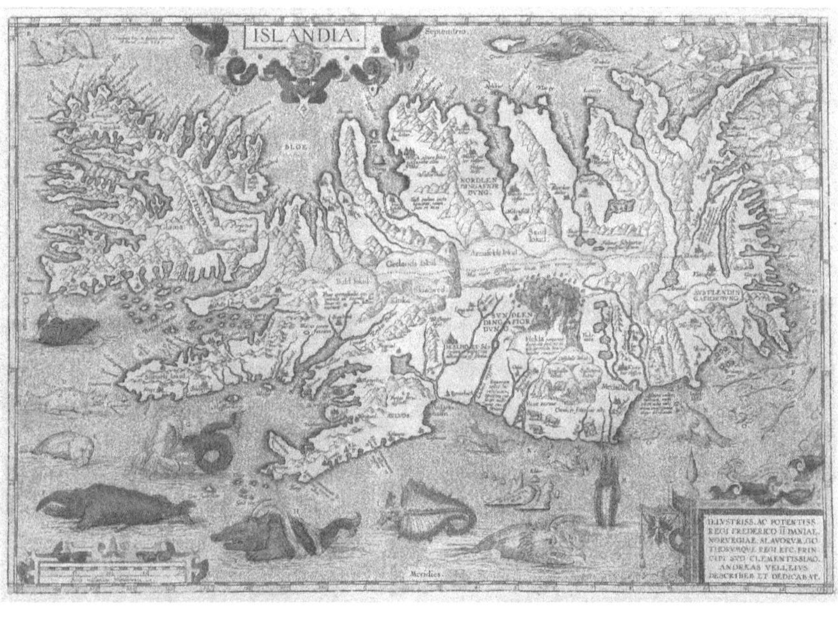

Figure 1.2 The Ortelius map of Iceland (*c*.1590 original version). This map was almost certainly based on one drawn by Bishop Guðbrandur Þorláksson. It is known that sometime before 1575 Guðbrandur had constructed a celestial globe accommodated to the latitude of Iceland (Hermannsson, 1926).

Guðbrandur. If he had chosen to show sea ice he would probably have placed it in the north and northeast where it occurred more frequently. Although Björn Jónsson, the author of *Skarðsárannáll*, did not enjoy the support of Bishop Guðbrandur, he was helped and encouraged by Guðbrandur's grandson, another bishop, Þorlákur Skúlason (1597–1656) who was Bishop of Hólar from 1628 until his death, and also by Brynjólfur Sveinsson (1605–75) who was Bishop of Skálholt from 1639 to 1674.

Following the scholastic tradition of the day, Arngrímur Jónsson, Oddur Einarsson, and Gísli Oddsson all spent time at the University of Copenhagen.[7] During the years that Arngrímur was there (1585–89; 1592–93; 1602–03) he got to know several of the leading Danish historians and intellectuals (as described by Benediktsson, 1957a) who encouraged Arngrímur in his intellectual endeavours. However, their interactions were of mutual benefit. As far as mainland Scandinavia was concerned, most of the knowledge regarding Iceland's great literary and historical heritage had been lost. These historians were thus keen to learn what information might be available, not least to supplement their own historical writings. There was particular interest in what had happened to the Norse settlements in Greenland, and in this regard expeditions to Greenland were sent from Denmark in 1605, 1606, and 1607. In response to the enquiries of the historians, Arngrímur wrote the work on the history of Greenland, *Gronlandia*.[8]

Arngrímur Jónsson *lærði* (the Learned) (1568–1648): *Brevis commentarius de Islandia* ('A Brief Commentary on Iceland')

Arngrímur Jónsson was born at a farm named Auðunarstaðir in Víðidal in the north of Iceland but, at the age of eight, went to live at Hólar with his close relative, Bishop Guðbrandur Þorláksson. Guðbrandur made himself responsible for Arngrímur's education and played an important role in Arngrímur's life. As noted, Arngrímur's first period of study at the University of Copenhagen was during 1585–89. Subsequently he held a number of positions as a minister in northern Iceland, but was also closely associated with the school at Hólar, serving variously as headmaster (*rektor*) and as Guðbrandur's assistant. In 1592 he travelled again to Copenhagen, remaining until 1593. The prime reason for this was an errand for Bishop Guðbrandur, but it marked the beginning of his close acquaintanship with the Danish historians noted above, and the publication of his first work, the *Brevis commentarius de Islandia* ('A Brief Commentary on Iceland'). He cemented his collaboration with the Danish scholars on a third visit to Copenhagen in 1602–03. While still relatively young, Arngrímur came to be known as Arngrímur *lærði* or 'Arngrímur the Learned' in recognition of his scholarly talents. Certainly his writings were prodigious. As well as the *Brevius Commentarius* he wrote another description of Iceland entitled *Crymogea* (a Latinized name for Iceland). The only reference to sea ice in this work concerns an account of one of the first would-be settlers to Iceland, Flóki Vilgerðarson, who was said to have given Iceland its name for the sea ice. Arngrímur's scholarly output includes works of a religious nature and translations of some of the Icelandic sagas (Benediktsson, 1957a).

The *Brevis commentarius* may be seen in the context of the intellectual activity in Europe that accompanied the Reformation and the Renaissance, including the production of a number of early geographical accounts. It is unlikely that the foreign writers who produced them had ever been anywhere near Iceland, and much of what they wrote owed its origins partly to their imaginations and partly to earlier works which were equally uninformed. In consequence, these works were full of misconceptions. Regarding Iceland as lying far beyond the boundaries of the civilized world they propagated the idea that hell itself was to be found in Iceland, in Mount Hekla, or in the sea ice lying off the coasts. The poverty, both material and spiritual, of the Icelanders was also emphasized. Arngrímur specifically mentions the writers Munster (whose work from 1544 survives in an edition of 1598) and Frisius (1582).

Guðbrandur Þorláksson felt that a written refutation of such scurrilous works was necessary. After his appointment as Bishop of Hólar, the time he could devote to scholarly pursuits was cut down by his official duties (and poor health). Rather than attempting a refutation himself he therefore looked for a younger scholar who might undertake this task. His choice eventually fell on his relative, Arngrímur Jónsson, and by the spring of 1592, Arngrímur had prepared the manuscript of his *Brevis commentarius*. He brought the manuscript with him to Copenhagen and it was published there in 1593. The references to medieval manuscripts contained in it were of great importance for northern history and literature, and gave scholars in Scandinavia, especially those whom Arngrímur met in Denmark, the first hints that such material existed at all. The *Brevis commentarius* was considered sufficiently important at the time of its writing that an English translation was included in Hakluyt's *The Principal Navigations* 1 (Hakluyt and Goldsmid, 1598). Arngrímur's works are published in Benediktsson (1950, 1957b), in the original Latin. His description of sea ice includes the following:

> Regarding the notion that sea ice is always land-fast to Iceland or, as Munster asserts shortly after, that it is fast for eight consecutive months; neither of them are true. For the most part the ice melts in April or May and is driven towards the west. It does not then return before January or February, and very often even later. It would be possible to count up many years in which

this ice, the harsh scourge of our nation, has not been seen at all around Iceland. This was the case in this year, 1592. Thus, we can see the truth of what Frisius wrote: 'navigation to this island is only possible for four months because the ice and cold close up the passage' when English ships come to us every year, sometimes in March, sometimes in April and some in May. German and Danish ships usually come in May and June and some do not leave again until August. Last year, 1591, a certain German ship laden with copper stayed in the port of Vopnafjord in Iceland for about fourteen days in November after which it left without difficulty. Thus, since the ice of Iceland is neither perpetual, nor remains for eight months, Munster and Frisius can be seen to be wrong.[9]

What is interesting in this passage is that it seems clear that, despite Arngrímur's assertion that ice was not present off the coast of Iceland for seven to eight months in a year, it is evident that, around the time when he was writing, sea ice was a frequent and unwelcome visitor to Iceland. There is no question of the ice *not* being present. He says, for example, that 'for the most part the ice melts in April or May and is driven towards the west. It does not then return before January or February and very often even later.' The ice clearly came 'very often' to Iceland around this time (the late sixteenth century). Arngrímur is also aware of the variability of the presence and absence of the ice off Iceland, and states that the ice does not appear *every year* and, furthermore, that this was the case during the year in which he wrote *Brevis commentarius*: 1592. The fact that he specifically mentions this as an ice-free year suggests that this was the exception rather than the rule during his lifetime and up to the time of writing.[10]

Oddur Einarsson (1559–1630): *Qualiscunque descriptio Islandi* ('A Draft Description of Iceland')

Oddur Einarsson was born at Möðruvellir in Hörgárdalur, not far from the northern capital of Akureyri. He attended the school at Hólar and then the University of Copenhagen from 1583 to 1585. He was particularly interested in mathematics and astronomy and studied at that time with Tycho Brahe (1546–1601) the nobleman astronomer who undertook meteorological observations on the island of Hven in the sound between Denmark and Sweden during the years 1582–97 (Oddsson, 1942, p. 55). Subsequently, in 1586, Oddur became headmaster (*skólameistari*) at the Hólar school. In 1588 a new bishop was to be elected to the Skálholt see. Bishop Guðbrandur was unable to attend, but sent letters recommending Oddur (although he was not specifically named) stating, among other things, that it was necessary to have as accomplished a scholar as possible in the office (Benediktsson,

1957a: 33). Oddur travelled again to Copenhagen in the autumn of 1588, returning in the spring of 1589 to Iceland.

Clearly, Guðbrandur had a very high opinion of Oddur. His contemporaries also regarded him as a gifted scholar and it is known that he produced a variety of works, in particular records of the Reformation period. It would appear that it was Oddur whom Guðbrandur first had in mind when he was seeking a young scholar to produce a refutation of the foreign accounts of Iceland, and it is unclear why Oddur did not do this (Hermannsson, 1917). Certainly it must have been his intention to write something along these lines. According to Benediktsson, 'it is almost certain that the anonymous work, *Qualiscunque descriptio Islandiæ*, was drafted by him in Copenhagen in the winter of 1588–89 but not finished' (Benediktsson, 1957a: 33). Oddur was consecrated as Bishop of Skálholt in 1589, and remained in office until his death in 1630. Arngrímur does not appear to have known Oddur's work (Benediktsson, 1957a: 46). While Arngrímur's *Brevis commentarius* was published and enjoyed a certain measure of publicity, Oddur's work vanished into obscurity and was not found until the early twentieth century in the National Library in Hamburg. It was then edited by Fritz Burg (1928) and ascribed by Burg to Sigurður Stefánsson (*c.*1570–94).[11] It fell to the Icelandic scholar Jakob Benediktsson (1907–99) to deduce that the author of the *Qualiscunque* was Oddur Einarsson (Benediktsson, 1956). As an early geographical work the *Qualiscunque* is outstanding for its detail and accuracy, and is an excellent description of Iceland, both with regard to its people and the country itself. Why Oddur never sought to have it published in his lifetime remains an enigma. Also, given the small number of Icelandic scholars at home and abroad in Denmark around the time in question, it also seems strange that Arngrímur does not seem to have been aware of the work that was written some four or so years prior to his.

Given that Oddur's work was not published, and subsequently lost until the early twentieth century, it is ironic that it has been said of it that it is 'the foremost amongst all the learned writings on Iceland and the Icelanders'.[12] In addition to giving accounts of the Viking Age, the discovery of Iceland, Greenland and Vínland (North America), and information on such diverse subjects as birds, fish, food, and amusements in Iceland and the Icelandic language, Oddur discusses Iceland's geographical position and climate, as well as natural phenomena such as volcanic eruptions and sea ice.

> If it so happens that someone drifts north past the island (Iceland) he will see before him the blue-green ice of immense breadth and horrible extent which reaches out over all the oceans surrounding the northern coasts of Greenland. And some people think that it was the proximity of the ice that caused those who first discovered Iceland to give it the name they did, either because they first encountered it near Iceland off the coast or because great quantities of it

reached the shores of the land itself. The Icelanders who have settled on the northern coasts are never safe from this most terrible visitor. The ice is always to be found between Iceland and Greenland although sometimes it is absent from the shore of Iceland for many years at a time, by the grace of God. Certainly, those who live in these regions suffer from great hardship caused by its presence when it has been evident for long periods of the time. In particular, on account of the barrenness of the fields which it caused. This occurs because the vitality goes out of the Earth and the sap which gives fertility is wasted as soon as the ice has become land-fast and the damaging cold has touched the fields. This island could not be inhabited by men for long if such an unwelcome guest came to trouble it every year, but divine providence assures us that this ice is held within certain bounds so that it does not approach the island of Iceland, except when God has decided to punish our people, and this occurs at varying intervals. Sometimes it is scarcely to be seen for a whole decade or longer, sometimes it comes after five years' absence, sometimes it occurs almost every other year, sometimes twice or three times in the same year. It moves with such speed and strength ... and although on one day the ice can scarcely be seen even from the high mountains, it has often occurred that on the next it fills up all harbours and fjords to such an extent that to observers it seems as if all the ice in the sea has suddenly arrived at the coast of Iceland ... It makes a great deal of difference at what time of year the ice comes. In the autumn and at the time of the winter solstice when the frost has already got into the ground and there is snow cover, its presence does less damage but, during the spring and summer when the weather is becoming milder, the ice invariably brings disaster with it, because that is when it has the greatest power and the grass is most adversely affected. The northerners are therefore far worse off than the southerners who never see this ice as, whenever it starts to drift to east or west, it is swept back by the strong currents of the great sea.[13]

Oddur, like Arngrímur, is aware of the variability of the ice, but he is even more informative. It is extremely interesting that he knew, for example, not just that the ice has a damaging effect, but that it makes a difference at what time of year the ice comes, and that its presence is particularly disastrous during the spring and summer. That this is the case is well-known from modern research (see, e.g., Friðriksson, 1969). It is also clear that Oddur had knowledge of the existence and effects of ocean currents.

It is noteworthy that Oddur says that the ice never reaches the south coast. The ice did occasionally reach the south when it drifted around from the east coast, but only very rarely. When considering the period up to the late nineteenth century, this happened perhaps once or twice a century. For the period discussed here this occurred in 1605 and 1615.[14] However, this was some time after Oddur wrote his work *Qualiscunque*, which was almost certainly drafted during the winter of 1588–89 as noted earlier. Given that he would have experienced the sea-ice events of 1605 and 1615 he would clearly have changed his mind in later life.[15]

Gísli Oddsson (1593–1638): *Annalium in Islandia farrago* ('Annal Collection for Iceland'); *De mirabilibus Islandiae* ('The Marvels of Iceland')

Gísli Oddsson was the son of Oddur Einarsson. He followed a similar path to his father, in that he attended the school at Skálholt and then the University of Copenhagen (the latter from 1613 to 1616). After that he became a minister at Skálholt and in 1621 headmaster or rector (*rektor*) of the Skálholt school. In 1622 he was appointed minister at Holt by Eyjafjöll in the southeast of Iceland. Following his father's death in office in 1630, he was ordained Bishop of Skálholt in 1632. Oddur himself died not long afterwards, in 1638, at the age of 45. It is clear that he inherited his father's love of learning, if not his aptitude for mathematics. An interesting aside is that he left an unfinished translation (from Latin into Icelandic) of Arngrímur Jónsson's *Crymogea* (Hermansson, 1917: iii). Gísli produced two works written in Latin (published in Hermannsson, 1917). Icelandic translations are available by Jónas Rafnar (see Oddsson, 1942). Gísli's works, written at Skálholt, are *Annalium in Islandia farrago* ('Annal collection for Iceland') finished in 1637 and *De mirabilibus Islandia* ('The Marvels of Iceland') written in 1638, and completed shortly before his death. As the title of the *Annalium* suggests, this work takes an annalistic form with yearly entries from 1106 to 1637. The early part to 1402 has been copied from other works. The rest is compiled from contemporary sources, now lost, and the later accounts are from Gísli's own observations (Hermansson, 1917: vii). The *Annalium* corroborates other accounts, notably *Skarðsárannáll*, regarding the severity of the early seventeenth century. The *De mirabilibus* has parallels with both the *Qualiscunque descriptio Islandia* and *Brevis commentarius* in that it devotes its forty short chapters to different topics relating primarily to the natural history (including birds, fish, monsters, and wild and domestic animals, geology) and general life in Iceland. The second chapter concerns sea ice. The most relevant parts are translated below.

> Every time the sea ice reaches our shores, especially to the north, albeit rarely to the greater part of the country, it brings the greatest calamity. However, this does not occur every year, rather at certain intervals and single years, decided by Almighty God. The calamity causes dire effects to the health equally of human beings, livestock, and all animals that live here and leads directly to dearth. Except when it sometimes brings with it great quantities of fish, seals, whales and these kinds of creature, just as tithes were collected,[16] but

> it seldom benefits the entire country.[17] In addition to this it is not unusual to meet polar bears and falcons that are brought with it and are left behind by chance, when the ice leaves. The ice is perhaps most noteworthy because of an amazing wonder that is often brought along with it. This is a great quantity of driftwood. From this dividend almost all houses in this country are very well maintained and if we did not have this, our settlements would perish completely. The trees are of varying kinds: spruce, beech, elm, ash and other kinds of tree which do not need to be mentioned here. While some of these trees are slender and thin, others, on the other hand, have such big trunks that they have reached over 48 ells[18] long, as I recall, with corresponding width around. People are not sure where this timber comes from … .[19]

Here Gísli emphasizes the great variability of the visits of sea ice to Iceland as well as the mainly negative, but also positive impacts, that have been noted in the accounts of Arngrímur and Oddur. It is interesting to speculate whether Gísli had the *Qualiscunque descriptio Islandiae* to hand. One would assume so, but although his work is very similar in that it aims to be a description of Iceland, it is original and does not show any signs of copying from that work. Although it has to be said that, of the two, Oddur's is the more polished account, Gísli does add information not to be found in his father's work, such as the comments on driftwood. Perhaps it was his intention to augment the earlier work.

Björn Jónsson (1574–1655): *Skarðsárannáll* (the Annals of Skarðsá)

The medieval Icelandic annals form an invaluable contribution to Icelandic historiography; however, they come to an end in 1430 (Ogilvie, 1991).[20] In the early seventeenth century the resumption of annal-writing was pioneered by Björn Jónsson (1574–1655). Subsequently, around forty such annals were produced. These are published in the collection *Annálar 1400–1800* (1922–27). It appears that Björn was encouraged to begin his annal-writing task by Bishop Þorlákur Skúlason of Hólar. Björn grew up at Reynistaður in the county of Skagafjord in northern Iceland, a place known for scholarship, and it is likely that he was introduced to literary and Latin studies there.[21] It would have been natural for him to attend the school at Hólar, but for a variety of reasons it seems Björn did not find favour with Bishop Guðbrandur (Þorkelsson, 1887: 42). This was unfortunate for Björn who, although he did not have a formal education like Arngrímur *lærði*, Oddur Einarsson, and Gísli Oddsson, he was clearly attracted to scholarship from an early age. After Þorlákur became Bishop in 1628, Björn's situation changed for the better as he then had access to the Hólar archives.

The later annals were generally named according to the place where they were composed, and thus the annal that Björn compiled is known as *Skarðsárannáll* from Skarðsá in the Skagafjord district in the north, where he lived from 1605 until his death. *Skarðsárannáll* covers the period 1400 to 1640 but, as most of the work on the annal was done during the years 1636 to 1639, its contemporary span is short. Nevertheless, there is evidence to suggest that Björn's work is reliable before 1636, at least from *c.*1580. His work must have been based partly on oral tradition, but he also had many documents at his disposal after he had access to these at Hólar, and he often worked there for long periods (Þorkelsson, 1887). Björn did not write specific narratives on the nature of sea ice but wrote accounts of sea-ice events for the years 1602 to 1639.[22] The example below is for the extreme ice year of 1615 when the ice reached the east and south coasts.

> A severe winter with much lack of pasture. All over Iceland, nearly all outdoor livestock died that did not have hay. Ice drifted in to the north at *þorri*.[23] Then the sea ice surrounded all of Iceland and stayed at the north coast until the moving days.[24] Ice penetrated to the southeast to Grindavík. It reached Reykjanessröst and Vog and all Suðurnes.[25] No one remembered that the ice had reached as far as Reykjanessröst before. There was seal-hunting on the ice off Suðurnes ... There were widespread shipwrecks in the ice ... There were many polar bears in the country but they did not do any harm ... and were killed in the east, south and the north ...[26]

Here Björn emphasizes both the negative and positive aspects of the ice. The cold weather that accompanied the ice caused lack of pasture which in turn caused livestock to die, shipwrecks occurred in the ice, and polar bears were brought on land, transported by the ice. However, there was also 'seal-hunting' on the ice which would have augmented food supplies. A summary of the descriptions of the occurrence of sea ice for the period *c.*1580 to 1639, with information coming primarily from *Skarðsárannáll*, is given in Table 1.1.

Conclusions

The writings of the four 'sea-ice' scholars discussed here form a very interesting chapter in sea-ice historiography. Their dedication to learning and scholarship is striking. Of the four, it is Arngrímur Jónsson who undoubtedly contributed most to general Icelandic scholarship, but with regard to what may be called 'ice scholarship' he has more than an equal in Oddur Einarsson. Oddur's son, Gísli Oddsson, perhaps did not have quite the same impact in scholarly terms, but is well worthy of inclusion here, not least because of his brief but highly interesting accounts of 'the marvels of Iceland' that included sea

Table 1.1 Sea ice off Iceland's coasts *c*.1580–1639

Year	Comments
1580s	It seems likely that there was much sea ice off Iceland in the 1580s (QD, BC).
1592	No sea ice this year (BC).
1590s	Much sea ice in the early 1590s (QD, BC).
1602	Severe sea-ice year and ice until far into summer. Unusually severe year (*Skarðsárannáll*; *Annalium farrago*).
1604	Sea ice brought whales. Much seal hunting (*Skarðsárannáll*).
1605	Severe sea-ice year. Ice reached the south coast (*Skarðsárannáll*).
1608	Sea ice until summer (*Skarðsárannáll*).
1612	Sea ice present (*Skarðsárannáll*).
1615	Very severe sea-ice year. Stayed to summer and reached southwest coasts.
	Many bears and seals on the ice (*Skarðsárannáll*; *Annalium farrago*).
1618	Sea ice came in the spring (*Skarðsárannáll*).
1622	Sea ice present. Much seal hunting (*Skarðsárannáll*).
1624	Sea ice came to the north in the spring (*Skarðsárannáll*).
1625	Severe sea-ice year. Stayed to beginning July. Beached whales (*Skarðsárannáll*).
1626	Sea ice. Seal hunting on the ice (*Skarðsárannáll*).
1628	Severe sea-ice year. Stayed to around mid-August (*Skarðsárannáll*).
1629	Some ice to the north. Did not stay. Seals caught. Polar bear sighted (*Skarðsárannáll*).
1633	Severe sea-ice year with ice from ca. February to late June (*Skarðsárannáll*).
1636	Some ice. Mentioned for Skagafjord. Seal catching (*Skarðsárannáll*).
1638	Sea ice in the north in autumn. Useless (*Skarðsárannáll*).
1639	Severe sea-ice year. Came to the south (*Skarðsárannáll*).

ice. In some regards, Björn Jónsson has conveyed the most useful information in that he provided annual accounts of sea ice from 1602 to 1639. The purpose of this chapter was to focus on these men and not to provide an analysis of the climate and sea-ice variations off the coasts of Iceland during the period in question. That has been done elsewhere (Ogilvie, 1982 *et seq.*). However, it seems appropriate nevertheless to emphasize that, during the period when our sea-ice scholars were writing, there was frequent and extensive sea ice reaching the coasts of Iceland. According to the sources it seems likely that there was much sea ice during the 1580s and the early 1590s. Particularly severe years with long-lasting sea ice occurred in 1602, 1604, 1605, 1608, 1612, 1615, 1618, 1622, 1624, 1625, 1626, 1628, 1629,

1633, 1636, 1638, and 1639. Clearly some years were more severe than others, but there can be little doubt that this period was one of very high sea-ice occurrences, see Table 1.1.

The table shows a summary of information regarding sea ice covering the general period of the 1580s and 1590s from the *Qualiscunque descriptio Islandiæ* by Oddur Einarsson (QD) and *Brevis Commentarius de Islandia* by Arngrímur Jónsson (BC). From 1602 onwards the summaries are mainly from *Skarðsárannáll* by Björn Jónsson, with two from *Annalium farrago* by Gísli Oddsson.

What is known regarding the variability in the incidence of sea ice from recent analyses (Miles *et al.*, 2014, 2020) accords well with the statements by our scholars, who are also in agreement on this topic: 'It would be possible to count up many years in which this ice, the harsh scourge of our nation, has not been seen at all around Iceland. This was the case in this year, 1592' (Benediktsson, 1950: 16); 'The ice is always to be found between Iceland and Greenland although sometimes it is absent from the shore of Iceland for many years at a time …' (Einarsson, 1971: 34–35); 'Every time the sea ice reaches our shores … it brings the greatest calamity. However, this does not occur every year, rather at certain intervals and single years …' (Oddsson, 1942: 5).

Although all these scholars focus on the detrimental aspects of the ice, Gísli Oddsson is unique in that he also places much emphasis on its positive benefits, primarily in the form of the marine mammals that supplemented the food supply, and driftwood that the ice brought. Björn Jónsson has a different approach in his writings as he does not directly comment but just soberly describes the presence of the ice and its effects, but he adds the detail that polar bears would sometimes accompany the ice and notes the fact that seal hunting took place on the ice (in 1604, 1615, 1622, 1629, and 1636). In his annalistic work, *Annalium farrago*, Gísli Oddsson also comments on the seal hunting and polar bears for the year 1615 (see Table 1.1).

Though much is known regarding sea-ice variations in recent times, in particular since the beginning of the era of satellite data in the 1970s, clearly far less is known regarding variations in the past. The careful observers of sea ice, the 'sea-ice scholars' discussed here, wrote what they did in large degree purely for the sake of scholarship and learning. Modern scholars reap the benefits of their work and erudition. It is clear that the scholars were well aware of the dual nature of the sea ice both as destroyer and as the bringer of certain benefits, such as sustenance in the form of marine mammals and the 'amazing wonder' that was the driftwood that it brought with it. However, that the ice was seen primarily as a 'calamity', a mighty force that brought with it cold, mists, and death, reminiscent of the underworld as conceived in Norse mythology, cannot be in doubt. But was there more to the sea-ice story than this? Did the ice (and indeed the fire, in the form of frequent and often devastating volcanic eruptions) inure Icelanders to a life of living with environmental change? Did the sea ice in fact give an unusual gift – that of resilience in the face of the unpredictability of the natural world?

Notes

The work in this chapter is supported by, and contributes to, the NordForsk-funded Nordic Centre of Excellence project (Award 766654) Arctic Climate Predictions: Pathways to Resilient, Sustainable Societies (ARCPATH). Support from the National Science Foundation (Award 2127861) Synthesizing Historical Sea-Ice Records to Constrain and Understand Great Sea-Ice Anomalies (ICEHIST) is also acknowledged. I also received a grant from the Icelandic Centre for Research (RANNÍS) *Rannsóknarsjóður Sigrúnar Á. Sigurðardóttur og Haralds Sigurðssonar* Award 170191–4001. *Upphaf kortagerðar og saga íslenskrar landfræði* ('The Foundations of Map-Making and Geography in Iceland' also gave inspiration for this chapter. I am grateful to Viðar Hreinsson for assistance in the translation of the account on sea ice from Gísli Oddsson's work. He also procured for me the work by Jakob Benediktsson on Arngímur Jónsson. I am also very grateful to Árni Daníel Júlíusson for providing me with his article of 2002 on *Icelandic Sagas Around 1600* that contains much relevant background information for this chapter (see references). My colleague Martin Miles very kindly provided Figure 1.1 showing North Atlantic locations and ocean currents. I am grateful to my 'sea-ice colleagues' Ingibjörg Jónsdóttir, Trausti Jónsson, Þór Jakobsson, and Páll Bergþórsson for inspiring collaboration over many years.

1 Translated by the author from the original Norwegian.
2 As regards the actual writing, all these scholars would have written their accounts primarily, if not entirely, on vellum. Although paper appears to have been introduced to Iceland as early as 1423, it continued to remain scarce for a considerable time. See https://papertrailsiceland.wixsite.com/papertrails/post/the-oldest-icelandic-paper-documents-and-manuscripts).
3 Iceland is said to have been given its name by an early would-be settler, Flóki Vilgerðarson, who visited Iceland in *c.*865. He and his companions spent the summer fishing and neglected to gather enough hay to feed the livestock over the winter. His companions subsequently died, and, in his vexation, Flóki left, but not before deciding to call this new land 'Iceland' from the sea ice he saw in the fjords. See the early work *Landnámabók* ('The Book of Settlements'), edited by Jakob Benediktsson, 1968, pp. 36–39.
4 To this day, the Icelandic expression for a 'windfall' is *hvalreki*, meaning a whale that has been stranded on shore.

5 A notable example is Björn Jónsson who wrote, among other things, a poem about the severe ice winter in 1625 entitled 'Svellavetur' ('Winter of frozen ground'). See Jónsson, 1913.

6 For recent work on this topic, see e.g., Ebenesersdóttir *et al.*, 2018 and Plomp *et al.*, 2021.

7 Iceland's first university, the University of Iceland in Reykjavík, was not established until 1911. Until that time, and even for some time after, for higher learning Icelanders usually attended the University of Copenhagen (founded in 1479). Björn Jónsson was a self-educated scholar, and did not attend the university.

8 Among other things, *Gronlandia* contains the so-called *Vínland Sagas*, the *Saga of the Greenlanders*, and the *Saga of Eirík the Red*, that describe the settlement of Greenland by the Norse, and voyages to the mainland of North America.

9 This quotation from *Brevis commentarius* has been translated by the author from the original Latin in the edition by Benediktsson, 1950, p. 16). When Arngrímur writes, 'Thus we can see the truth of what Frisius wrote' in regard to navigation only being possible for four months of the year, he is, of course, being ironic.

10 This discussion on Arngrímur draws on Thoroddsen, 1892–1904, vol. I and 1908–22, as well as the thesis of Jakob Benediktsson on Arngrímur (Benediktsson, 1957a).

11 This was not an unreasonable assumption as Sigurður was clearly a man of many talents: a Latin scholar, a poet, artist (a painter), and a musician (*Norðanfari* 59–60, 1880: 126–127). He attended the University of Copenhagen in 1593 and became headmaster of the school at Skálholt in 1594 on his return. He died shortly after in a tragic drowning accident (Ólason, 1919).

12 Benediktsson, in his preface to the Icelandic translation – *Íslandslýsing* – 'Description of Iceland' of Oddur's work, 1971, pp. 34–35).

13 Translated into English by the author from Burg, ed., *Qualiscunque descriptio Islandiæ*, 1928, pp. 4–5; and Einarsson, 1971, pp. 34–35.

14 Sea ice also reached the south later in the seventeenth century in 1694 and 1695 (Vilmundarson, 1972; Ogilvie, 2020). Interestingly, a period of mild climate with little sea ice appears to have occurred during *c.*1649 to 1682 (Ogilvie, 2020: 18).

15 Sea ice may also have reached the south coast in *c.*1552, and it seems that this information came from Oddur. He encouraged another scholar, Jón Egilsson (1548–1636), to write an historical work entitled the *Biskupa-Annálar* ('History of Icelandic Bishops'). This was completed in 1605. Although this work does contain some weather and climate information, the only mention of sea ice is an interpolation which may refer to 1552. The editor of the currently available publication (1856), Jón Sigfússon, suggests that this interpolation should be attributed to Oddur. It is interesting that Jón Egilsson's work was completed in 1605, a year when the sea ice did reach the south. It is possible to conjecture that the event piqued Oddur's interest and that he found an account referrring to the event of *c.*1552 in the archives at Skálholt. See Egilsson, *Biskupa-Annálar*, p. 100, note 2 and Ogilvie, 2020.

16 Presumably this slightly odd phrasing is meant to suggest that the benefits of the ice, such as marine mammals etc., were as if tithes were being collected (the one tenth value of an item that was typically given to the church).
17 This echoes his father's statement that sea ice did not reach the south coast.
18 An 'ell' was approximately 114 centimetres. In other words, the tree trunks are said to be some 54.72 metres long.
19 Translated from Oddsson, 1942, pp. 57–58 by the author with input from Viðar Hreinsson. See also the paper by Ólafur Eggertsson (1993) on the origins of driftwood on the coasts of Iceland.
20 The medieval Icelandic annals are published in Storm, 1977.
21 Björn was born at Ingveldarstaðir, Skagafjörður, but his father died when he was 8 years old and he then went to live at nearby Reynistaður.
22 No sea ice is recorded for the last year he covered, 1640. *Skarðsárannáll* was continued for the years 1641–45 by Bishop Brynjólfur Sveinsson (1605–75), but no sea ice was recorded for these years.
23 The old Icelandic month of *þorri* began on a Friday between 9 and 16 January and lasted until between 8 and 15 February (according to the Old Style or Julian calendar).
24 The 'Moving Days' began in early May and lasted approximately a week.
25 Suðurnes encompasses the southern peninsula region of Iceland.
26 Translated from the Icelandic by the author from *Annálar 1400–1800*, *Skarðsárannáll*, I, pp. 202–203.

References

Annálar 1400–1800, I (1922–27) *Skarðsárannáll*. Reykjavík: Hinu Islenzka Bókmentafélagi, 28–265.

Benediktsson, J., ed. (1950) *Arngrimi Jonæ Opera Latine Conscripta* Vol. I, in J. Helgason, ed., *Bibliotheca Arnamagnæana*, Vol IX. Copenhagen: Ejnar Munksgaard.

Benediktsson, J. (1956) 'Hver Samdi Qualisqunque descriptio Islandiæ?', in *Nordæla: Afmæliskevðja til Sigurðar Nordals*. Reykjavík: Helgafell, 97–109.

Benediktsson, J., ed. (1957a) *Arngrímur Jónsson and His Works*. Copenhagen: Ejnar Munksgaard.

Benediktsson, J., ed. (1957b) '*Arngrimi Jonæ Opera Latine Conscripta* Vol. IV, Introduction and Notes', xv–116, in J. Helgason, ed., *Bibliotheca Arnamagnæana* Vol XII. Copenhagen: Ejnar Munksgaard, xvi–593.

Benediktsson, J., ed. (1968) *Íslendingabók. Landnámabók*. Íslenzk Fornrit I. Reykjavík: Hið Íslenzka fornritafélag.

Burg, F., ed. (1928) *Qualiscunque descriptio Islandiæ*. Veröffentlichungen aus der Hamburger Staats und Universitäts Bibliothek, Band I. Hamburg.

Ebenesersdóttir, S. S., M. Sandoval-Velasco, E. D. Gunnarsdóttir *et al.* (2018) 'Ancient Genomes from Iceland Reveal the Making of a Human Population', *Science* 360(6392): 1028–1032. doi: 10.1126/science.aar2625.

Eggertsson, Ó. (1993) 'Origin of the Driftwood on the Coasts of Iceland: A Dendrochronological Study', *Jökull* 43: 15–32.
Egilsson, J. (1856) Biskupa-annálar Jóns Egilssonar með formála, athugagreinum og fylgiskjölum eptir Jón Sigurðsson, *Safn til Sögu Íslands og Íslenskra Bókmenta Að Fornu og Nýju* 1, Copenhagen.
Einarsson, O. (1971) *Íslandslýsing Qualiscunque descriptio Islandiæ*. Reykjavík: Bókaútgáfa Menningarsjóðs.
Friðriksson, S. (1969) 'The Effects of Sea Ice on Flora, Fauna and Agriculture', *Jökull* 19: 146–157.
Frisius, G. (1582) *Opera collecta*. Frankfurt, 8vo.
Hakluyt, R. and Goldsmid, E., eds 1598. *The Principal Navigations, Voyages, Traffiques, and Discoveries of the English Nation*. London: George Bishop Ralph Newberie and Robert Barker. Reprinted 1928, London, Toronto, New York.
Hartman, S., A. E. J. Ogilvie, J. H. Ingimundarson, A. J. Dugmore, G. Hambrecht, and T. H. McGovern (2017) 'Medieval Iceland, Greenland, and the New Human Condition: A Case Study in Integrated Environmental Humanities', *Global and Planetary Change* 156(September): 123–139. https://doi.org/10.1016/j.gloplacha.2017.04.007.
Hermannsson, H., ed. (1917) *Annalium in Islandia farrago* and *De mirabilibus Islandiæ* by Gísli Oddsson, in *Islandica* X. Ithaca, NY: Cornell University Library.
Hermannsson, H., ed. (1926) Two Cartographers Guðbrandur Thorláksson and Thórður Thorláksson, *Islandica* XV. Ithaca, NY: Cornell University Library.
Jakobsson, M., L. A. Mayer, B. Coakly, and J. A. Dowdeswell (2012) The International Bathymetric Chart of the Arctic Ocean (IBCAO) Version 3.0, *Geophysical Research Letters* 39(12). doi: 10.1029/2012GL052219.
Jochumsson, M. (1988) *Hafísinn*. Reykjavík.
Jónsson, B. (1913) 'Svellavetur', *Andvari* 38(1): 104–110. Available at: http://timarit.is/view_page_init.jsp?gegnirId=000509578 (accessed 21 March 2022).
Júlíusson, Á. D. (2002) 'Icelandic Sagas Around 1600. Reception, Interpretation and Context'. Conference proceedings for *Sagas and Societies*, Borgarfjord 2002. Available at: https://publikationen.uni-tuebingen.de/xmlui/handle/10900/46199 (accessed 21 March 2022).
Miles, M. W., C. S. Andresen, and C. V. Dylmer. (2020) 'Evidence for Extreme Export of Arctic Sea Ice Leading the Abrupt Onset of the Little Ice Age', *Science Advances* 6(38): 1–6. Available at: http://advances.sciencemag.org/ (accessed 21 March 2022).
Miles, M. W., D. V. Divine, T. Furevik, E. Jansen, M. Moros, and A. E. J. Ogilvie (2014) 'A Signal of Persistent Atlantic Multidecadal Variability in Arctic Sea Ice', *Geophysical Research Letters* 41(2): 463–469. doi: 10.1002/2013GL058084.
Munster, S. I. (1598) *Cosmographey*. Basel.
Nicholls, N. (2010) 'Why do We Care About Past Climates? An Editorial Essay', *WIREs Climate Change* 1(2): 155–157. doi:10.1002/wcc.4.
Norðanfari (1880) Skólameistararöð í Skálholti, tölublað (a list of schoolmasters/rectors at Skálholt by Oddur Einarsson, printed in this newspaper). Available at: https://timarit.is/page/2039236#page/n0/mode/2up (accessed 2 April 2022).
Oddsson, G. (1942) *Íslenzk Annálabrot* (*Annalium in Islandia farrago*) *og Undur Íslands* (*De mirabilibus Islandiæ*), trans J. Rafnar. Akureyri: Þorsteinn M. Jónsson.

Ogilvie, A. E. J. (1982) 'Climate and Society in Iceland from the Medieval Period to the Late Eighteenth Century'. Unpublished PhD thesis, School of Environmental Sciences, University of East Anglia, Norwich.

Ogilvie, A. E. J. (1991) 'Climatic Changes in Iceland AD *c.*865 to 1598', in G. F. Bigelow, ed., *The Norse of the North Atlantic, Acta Archaeologica Vol. 61–1900* Munksgaard, Copenhagen: 233–251.

Ogilvie, A. E. J. (2005) 'Local Knowledge and Travellers' Tales: A Selection of Climatic Observations in Iceland', in C. Caseldine, A. Russell, J. Harðardóttir, and O. Knudsen, eds, *Iceland – Modern Processes and Past Environments.* Developments in Quaternary Science 5, Amsterdam, Boston, Heidelberg, London: Elsevier, 257–287.

Ogilvie, A. E. J. (2010) 'Historical Climatology, Climatic Change, and Implications for Climate Science in the 21st Century', *Climatic Change* 100(1): 33–47.

Ogilvie, A. E. J. (2015) 'An Ancient Enemy Observed: Images of Sea Ice in Selected Narratives of Iceland from the Settlement to the Late Nineteenth Century', in M. Gustavsson and D. Retsö, eds, *Långa linjer och många fält: Festskrift till Johan Söderberg.* Stockholm Studies in Economic History 65. Stockholm: Acta Universitatis Stockholmensis, 137–155.

Ogilvie, A. E. J. (2017) 'A Brief Description of Sea Ice', in: E. Ogilvie, *Out of Ice.* London: Black Dog Publishing, 88–90.

Ogilvie, A. E. J. (2020) 'Famines, Mortality, Livestock Deaths and Scholarship: Environmental Stress in Iceland *c.*1500–1700', in A. Kiss and K. Prybil, eds, *The Dance of Death: Environmental Stress, Mortality and Social Response in Late Medieval and Renaissance Europe.* London: Routledge, 9–24. https://doi.org/10.4324/9780429491085.

Ogilvie, A. E. J., B. T. Hill, and G. R. Demarée (2021) 'A Fleet of Silver: Local Knowledge Perceptions of Sea Ice from Iceland and Labrador/Nunatsiavut', in E. Panagiotakopulu and J. P. Sadler, eds, *Biogeography in the Sub-Arctic: The Past and Future of North Atlantic Biota.* Oxford: John Wiley & Sons Ltd, 273–291. doi: 10.1002/9781118561461.

Ogilvie, A. E. J. and T. Jónsson (2001) '"Little Ice Age" Research: A Perspective from Iceland', *Climatic Change* 48(1): 9–52.

Ólason, P. E. (1919) 'Sigurður Stefánsson', in *Menn og Menntir 1.* Reykjavík: Bókaverslun Ársæls Árnasonar.

Pfister C., R. Brázdil, J. Luterbacher, A. E. J. Ogilvie, and S. White (2018) 'Early Modern Europe', in S. White, C. Pfister, and F. Mauelshagen, eds, *The Palgrave Handbook of Climate History.* London: Palgrave Macmillan, 265–295. https://doi.org/10.1057/978-1-137-43020-5_23.

Plomp, K. A., H. Gestsdóttir, K. Dobney, N. Price, and M. Collard (2021) 'The Composition of the Founding Population of Iceland: A New Perspective from 3D Analyses of Basicranial Shape', *PLOS one* 16(3): e0249393. https://doi.org/10.1371/journal.pone.0249393.

Sigurðsson, H. (1971) *Kortasaga Íslands: frá öndverðu til loka 16. aldar.* Reykjavík: Bókaútgáfa Menningarsjóðs og þjóðvinafélagsins.

Sigurðsson, H. (1978) *Kortasaga Íslands: frá lokum 16. aldar til 1848.* Reykjavík: Bókaútgáfa Menningarsjóðs og þjóðvinafélagsins.

Storm, G., ed. (1888) *Islandske Annaler Intil 1578*. Christiania: Udgivne for det norske historiske kildeskriftfond (reprinted Oslo, 1977).

Thorarinsson, S. (1956) *The Thousand Years Struggle against Ice and Fire*. Museum of Natural History, Department of Geology and Geography, Reykjavík, Miscellaneous Papers, No. 14. Reykjavík: Bókaútgáfa Menningarsjóðs.

Thoroddsen, Þ. (1892–1904) *Landfræðissaga Íslands. Hugmyndir manna um Ísland, náttúruskoðun og rannsóknir fyrr og síðar I–IV*. Reykjavík, 1892–96; Kaupmannahöfn, 1898–1904: Gefin út af Hinu Íslenzka bókmenntafjelagi.

Thoroddsen, Þ. (1908–22) *Lýsing Íslands I–IV*. Copenhagen: Gefin út af Hínu íslenzka bókmenntafélagi.

Thoroddsen, Þ. (1914) 'Islands klima i Oldtiden', *Geografisk Tidskrift* 22: 204–216.

Vilmundarson, Þ. (1972) 'Evaluation of Historical Sources on Sea Ice Near Iceland', in Thorbjörn Karlsson, ed., *Hafísinn*. Reykjavík: Almenna Bókafélagið, 313–332.

Þorkelsson, Jón. (1887) 'Þáttur af Birni Jónssýni á Skarðsá', *Tímarit Hins íslenzka bókmentafélags* 8: 34–96. Available at: https://timarit.is/page/2317330#page/n33/mode/2up (accessed 21 March 2022).

2

A moving element: ice, culture, and economy in northern and northwestern Russia

Alexei Kraikovski

It is unsurprising that Russia enjoys a 'frosty climate'.[1] Many, if not all the works offering a long-term perspective of Russian history contain various elements of environmental determinism: 'All of the Russian states shared essential features and faced common challenges. Geographic location and climate shaped all of them … poor soils, northern location … and continental climate … made production of edible grains difficult' (Engel and Martin, 2015: x). However, the links between the inhabitants of the Russian plain and frost suffers from being too generic. Frost is not a uniform phenomenon. Every member of northern communities in Russia knows that dealing with a cold climate means interaction with many and varied phenomena. Moreover, the experience of snow is not a uniform one, depending on quantity, thickness of the layer, structure, and air and surface temperatures. Therefore, the story of the interaction of Russians with their 'cold nature' deserves more detailed examination and introspection. The story of human interaction with frost, ice, frozen ground, and snow offers insights into the practices of living in a 'severe' climate. Frozen ground can and does vary in extent and thickness, and this has generated new studies on the place of permafrost within Russian sciences and society. Seasonally speaking, Russians have long understood that permafrost has an active layer that thaws over the summer season (for a recent study of Russian permafrost, see Crate, 2021).

This chapter is about the coastal areas of Russia. It considers the dynamic encounters of sea and land as a prerequisite to make sense of the significant role of ice in the life of local communities. Contact with ice forms an important part of the littoral, and literal, experience of Russians living in the north and northwest of the enormous Eurasian Empire, in the White Sea basin and around the Gulf of Finland. I will use this concept of the littoral, provided by the 'New Coastal History' (Worthington, 2017), to approach ice as an integral part of life on the Russian coastline. Considering the baseline not as a legal border or the tidal mark as a natural geographical barrier, but rather as the central axis of the human activities in the coastal areas, the New Coastal History offers an instrument to unlock the societal mechanisms shaping the opportunities and limits of nature use and control that are widely understood. Henceforth, I address the various aspects of ice experiences for the coastal dwellers and argue that this natural phenomenon shapes both individual and collective experiences of time and space. Ice, as the chapter reveals, occupies a significant place in the social and economic imagination of the coastal community members. Thinking with ice as timescale and then the spatiality of ice helps to tease out further these dynamics and imaginaries. Finally, the chapter considers the social imaginaries of ice that were and continue to remain important for the planning of local and regional political and economic life.

Prosperity and piety

The vast area under study, situated in the northwestern corner of Russia, is covered with marshy taiga forest.

It has the Gulf of Finland in the southwest and includes the huge lakes of Ladoga and Onega, the powerful rivers of the Neva, the Northern Dvina, and the Onega, and the White Sea at its northern edge. The area is characterized by the onset of early autumnal frost and long winters, significantly complicated in the early modern period by the 'Little Ice Age' (Borisenkov and Pasetskii, 1983: 112). Local inhabitants, through the centuries, accumulated great experience of dealing with river and sea ice due to the abundance of open water combined with the severe climate. The complicated historical development of the area also contributed to the diversity of traditional ways of dealing with the surrounding nature, including frozen water. Since the early Middle Ages, the major parts of this vast area remained under the control of Great Novgorod. This famous metropolis of the north and northwest of medieval Rus' shaped the spatial structure of the area with quite diverse types of settlements, such as cities and towns, manors, fortresses, monasteries, and villages. In 1478 all the possessions of Great Novgorod were incorporated into the Muscovian State (Birnbaum, 1981).

The sixteenth and seventeenth centuries witnessed several significant transformations in the spatial organization of the region. First, Russians gradually lost political and military control over the Eastern Baltic. After the Time of Troubles (*Smutnoe vremya*) in the early seventeenth century, the entire coast of the Gulf of Finland came into the hands of Sweden. As a result, the Russian merchants had to deal with the Swedish authorities in order to secure access to the Baltic ports (Kotilaine 2005: 294–375). Therefore, the prevailing geopolitics of the region was a great deal more turbulent.

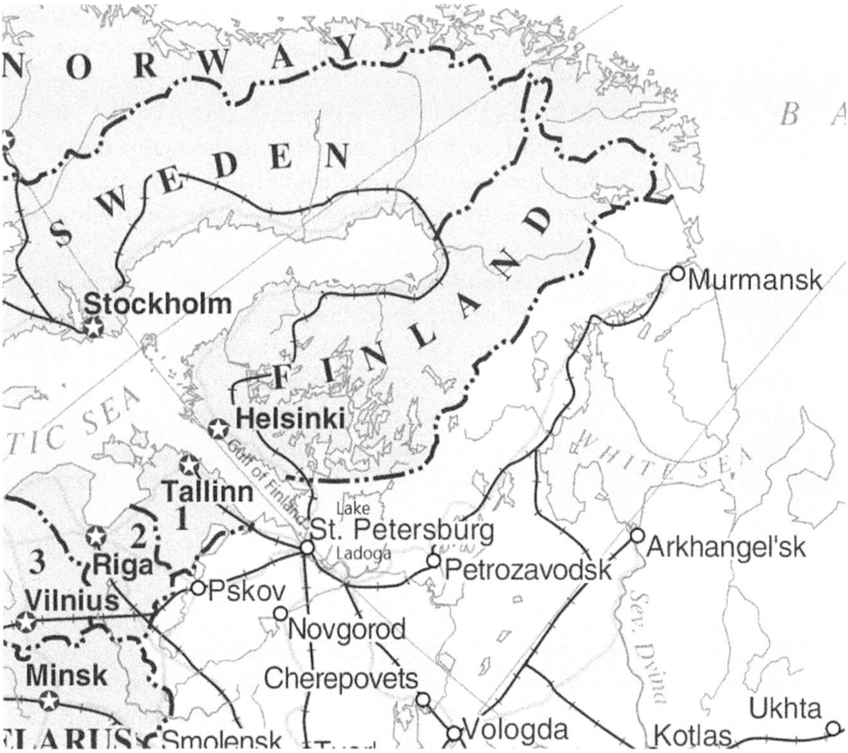

Figure 2.1 Map of northwest Russia and wider region.

Second, the economic, social, and cultural development area of the Russian North received significant impact via a process known as monastic colonization. Orthodox monks, in the fifteenth and sixteenth centuries, moved to the poorly populated areas of the Russian North in order to live the ideal Christian life full of hard work. The Russian government supported them with economic privileges. As a result, the newly founded monasteries became important economic and cultural centres and formed the base for the specific area later called 'the Russian Thebaide'. This name described the north as the place where monks and local peasants formed a sort of unity in their successful transformation of a severe nature, to achieve a combination of prosperity and piety (Kraikovski *et al.*, 2020).

Third, after the port of Archangelsk was founded in 1584, in the downstream of the Northern Dvina River, about 35 kilometres upstream from the White Sea, it soon became the major centre of Russia's international trade. As a result, the river route along the Sukhona and the Dvina, connecting

Archangelsk with the town of Vologda and further with central Russia, became one of the most important communication lines in the country (Kotilaine, 2005: 234–293).

The situation changed radically in the eighteenth century. The victory over Sweden in the Great Northern War made Russia the major power on the Baltic Sea, while St Petersburg, the new capital of the young empire, eventually became the new Baltic metropolis, completely controlling the Gulf of Finland from 1809 onwards (Kraikovski and Lajus, 2019). The port of Archangelsk lost its dominant position to St Petersburg, which over the course of time put an end to the flourishing commercial life along the Sukhona and Dvina Rivers (Kahan, 1985: 254–256). Following these changes, the local population in the north faced many geopolitical, social, and economic transformations, which meant adjusting everyday strategies and practices to an emerging novel situation. What has been less well understood is the longer-term societal interaction with ice and its connection to social and economic transformation.

Ice and time

Seasonality is perhaps the most visible peculiarity of ice. To a very great extent, ice shaped the seasons of the year for local inhabitants. The time of freezing and melting of ice in the major rivers was often recorded as the most important indicator of influence for the life of communities during the entire year. For instance, the unusually strong ice flow on the Dvina in 1621 was recorded in the local chronicles as highly disruptive because of its capacity to block human movement and the associated dangers of late spring flooding (Borisenkov and Pasetskii, 1983: 206).

St Petersburg, as the emerging centre of Russian modernity, became the place where the monitoring of ice cycles became routine. The inhabitants of the modernized empire needed to know the seasonal cycle and recorded natural events as a strategy for regularizing the annual and seasonal flow of time (Kraikovski and Lajus, 2017). One could say, based on the observations of the ice-shaped seasonality of St Petersburg's urban life, that the capital of the Russian Empire had three seasons rather than four (winter, spring, summer, and autumn), namely an ice-covered Neva, a time of no ice, and a time of unstable or moving ice.

The influence of this seasonality was quite predictably very visible in the economic activities of the ports, both on the seashore and on the navigable rivers. The two geographical regions were inseparable in the cryospheric imaginaries of residents and traders. Margarita Dadykina recently studied the production and transportation practices on the salt-boiling enterprises situated in the possessions of the Spaso-Prilutsky monastery. She demonstrated that the ice of the White Sea and the inflowing rivers shaped the cycles of production and the temporality of commerce. Indeed, in the town of Sol Vychegodskaya, the famous centre of salt production, the pattern of ice covering for the river Vychegda separated the period of high productivity from the time of weak salt-boiling processes. The ice flow combined with the spring floods led to the rapid decrease of salinity of the underground brines used as a raw material for the salt-boiling premises, and therefore compromised the most intensive time of production. The monks had to stop their activities and wait until salinity returned to a commercially valuable 'normal'. Instead, however, the ice flow opened a period of intensive communication between the scattered monastic possessions. The period between ice flow and freezing included the delivery of salt to the major regional markets and the transportation of timber used as a fuel in the ovens of the salt-boiling factories. The freezing of the rivers, therefore, marked the start of a new cryo-economic cycle, when transportation costs became much higher (Dadykina, 2016, 2019).

Further north this seasonality could be even more pronounced. Official documents kept by the Russian authorities related to the commercial shipping in the White Sea and Barents Sea basin confirm that the ice fundamentally shaped the seasonality of sea-related activities. Even the areas situated rather far from the zone of freezing or melting were affected. Indeed, as we see in the history of eighteenth- and nineteenth-century Pomor hunting on Spitsbergen, ice in the White Sea Throat, the strait between the White Sea and the Barents Sea, regulated the seasonality of the White Sea ports, including the major one in Archangelsk. As a result, local hunters and foreign merchants involved in the commodification of marine mammals' fat faced significant challenges in planning their time to and from Spitsbergen. Traders and hunters were acutely aware that being ice free, near ice free, and ice-bound was intimately related, depending on seasonality, commercial and trading networks, and the intersecting physical geographies of water, ice, and multiple landforms.

Sea ice prevented outgoing shipping, so the hunters could not leave the White Sea harbours for the grounds situated in the Arctic, be that Spitsbergen or Novaya Zemlya, before summer (Kraikovski, 2015). Being trapped on the White Sea was not entirely disadvantageous, either. Merchants sometimes used the winter, when their vessels were frozen in the ice, as the opportunity to develop illegal operations, probably relying on the severe weather conditions to restrict the normal practices of surveillance and control. For instance, in 1712 the city authorities of Archangelsk had to investigate illegal commercial operations between the local dweller Semen Chudakov and the well-known English merchant Samuel Gartside. Chudakov, according to the results of investigation, illegally delivered a consignment of blubber to Gartside's vessel frozen in the ice in Maimaksa, in the relatively remote branch of the Northern Dvina. The scheme did not work that time, but the documents describe it as commonplace (Kraikovski, 2019).

In St Petersburg, the sea ice of the Neva and the Gulf of Finland shaped the seasons and determined many aspects of social and economic life of the imperial capital. The consumption patterns of local communities depended on the prevailing ice conditions of the gulf. The opening of international navigation made the importation of freshly delivered food possible with the oyster-eating season a good example of seasonal availability. According to eighteenth-century local newspapers, the ice determined the start of the oyster season in the imperial capital, and the time varied from late March to mid-May (Kraikovski, 2018).

The seasonality of port opening, and closure, was essential for the employment market of the city. Any early appearance of ice became problematic and even dangerous for the port and could cause reverberations for the entire economy of the Baltic metropolis, with ramifications for all social classes. The melting ice was a clear indication of spring, which meant first and foremost the opening of international navigation and therefore the onset of a new cycle of business activity in the city (Kraikovski and Lajus, 2017). Moreover, considering the active participation of the peasants from the Upper Volga area, known as the perfect experts in riverine shipping, in St Petersburg water transportation, we can note that the ice of the Neva and the gulf determined the life cycle of the economically active population far outside the city. These peasants left their houses in spring, spent seven months in the imperial capital, and were able to return home only when the Neva and the gulf were ice-covered again. They had to keep in mind the natural rhythm of the St Petersburg ice regime to plan their family and professional lives (Lurie, 2011: 205–209).

Given the direct correlation between sea ice, port access and maritime trade, a severe winter carried with it the potential for considerable socio-economic impact. For a long time, until the twentieth century, the local business community found ways to profit from an abundance of Neva ice. One notable development in St Petersburg was the onset of the ice delivery and storage industry. Before the invention of artificial freezing, this proved a lucrative business for many ice-rich suppliers in countries such as Norway and the United States (see the chapter by Jonathan Rees). The Neva water had an historically superb reputation in the city, and the same was true for the ice of the main river. The abundance of high-quality ice was important for urban dwellers. This industry, almost ignored by the city authorities in the eighteenth century, became a matter of state regulation after 1860, with the necessity to petition for the right to cut and sell the ice blocks. Therefore, the ability of the water to freeze in low temperatures became an important resource controlled by the government, like other natural resources. The value of this resource was inevitably linked to seasonal variations. Ice delivery to the numerous ice houses in the city region was extendable even during the summer season. This was an important advantage for city residents and businesses and helped manage the transition towards the introduction of artificial freezing (Dills, 2010: 22–36; Kraikovski and Lajus, 2010).

The dwellers of the coastal areas of the Russian North and Northwest had long and varied experience of what we may call an ice-shaped temporality. The presence or absence of ice on the sea or the river in every moment of the year was never cost neutral. Depending on the circumstances, it was normal or abnormal, predictable, or unexpected, desirable, or even catastrophic. Technological development and cycles of innovation disrupted what we might term the habitual seasonality of local ice. The icebreakers in the second half of the nineteenth century made the ports available all year round (Andrienko, 2009: 51–67), while the intensive construction of ground transportation routes, including railways, made summer season water transportation a thing of the past. While transport efficiencies no doubt emerged, water transportation routing in the summer nonetheless helped to bind scattered communities together. However, this turn towards enhanced connectivity was itself conditioned by the spatial distribution of ice.

Ice and space

The freezing of water in autumn and the melting of ice in spring radically changed coastal spaces, and therefore altered the working practices used by local dwellers for dealing with their habitual surroundings. Their active interaction with the ice surface was among the most important features of rural and urban life in north and northwestern Russia. Texts written by early modern foreign observers attest to these seasonal adjustments. Nicolaes Witsen described the day of 12 December 1664 in his diary of a visit to Russia as a member of the entourage of the Ambassadors of the General States of the United Provinces. He noted that the city of Pskov impressed him by its vast and very rich daily ice market, where one could buy onions, garlic, herring, and innumerable other commodities (Witsen, 1996: 60). This record confirms that the locals considered the ice of the Velikaia River an inseparable part of their urban trading space. In winter, the dwellers of the Russian North and Northwest enjoyed an additional 'territory' for buying, selling, and interacting with one another. The market, which occupied the major square in the centre of Pskov, moved to the river after freezing began. This proved useful for the sanitary conditions of trade, as the dirt, typical for the land-based market, could be just thrown under the ice and removed by the river flow. The frozen river performed an economic and ecological service – it provided additional space for trade in the winter, and it removed dirt and debris away from the city in the spring melt. Therefore, the freezing of the big river, with thick and stable ice reliably covering the water for several months, became an important asset of the climate, and the local dwellers used it actively and effectively. Ice was no longer framed as an inconvenience to trade.

With ice, St Petersburg opened a new chapter in the environmental history of Russian urbanization. Dealing with the frozen element was integral to urban space in the imperial capital. The numerous rivers and canals formed the skeleton of the city and in winter became a seamlessly integrated part of the everyday activities of the people living on the riverbanks. Grigorii Kaganov, in his research on the visual representations of St Petersburg's urban space, makes the point that river ice aided and abetted artistic interpretations of the city as a 'Venice of the North'. According to Kaganov's observations, Benjamin Patersen, the Swedish artist who worked in St Petersburg from 1787, represented the space of the Russian capital as a vast and open area for ice to gather and make its presence felt, representing in turn a severe and powerful north, and the sun, representing a warm and abundant south. Vasilii Sadovnikov, an artist of the first half of the nineteenth century, represented the river ice of the Moika near Palace Square in 1829 as a communal area, which belonged to the working populace, like watercarriers, peddlers, coachmen, and washerwomen. People of that kind were not considered desirable on the fashionable streets of this very aristocratic part of the city, but the ice of the small river in the wintertime 'belonged' to them (Kaganov, 1995: 56, 104–106). River ice, depending on its location and durability, was being imagined and represented as either indicative of St Petersburg's distinct aesthetic or as an ice-filled agora, enabling private use of a seasonal public space.

The powerful Neva, the major axis of the urban structure and, according to Dennis Shaw, the core of the St Petersburg 'space of modernity', had a distinct ice spatiality (Shaw, 2003). The vast space between the Peter and Paul Fortress, the Spit of Vasilievski Island, and the Winter Palace continued downstream the Great Neva between the Admiralty side and the university embankment in winter became the main representative square of the imperial capital. The ice accommodated sophisticated temporal infrastructure, including the hippodrome. Along the ice roads for crossing the Neva, as well as for those travelling along the river on sledges, there were strategically placed huts with drinking bowls for horses, and a slew of skating rinks for local dwellers. The ice also played an important role in the religious ceremony of the Epiphany, when the Neva represented Jordan, and a big hole in the ice prepared inside a wooden pavilion was considered the image of the place in Palestine where John the Baptist gave Jesus the power to reveal his divine nature. During Shrovetide (*Maslenitsa*), the festival before the Lent, which recalled pagan traditions of celebrating spring, the city dwellers erected on the ice of the Neva infrastructure for tobogganing, and transformed this vast area into a zone of transgressional fun for commoners (Kraikovski and Lajus, 2010).

These observations affirm that urban space in St Petersburg was transformed by ice, but the transformation was not homogenous. The cultural, economic, and social context of ice-related activities mattered. The ice surface was a matter of utilization, consideration, and contextualization when it came to social and economic life. There was no one ice surface. It was highly textured, physically and socially. At every moment the contact with a certain part of the ice could be appropriate or inappropriate for the representative of a certain social group. These spatial structures were temporal within one calendar year, but at the same time they took on a sense of permanence in a longer-term perspective, revived and repeated winter after winter for decades and even centuries.

Transformation of the coastal space significantly also altered the practices of dealing with the natural resources provided by watery habitats. The presence of ice restricted shipping and therefore fisheries and hunting of marine mammals because boats could not be used. Ice enabled direct access to the areas that were not available for fishermen in warmer seasons without vessels (Kraikovski, 2015). Foreign observers were impressed with the methods and approaches used by locals to fully enjoy this advantage of ice fishing. For instance, in his diary Nicolaes Witsen described in detail the winter fisheries on Lake Ilmen near Novgorod, that he considered 'very strange'. First, he argued, local fishermen appeared to be able to catch and take from under the ice 'millions' of small fishes looking like smelt. This was most probably *snetok*, the small smelt, which was quite abundant in the lakes of the Russian Northwest and occupied an important place in the local diet. Witsen continued, however, that the Russians impressed visitors with their specific winter fishing technique, involving a hut on the ice made from sledges, straw, and covered with their fur coats. Inside the fisher had a small fire, necessary for heating and for keeping the hole in the ice open. Through the hole, Witsen recorded, he caught the fish below the ice. [Witsen, 1996: 60). The ice therefore provided welcome opportunities that were not available in the warm season. While local practices could look exotic to outside observers, they fitted very well into routine life in the ice-covered coastal zone.

Ice and mobility

Frost, snow, and ice have long been recognized as decisive to historical patterns of mobility (Hastrup and Olwig, 2012). The environmental history of the Russian North and Northwest, however, provide deeper insight into the multifaceted opportunities for travel via freezing rivers and coastal waters.

The transformation of space after the freezing of water made the surface of rivers, lakes, and coastal parts of the sea suitable for ground transportation. Foreign observers paid attention to the clear link between the freezing and the opportunities for enhanced connectivity between the distant points of this enormous country. For instance, in 1700 Cornelis de Bruijn mentioned that the merchants coming for commerce to Archangelsk in summer normally had to idle for one month in the north after the end of the trade markets, and the departure of foreign vessels. de Bruijn argued that they had to wait for snow and ice to be able to travel between Archangelsk and Moscow (de Bruijn, 1988: 40).

Other observers provided more details on dealing with ice, while moving through the vast snow-covered area where travellers faced harsh frost. Balthazar Coyett, a Dutch traveller, grew up in the tropical southeast Asian possessions of the United Provinces. This could well have explained his attentiveness to the methods of dealing with the frost, snow, and ice he observed on the way to the embassy of Coenraad van Klenck from Archangelsk to Moscow in 1675. The frost on the Sukhona River was strong enough to create problems for foreigners with no experience of how to manage their mobility in the severe winter of the Russian North. The travellers, Coyett reported, found that all the beverages they carried on the sledges were completely frozen, so they had to drink their wine with ice, which was not comfortable. As a result, the foreigners with experience of living in the area advised others to use ice to protect the barrels of wine from freezing. On 18 November, Coyett reported, the ambassador ordered a hole to be created in the ice of the river and several times put barrels of wine into the water until a thick layer of ice appeared. Moving wine above and below the frozen surface was arguably a more subtle form of mobility. But the idea was to create a crust to keep the wine cold and yet prevent freezing (Koiet, 1900: 346).

The links between ice and mobility practices became more complex and multifaceted with the complication of urban life during the Europeanization and modernization of the eighteenth century. St Petersburg became a cradle for the new practices of urban life, including mobility inside and outside the city. The abundance of water in the new capital of the Russian Empire made it possible for water routes to connect further with the central regions of Russia, enabling further food supply transportation. In this way, ice completely transformed the transport situation for the dwellers of St Petersburg. On one hand, the winter infrastructure, before the introduction of railways, provided much less opportunities for the delivery of big consignments of commodities compared to the summer situation (Jones, 2017.) On the other, ice made the immediate vicinity of the capital more accessible through winterized transport. The island of Kotlin with the naval fortress of Kronstadt received a direct communication line with the capital. The seasonal road on the ice was well equipped and maintained, including the landmarks, lights, and even temporary buildings for rest (Kraikovski and Lajus, 2017: 248). In 1880, the shorter way from Kronstadt to Oranienbaum, the nearest point on the southern coast of the gulf, even became the arena of a risky experiment with a seasonal railway constructed on ice (Amirkhanov, 2018: 192).

The presence of ice was and remains ambivalent if we consider it through the perspective of immobility and motion. As is well known, ice makes the water surface solid and hard enough to afford terrestrial transportation. But ice was also an insistent and lively agent. The seasonality of river and sea in the area mattered. Ice flow could restrict almost all sorts of mobility. Moving ice was dangerous for vessels, and at the same time open water or thin ice was a hazard for those on sledges. Until permanent bridges appeared, even the big urban centres faced significant problems in both internal and external transportation. Foreign observers described moving ice as one of the most characteristic features of the Russian spring, bringing with it a sense of frisson. William Richardson, an observant and educated Englishman, reported from St Petersburg in 1777 that the ice on the Neva was very strong and broke quite unexpectedly with great noise. Moving ice was lively, truculent, and always dangerous. Moreover, he noted the link between the wind direction and the ice condition, when the eastern wind brought huge pieces of ice from Lake Ladoga making the weather in the city cold enough to be compared to winter (Richardson, 2013: 43–44), a vision similar to the one that German painter Caspar David Friedrich had of river ice on the Elbe where he made an extensive study for his 'The Sea of Ice' (*Das Eismeer*) in 1823–24.

The role of moving ice in marine harvesting in the area deserves further discussion. Fishermen in the Gulf of Finland and Lake Ladoga knew very well that in spring drowning was a perpetual danger during ice fishing. Tragedies of this kind were not unusual for St Petersburg (Kraikovski and Lajus, 2010). Arctic ice could be even more destructive when in motion. The history of the four Pomors, who in the 1740s had to spend six years isolated on one of the islands of Spitsbergen (Roberts, 2003), is almost completely based on the dangerous and enduring power of moving ice.

According to the report by P. Le Roy, their ship-based journey to the archipelago was made possible by moving ice. After the crew decided to spend winter on this shore, they had to go to the island looking for a place to stay. As they had to move across the surface of rather unstable ice, they took very little equipment. Next morning they could not find the vessel, which was most probably destroyed by the moving ice together with the rest of the crew. So the ice became the powerful more-than-human actor of the story, which then determined the enduring conditions for the four Pomors who remarkably survived an extremely severe ordeal (Le Roy, 1955: 9–10). The Pomors of the White Sea were quite capable of developing methods for hunting marine mammals and used their knowledge of moving ice in the White Sea Throat to good effect. Every spring the hunters used the confluence of spring ice flow from the White Sea to the Arctic Ocean with the seasonal migration of seals to profiteer. The hunting teams took small boats for crossing the glades and moved together with the ice, killing the animals and butchering the carcasses immediately. Despite the danger of being lost in the sea, their knowledge of the natural conditions of the area was normally sufficient to predict the arrival of the ice floes at certain points on the coast, where they then met the merchants interested in purchasing their haul (Kraikovski, 2015).

Ice and imaginaries

Ice was integral to the economy of the Russian North and Northwest. It provided a resource base for limited yet important local seasonal industry such as seal hunting and fishing. At the same time, the absence and presence of ice shaped the availability of commodities in local markets, decisively affecting the seasonality of commerce. It could influence the quality of commodities and therefore was important for variability in consumption patterns. Last but not least, the ice was an important actor on the transportation business, able to anchor, promote, and/or restrict mobility. However, the eighteenth-century transformation of the Russian economy provided further opportunities to conceptualize the economic agency of ice in the region.

Lissa Roberts has perceptively noted that, in the eighteenth century, turbulent social and political transformations influenced economic practices though mediation of socio-economic imaginaries (Roberts, 2014). For Russia, this meant that the officials intended to transform and regulate the economy of the enormous empire through bureaucratic interventions in St Petersburg. They based their decisions and further administrative efforts on a certain understanding of the local environment and socio-economic situation. For the Russian North this meant that the marine harvesting was transformed according to the vision of the St Petersburg elite. It contributed to the 'construction' of an imaginary Russian Arctic. It provided the ideological and economic base for governmental efforts to create a new branch of maritime industry in the Russian North – namely Spitsbergen whaling. Ice was enrolled in these emergent economic imaginaries. The sources demonstrate that the actors of this elemental story were quite attentive to the problem of ice on and below the vast area between Archangelsk and Spitsbergen, and this powerful agential force appeared frequently in the discussions concerning the prospects for the new international industry (Kraikovski, 2017). In other words, the severe natural conditions of the area shaped visions of how the Westernization and Europeanization of the Russian economy should unfold. Operating with the images of a harsh environment, the governmental officials and local practitioners had to include river and sea ice into their socio-economic imaginaries, making this resilient, diverse, and unpredictable natural phenomenon important for the making and remaking of social and economic transformation. As historians of the Soviet Union have noted, this encounter with ice, frost, frozen ground, and the cold was to be taken to new developmental heights during the Stalin era (McCannon, 1997; Chu, 2021).

Note

1 Support from the Basic Research Program of the National Research University Higher School of Economics is gratefully acknowledged.

References

Amirkhanov, L. (2018) *Kronshtadt. Gorod-krepost: Ot osnovaniia do nashikh dnei.* Tsentrpoligraf.

Andrienko, V. (2009) *Ledokolnyi flot Rossii 1860-e–1918 gg.* Evropeiskie izdaniia.

Birnbaum, H. (1981) *Lord Novgorod the Great: Essays in the History and Culture of a Medieval City-state.* Slavica Publishers.

Borisenkov, E. P. and V. M. Pasetskii (1983) *Ekstremal'nye prirodnye iavlenija v russkikh letopisiakh XI-XVII vv.* Gidrometeoizdat.

Chu, P.-Y. (2020) *The Life of Permafrost: A History of Frozen Earth in Russian and Soviet Science.* Toronto: University of Toronto Press.

Crate, S. (2021) *Once Upon the Permafrost: Knowing Culture and Climate Change in Siberia.* Tempe, AZ: University of Arizona Press.

Dadykina, M. (2016) 'Managing Space: The Structure and Organisation of the Communication System in the Spaso-Prilustky Monastery (16th–17th centuries)', *Quaestio Rossica* 4(3): 123–140.

Dadykina, M. (2019) 'Monastyrskoe khoziaistvo i praktiki obrashcheniia s prirodnymi resursami na Russkom Severe, XVI–XVII vv', *Rus, Rossiia: Srednevekove i Novoe vremia*, 6: 459–463.
de Bruijn, K. (1988) 'Puteshestviia v Moskoviiu', in *Rossiia XVIII v. glazami inostrantsev*. Lenizdat, 19–188.
Dills, R. (2010) 'The River Neva and the Imperial Façade: Culture and Environment in Nineteenth Century St Petersburg Russia'. PhD dissertation, University of Illinois at Urbana-Champaign. Available at: http://hdl.handle.net/2142/18391 (accessed 21 March 2022).
Engel, B. A. and J. Martin (2015) *Russia in World History*. Oxford University Press.
Hastrup, K. and K. F. Olwig (2012a) 'Climate Change and Human Mobility', in *Climate Change and Human Mobility: Challenges to the Social Sciences*. Cambridge: Cambridge University Press, 1–20.
Hastrup, K. and K. F. Olwig, eds (2012b) *Climate Change and Human Mobility: Challenges to the Social Sciences*. Cambridge: Cambridge University Press.
Jones, R. E. (2017) *Bread Upon the Waters: The St Petersburg Grain Trade and the Russian Economy, 1703–1811*. University of Pittsburgh Press.
Kaganov G. (1995) *Sankt-Peterburg: obrazy prostranstva*. Indrik.
Kahan, A. (1985) *The Plow, the Hammer, and the Knout: An Economic History of Eighteenth-Century Russia*. Chicago, IL: University of Chicago Press.
Koiet, B. (1900) *Posolstvo Kunraada fan-Klenka k tsariam Alekseiu Mikhailovichu i Feodoru Alekseevichu*. Arkeograficheskaia komissiia.
Kotilaine, J. T. (2005) *Russia's Foreign Trade and Economic Expansion in the Seventeenth Century: Windows on the World*. Brill.
Kraikovski, A. (2015) '"The Sea on One Side, Trouble on the Other": Russian Marine Resource Use before Peter the Great', *Slavonic and East European Review* 93(1): 39–65.
Kraikovski, A. (2017) 'Good Fisheries vs. Bad Fisheries: Ideological and Scientific Base for the Governmental Projects of Modernization of Russian System of Marine Harvesting in the Eighteenth Century', in *Environmental History in the Making: Environmental History*, vol 7. Springer, 49–70.
Kraikovski, A. V. (2018) '"Ustersy" dlia nashego moria. Vpechatleniia zagranichnykh puteshestvii i proekty evropeizatsii potrebleniia ryby i moreproduktov v Rossiiskoi imperii v XVIII veke', in *Evropeiskie marshruty Petra Velikogo: K 300-letiiu vizita Petra I vo Frantsiiu*. Evropeiskii dom, 272–279.
Kraikovski, A. (2019) 'Profits from Under the Water: The International Blubber Market, Russian Monopolistic Companies and the Idea of Whaling Development in the Eighteenth Century', *International Journal of Maritime History* 31(1): 34–49.
Kraikovski, A., M. Dadykina, Z. Dmitrieva, and J. Lajus (2020) 'Between Piety and Productivity: Monastic Fisheries of the White and Barents Sea in the 16th–18th centuries', *Journal of the North Atlantic* 41.
Kraikovski, A. and J. Lajus (2010) 'The Neva as a Metropolitan River of Russia: Environment, Economy and Culture', in *A History of Water. 2: Rivers and Society: From Early Civilizations to Modern Times*. IB Tauris, 339–364.
Kraikovski, A. and J. Lajus (2017) 'Living on the River Over the Year: The Significance of the Neva to Imperial St Petersburg', in M. Knoll, U. Lübken, and D. Schott,

eds, *Rivers Lost, Rivers Regained: Rethinking City–River Relations*. Pittsburgh University Press, 235–252.

Kraikovski, A. and J. Lajus (2019) 'The Metropolitan Bay: Spatial Imaginary of Imperial St Petersburg and Maritime Heritage of the Gulf of Finland', *Humanities* 8(37): 1–12.

Le Roy, P. L. (1955) *Prikliucheniia chetyrekh rossiiskikh matrosov, k ostrovu Shpitsbergenu bureiu prinesennykh*. Gosudarstvennoe Izdatelstvo Geograficheskoi literatury.

Lurie, L. (2011) *'Pitershchiki'. Russkii kapitalizm. Pervaia popytka*. BKhV-Peterburg.

McCannon, J. (1997) *Red Arctic: Polar Exploration and the Myth of the North in the Soviet Union,1932–1939*. Oxford: Oxford University Press.

Richardson, W. (2013) 'Pis'ma 1768–1771 gg', in *Ekaterininskii Peterburg glazami inostrantsev: Neizdannye pisma 1770-kh godov*. Paritet, 17–102.

Roberts, D. (2003) *Four Against the Arctic: Shipwrecked for Six Years at the Top of the World*. Simon and Schuster.

Roberts, L. (2014) 'Practicing Oeconomy During the Second Half of the Long Eighteenth Century: An Introduction', *History and Technology* 30(3): 133–148.

Shaw, D. J. B. (2003) 'St Petersburg and Geographies of Modernity in Eighteenth-Century Russia', in A. Cross, ed., *St Petersburg, 1703–1825*. Basingstoke: Palgrave Macmillan, 6–29.

Witsen, N. (1996) *Puteshestvie v Moskoviiu, 1664–1665*. Simposium.

Worthington, D. (2017) 'Introducing the New Coastal History: Cultural and Environmental Perspectives from Scotland and Beyond', in *The New Coastal History*. Cham, Switzerland: Palgrave Macmillan, 3–30.

3

Ever higher: the mountain cryosphere

Dani Inkpen

> I used to look forward to falling asleep with the glacier in the evening. I also looked forward to waking up to it in the mornings. (Pastor Jón, *Under the Glacier*)

The mountain cryosphere, often overshadowed in the scholarly literature by the Arctic, Antarctic, and Greenland, deserves a place in the ice humanities. If the ocean is valuable to environmental humanists because it provides 'a perspective that recognizes volume, matter, and emergence' (Steinberg and Peters, 2015: 248), reorienting our notions of space, the mountain cryosphere strongly recommends itself for the same reasons. Material objects with depth, breadth, compositional variation, and movement that force themselves into our stories and dreams, mountains and their cryospheric elements – snow, permafrost, and ice – offer abundant opportunities for investigating material and ideal relations among ice, humans, and nonhumans.

The mountain cryosphere is rapidly warming. In recent decades mountain surface air temperatures have risen 1.5 times faster than global rates (Hock *et al.*, 2019: 94). The focus of this chapter, mountain glaciers, are deemed by glaciologists to not have 'much of a future' (Haeberli *et al.*, 2019: 1). Such facts have made mountain glaciers highly amenable to narratives of global cryospheric collapse. Photographs of receding mountain ice have come to stand for global warming (Carey, 2007; Inkpen, 2018) (Figure 3.1). But the mountain cryosphere harbours more and different stories than just those of climate apocalypse. Mountains link ice intimately and immediately to human and other-than-human life.

Mountains and ice are co-constitutive: mountains provide habitats for ice, and ice reshapes mountains; when ice disappears, it causes rockslides. Found on every continent and covering roughly a quarter of dry land, mountains are also human habitats, home to 12 per cent of the world's population, with another 14 per cent living near them (Price, 2015: 3). They are domestic spaces. They have also long been popular travel destinations,

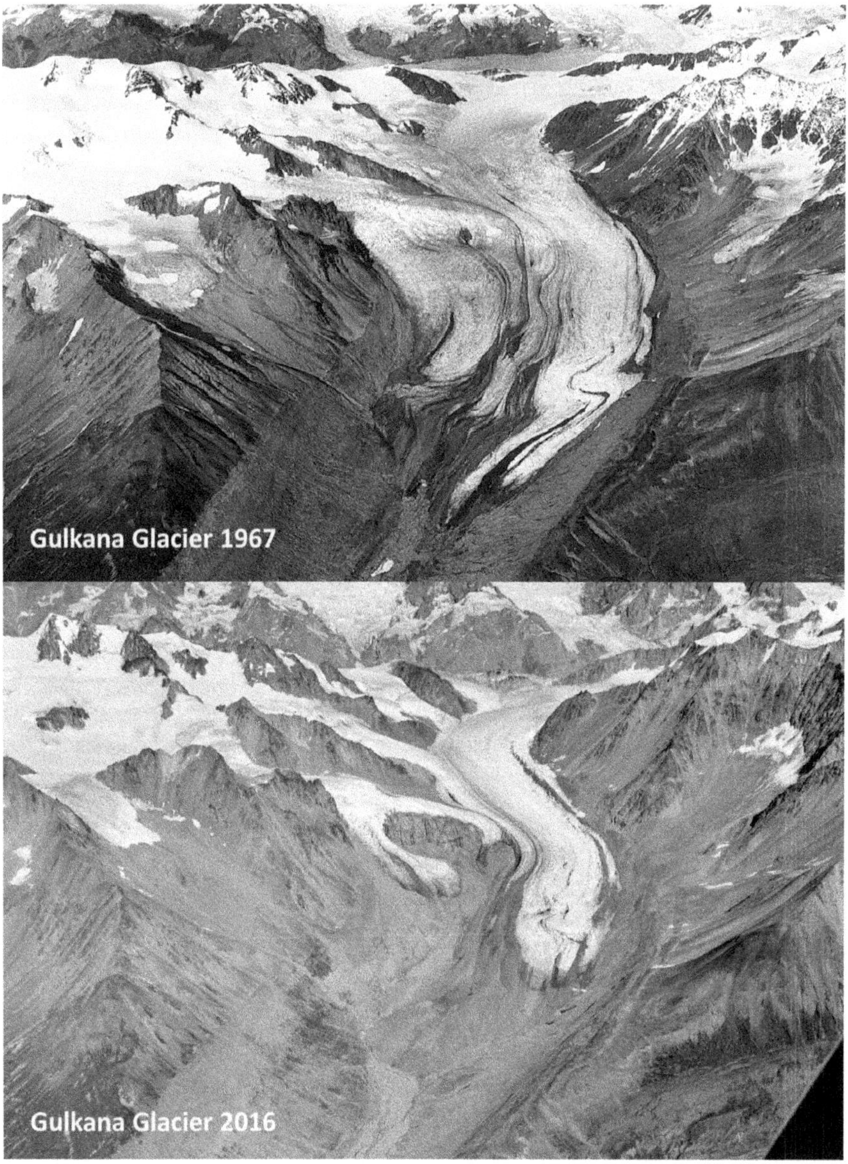

Figure 3.1 Gulkana Glacier, Alaska, 1967 and 2016.

be it for pilgrimage, leisure, research, health, adventure, or mixed motives. Mountains are sites of overlapping endeavours. Obstinate in their imposing materiality, these protuberances of rock and ice are also political, involving complex relations with human geopolitical schemes, sometimes frustrating, other times enabling (Scott, 2009; Debarbieux and Rudaz, 2015; Ferrari *et al.*, 2019). In the politics of knowledge-making mountain environs have been mobilized to bolster scientific credibility, rendering them also sites of epistemic authority (Hevly, 1996; Lane, 2009). Lofty calls to imagining, they are and have been experienced as sacred, sublime, romantic, and horrifying (Schama, 1995; Nicolson, 1997; Macfarlane, 2003). Havens of sacrosanctity for believers of all types, even the faithless find a potency in mountains as distillations of high ideals and aspirations reified (Bernbaum, 1998). Yet mountains are also vessels for the all-too-human. They are gendered and racialized, as when individual peaks such as Chomolungma (Everest) are deemed female, or when mountain folk are racially narrated as backward and primitive or, conversely, as paradigms of vigour and virtue (Debarbieux and Rudaz, 2015: 74–87). In short, mountains are thoroughly humanized geographies.

Mountains bring the cryosphere into close contact with human life. As a result, the mountain cryosphere is more directly and intentionally modified than other aspects of the cryosphere (Dodds, 2018: 148–150). For instance, residents of the Hindu Kush and Karakoram have for centuries practiced a form of icy matchmaking called glacier grafting that involves placing male and female chunks of ice in cool, shady spots above 4,000 metres before the onset of winter and then awaiting the birth of glacial progeny (Faraz, 2020). This tactic is rumoured to have halted Genghis Khan's advancing armies (Shaheen, 2016: 5). Less storied interventions tend to be more technocratic, economically oriented, and functionalist. The ski industry blankets underfed slopes in snow generated by fossil-fuelled engines and frees overfed ones of their precarious burdens with explosives dropped from helicopters; Swiss denizens protect cherished ice streams by draping them in white, shroud-like cloths; Andean glaciers are mined for ice. Artificial glaciers such as Stupas (Figure 3.2) functionally mimic, but little resemble, the real thing: contrivances that exploit variations in altitude and seasonal temperatures, they store and release water when it best suits human needs (though which humans is always a question). Artificial glaciers underscore an element common to many modifications of the mountain cryosphere. By reducing glaciers to a functional definition that focuses exclusively on their role as 'water towers' they underscore the promissory character of technological fixes to climate disruption.

This intimate and immediate entanglement of the mountain cryosphere with human life suggests that humanities scholars should pay attention. In this chapter, I seek to demonstrate the fecundity of the mountain cryosphere as an object of study by considering its unique ontological features (traits that mark it out as an object in the world): geographical dispersion; altitude; interiority; liminality; and uncanny temporality. While not all these features

Figure 3.2 Canadian Pacific advertisement featuring mountaineering, 1910. Note how the profile of the peak in the background, Mount Assiniboine, resembles that of the Matterhorn.

are exclusive to the mountain cryosphere, together they are constitutive of it. They structure the following sections where I use them to navigate recent work in history, anthropology, mountain studies, and literature that displays the potential of the mountain cryosphere to enrich our understanding of human–ice relations. Beyond narratives of global collapse, the mountain cryosphere harbours stories of race, gender, the phenomenology of ice, border politics, and time travel. Finally, in intimate and immediate contact with many different forms of human life, the mountain cryosphere invites us to 'listen for different stories' of life and beyond (Cruikshank, 2005: 74).

Lofty ontologies

At first pass, the mountain cryosphere is easy enough to define. It is that portion of the cryosphere that occurs in mountainous regions. Such conceptual partitioning of the cryosphere is uncontroversial. Scientists often carve out either Arctic sea ice and/or the Antarctic or Greenland ice sheets as special objects for study, and humanists follow suit. Yet, unlike these more well-known partitions (but rather like permafrost and snow), the mountain cryosphere lacks geographical continuity or even regional proximity. Mountains are found on every continent and display a vast range of conditions. Eighteenth-century Europeans took the Alps as an ideal standard, unjustifiably assuming a uniformity of mountain regions and the prototypicality of Europe (Debarbieux and Rudaz, 2015: 35); they saw Alpine glaciated peaks mirrored in the New Zealand Alps, the 'Canadian Alps' (Figure 3), and the 'American Alps' of Colorado. These assumptions buckled in the twentieth century as geographers and geologists came to appreciate the variation in mountain forms, formation, and character. For much of the century, it was unclear what even counted as a mountain (ibid.: 39).

In the 1990s, the United States Geological Survey pushed for a scientific consensus on criteria for adjudicating mountains. Based on considerations of slope, altitude, and topographic relief they deemed places over 2,500 metres above sea level to be mountainous; high plateaux and plains were ruled out by a minimum slope threshold (Price, 2015: 3). The mountain cryosphere, then, is that part of the cryosphere where frozen water forms around 2,500 metres, given certain conditions of climate and latitude. The actual distribution of ice and snow comprising it fluctuates according to altitude and latitude and the vicissitudes of seasons and years. It is spatially and temporally mercurial. Thankfully, such definitional porosity does not pose an insurmountable challenge to the ice humanist. Indeed, it invites investigation for those who understand critique to be a task for the humanities.

If the ontology of the mountain cryosphere marks it as diffuse, disconnected, porous, and fluctuating, its specific ontology is determined by verticality. It is the lofty analogue to the voluminous 'wet ontologies' of the oceans theorized by blue humanities scholars (Steinberg and Peters, 2015). As with wet ontologies, the vertical ontology of the mountain cryosphere offers opportunities for rethinking planar notions of territory and for cultivating a politics of verticality that accounts for 'reach, instability, force, resistance, incline, depth, and matter alongside the simply vertical' (Elden, 2013: 45). Eyal Weizman (2002) and James C. Scott (2009) have highlighted the importance of verticality for politics in arid contexts. Adding ice with its dynamism and vitality to the equation enhances the potential for verticality to enhance critical understandings of space. Thinking with the mountain cryosphere, then, brings fluidity, dynamism, and the material realities of frozen water into association with the physical and imaginative features of verticality.

Altitude: all-too-human verticality

The vertical dimension of mountains is a specific type: altitude. The idea of altitude as a universal topographical feature emerged in the seventeenth century, but was not commonplace until the eighteenth century. Altitude is not merely an axis on a three-dimensional field. Changes in altitude entail changes in air pressure, UV exposure, climate – the latter of course being necessary for the formation and maintenance of the mountain cryosphere. Alexander von Humboldt's *Tableaux* of Chimborazo and other mountains made beautifully clear that altitude also entails biological changes. High places engender unique adaptations in organisms. The Tibetan yak, for example, is so suited to living above 3,000 metres that it takes ill below that threshold (Della Dora, 2016: 78). The heights also prompt physiological reactions in unconditioned human bodies, which may include increased haemoglobin counts, swelling of tissues in the lungs and brain, and fatigue. As mountaineers well know, snow and ice exacerbate these physiological challenges by posing complex terrain to be navigated on less aerobic fuel and by reflecting burning UV rays into tender cavities like nostrils or the roof of the mouth.

The history of physiology offers a window into how race intersects altitude. How Sherpa, Gaddi, Gujari, and other Indigenous mountain people live, work, and play at heights that sicken shore dwellers has been a recurring obsession for Western scientists, and this research has rarely been divorced from racist categorizations and bioprospecting. As Vanessa Heggie has shown, adaptation to cold latitudes has generally been conceived as technological, while that to altitude has been considered biological and often racially specific. White polar explorers readily adopted, modified, and erased the origins of igloos, dog sleds, and pemmican in their races to the poles. Mountain dwellers, on the other hand, posed a trickier puzzle. Did they thrive at altitude because of short-term acclimatization or because of racially exclusive evolutionary adaptations (2019: 134–140)? Even without knowing the cause, turn-of-the-century British mountaineers ranked Himalayan peoples according to altitude and usefulness to British expeditions. At a time when climbing techniques and equipment were adapted to snow and ice routes, this meant that certain peoples were believed to be better suited to the mountain cryosphere than others. Gurkhas 'never failed us,' wrote mountaineer Tom Longstaff in 1908, 'they never complained, and they never lost their cheerfulness [...] They were superior to the best Garhwalis I have met, and even to the Bhotias, so I need hardly add that they bore no resemblance whatever to the Kumaoni or the down-country native' (364). In the early twentieth century, the Sherpa, partly as the result of their own efforts, emerged as the most desirable porters for technical, high-altitude work (largely for the characteristics Longstaff attributed to the Gurkhas) (Ortner, 1999: 58). The 1924 British Mount Everest Expedition sorted porters into racialized categories of low altitude or 'local' porters (mostly Tibetans) and high-altitude porters (mostly Sherpa). These vertical racial hierarchies held through much of the rest of the century, with ethnic Sherpa dominating high-altitude work in the Himalaya. Elsewhere, local Indigenous guides were often deemed inferior to white guides when it came to snow and ice work. At the end of the nineteenth century Swiss guides were imported to support the nascent climbing scene in the 'Canadian Alps'. Altitude served as a ladder onto which European mountaineers mapped racialized bodies, with the male, white body serving as the 'norm', and the highest, glaciated spaces reserved for the supposedly strongest and most virtuous.

Race is not the only human category to be mapped onto altitude. In many cultures, mountains are gendered. Chomolungma has already been mentioned. For Ladakhis, glaciers are male and lakes are female (Gagné, 2018: 110); and in the folklore of Gilgit Baltistan, grey, debris-covered, slow-moving glaciers are *po-gang* (male) and white or blue, fast-moving ones are *mo-gang* (female) (Faraz, 2020). In these instances, gender is a social category, marking mountains and glaciers as fellows in human-nonhuman societies that suggest ways of conceiving of mountain glaciers other than as (merely) sentinels of global warming – be it according to an ethics of care (Gagné, 2018) or in terms of kincentric thinking (Whyte, 2020: 4).

The mountain cryosphere was, and remains, gendered in less literal ways as well. In the nineteenth century, Europeans believed the heights were no place for ladies, who were thought to be physiologically suited to valleys and mid-slope sallies. Women's bodies were deemed less stable than men's and therefore less fit for the stresses of altitude and technical terrain. Michael Reidy (2015) has demonstrated the lengths to which this idea was promulgated by British physicist and mountaineer John Tyndall. Just as Humboldt carved the mountainside into biogeographical steps and British mountaineers ranked Himalayan indigenes according to elevation, Tyndall drew vertical hierarchies of gender. The high, snowiest regions excluded all things feminine; they were where Tyndall could 'feel in all [his] fibres the blessedness of perfect manhood' (quoted in Reidy, 2015: 163). Such notions held sway in male climbing cadres into the twentieth century to the extent that the presence of women on high, glacier-ringed peaks negated their existence. 'The [Aiguille du] Grépon has disappeared', one male climber proclaimed in 1929 after it was summited by Miriam O'Brien Underhill and Alice Damesme, 'as a climb it no longer exists. Now that it has been done by two women alone, no self-respecting man can undertake it' (quoted in Underhill, 1971: 158). Poet Helen Mort has recently toyed with this idea:

> When we climb alone
> en cordée feminine,
> we are magicians of the Alps –
> we make the routes we follow
> disappear. (2016: 2)

Mort's cheeky verse rightly implies that there were women climbers prepared to 'magically' prove physiology and convention wrong. Clare Roche finds that the climbs of middle-class Europeans such as Katherine Richardson, Lucy Walker, Margaret Jackson, and Meta Breevort 'contest the notion that mountaineering was a uniquely male activity' in the late nineteenth century (Roche, 2013: 2). All-woman and mixed parties were to be found on the crevassed slopes of the Mer de Glace, doing difficult, technical climbing. The glaciers of the high peaks and passes, though culturally coded as masculine, were stomping grounds for women too; the complex social dynamics of this situation are only just being probed by scholars.

Gender in the high alpine never operates alone: it is always situational and intersectional. In 1874, an ostensibly all-women's expedition lauded for their 'pluck' commented that 'amongst the mountains, particularly the glaciers, guides were much more useful than gentlemen' (ibid.: 11). Working-class guides and porters, who hauled equipment and cut steps in the ice, did not really count as men. At altitude, gender was conditioned by class and occupation, and this mattered for who was memorialized as an ascensionist. Like countless women technicians in the history of science, mountain labourers were rendered doubly invisible by their gender and class: invisible in their time and invisible to history (Oreskes, 1996: 101). Intersectional studies of gender at altitude beckon ice humanists. The European cultural-coding of mountain glaciers as masculine spaces and the presence of women 'trespassers' have received scholarly attention, as have the colonial aspects of mountaineering. Yet, how specific nexuses of class, race, colonial ambition, alternative forms of femininity and masculinity operated (and operate) in the mountain cryosphere remains a relatively untilled field.

Glacial deeps

Altitude is not the only form of verticality to mark the mountain cryosphere; glaciers themselves have a vertical dimension. Glaciers, though often depicted

Figure 3.3 Ice stupa in the Zanskar region of Ladakh.

from above or from the terminus (Figure 3.1), are places of interiority and depth. They may be entered, and mountain glaciers frequently are. Doing so gives them a profound placeful-ness of their own. Crevasses offer one way in. Often considered obstacles, these surface cracks are also portals to its dark interior, as we'll see, itself a portal – a time-travelling machine of sorts.

Mountaineers viscerally appreciate the transportive powers of crevasses. Most climbers don't enter willingly – though many have used them to traverse the ice or to study it – more often, they fall in. Then they are rarely in a position to enjoy their unique surroundings. Yet glacial interiors are capable of inspiring reflection and admiration. Perhaps the most well-known account of a glacier's interior is Joe Simpson's in *Touching the Void* (1997). After his climbing partner cuts the rope on which he dangles hopelessly, Simpson plummets onto an Andean glacier, shatters a snow bridge, and lands on an ice shelf midway down an enormous crevasse. Initially, he experienced the dark, silent, seemingly bottomless crevasse as a space of existential horror: 'the darkness beyond the light gripped my attention. I could guess what it hid, and I was filled with dread' (112). Unable to ascend, Simpson descends into the void. This route becomes his salvation when he finds a snow slope leading to the glacier's surface. Once he glimpses a way out his feelings about the place shift:

> For all its hushed cold menace, there was a feeling of sacredness about the chamber, with its magnificent, vaulted crystal ceiling, is gleaming walls encrusted with a myriad fallen stones. (135)

Sometimes the way to the light is through darkness. And sometimes ethereal sublimity can be appreciated only when there is a chance of survival.

Simpson learned the hard way what mountaineers know more commonly: glacier interiors are transportive. Yet, if on one level mountaineers appreciate the transportive qualities of glacier interiors, on another they seem heedless. Since 1971, the 34,000 people who have attempted Denali's West Buttress Route have left approximately 66 tonnes of faeces in the Kahiltna Glacier (Goodwin *et al.*, 2012: 434). In 2019, a Sherpa clean-up team removed 11 tonnes of trash and four bodies from Chomolungma, two from the dangerously shifting Khumbu Icefall (Sharma 2019; Khadka, 2019). They barely made a dent in the litter or the body count. The interiors of mountain glaciers, it seems, house more than just stories of existential dread and sublimity; they contain less elevated matters too.

Glacial interiors transport the matter they harbour, often on timescales that make them seem time-travelling machines. In 1991, the remains of a Neolithic Alpine traveller clothed in fur trousers and a woven shawl appeared on the surface of the Similaun Glacier and was christened Ötzi; in 1999, Kwäday Dän Ts'inchi (Long Ago Person Found), the body of a traveller from 300–500 years ago, emerged from a glacier in the Wrangell–St Elias Range and was claimed by the Tlingit; glacial archaeologists uncover medieval artefacts as Norwegian ice retreats. These ancient glacier revenants seem to compress time. As Peter Hansen has shown, museum curators, journalists, and publics seek to reconstruct recognizable people from glacier-expelled bodies. Ötzi's remains were used to build an archetypal model of Neolithic man. The model, Hansen writes, is 'fit, muscular, and bare-chested […] has a grizzled face and piercing gaze, caught in midstride as he glances to his left' (Hansen, 2013: 302) – a muscled older brother to modern heteronormative 'Man'. The director of the South Tyrol Museum where Ötzi resides exclaimed, 'we can finally look him in the eyes and recognize, to our amazement, that he is really one of us!' (ibid.). Here, time travel through glacial interiors is rendered evidence of a transhistorical universal humanity. Yet, the treasures revealed by retreating ice may also take on specific ethnic or national significances as attested to by Kwäday Dän Ts'inchi and the Viking finds of Norway.

The interiors of mountain glaciers are portals – metaphorical and literal. Contemplating the interiority of ice produces stories on different registers than that of global disaster. It brings their active materiality into relief, compelling us to recognize glaciers as all-encompassing environments in their own right, with the power to generate myriad experiences and stories.

Liminal spaces

Altitude and depth are not the only ontological features of the mountain cryosphere that make it a fecund object for humanist investigation. Mountain geographies are liminal. 'Conceptualized as natural barriers and [...] metaphorically associated with obstacles and challenges' (Della Dora, 2016: 12), mountains have often anchored borders, national or otherwise. Yet for a long time, mountain spaces, particularly the highest reaches, were not deemed politically valuable. Home to gods or demons, the frozen heights were alien to mundane politics. With the rise of nation states in the eighteenth century, natural barriers became national borders: the presumed fixity and self-evidence of the former bolstered the political legitimacy of the latter. At the same time the 'conquest' of mountains became a way of expressing national and imperial sentiments. Waving handkerchiefs from the summit of 'les Glacières' was part of the making of modern European statehood (Hansen, 2013). Since then, the mountain cryosphere has been intensely political, a space of conflict and contestation, and new possibilities; a place of oropolitics (Sircar, 1984).

Territories defined by mountain glaciers and watersheds have long undergirded claims to national and ethnic unity (Hansen, 2013; Debarbieux and Rudaz, 2015; Ferrari *et al.*, 2019). In Europe, the Treaty of Utrecht (1713) marked the formal beginning of a European tradition in which 'natural' borders of nations were defined by major mountain watersheds. Recent changeability in the mountain cryosphere has complicated such assumptions of territorial fixity. Cartographers, once content to sketch lines along seemingly empty white expanses on the map, face new challenges in a world of cryospheric transformation and increasingly precise surveillance. Territorial demarcations once believed solid are flowing away in melt streams whose courses shift with newly revealed topographies. And as glaciers melt, they 'pirate' rivers away onto new courses in a geological blink of an eye (Shugar, 2017).

Italy offers a striking example of how nations have responded to a changing mountain cryosphere. In 2005 and 2009 the Italian government signed bilateral agreements with Austria and Switzerland that redefined their Alpine borders as 'moving borders' – 'no longer permanently fixed, but [...] depend[ing] on the gradual changes caused by the erosion and reduction of the glaciers, up the extreme instance of their disappearance' (quoted in

Ferrari *et al.*, 2019: 61). Receding ice has joined war, annexation, and treaties as a means of altering borders. Between 2013 and 2016, the Italian Limes Project used cartography and art to explore the implications of this arrangement by allowing the pulsating Gräfferner Glacier on the Italian–Austrian border to 'speak for itself'. Using an elaborate translation apparatus of sensors, solar panels, pantograph, helicopters, software, data, and human bodies and minds, they managed to transcribe the micromovements of snow and ice in real time: to translate 'the moving border legislation into the manufactured image of a "natural" process' (ibid.: 178). In the end, however, the imprecision introduced by the thickness of the pen drawing the glacier's movements 'dissolved' the 'obsessive nature of measurement' (ibid.). The project's ultimately futile attempts to catch a glacier and pin it down reveal how the mountain cryosphere, though liminal, frustrates territorial ambitions. Territories always were, and remain, the authors contend, constructed and imprecise.

As a border zone to be guarded and fought for, the mountain cryosphere radically alters the practice of war, rendering it especially lethal by introducing 'strong gusts of wind, lightning, falling rocks, avalanches, crevasses, blizzards, extreme cold, and in some cases, lack of oxygen' to the usual dangers of combat (Della Dora, 2016: 94). The most infamous war-torn glacier is the Siachen, resting between 5,425m and 7,773m on the Pakistani–Indian border in the Karakoram Himalaya. Since the 1947 partitioning of India and the Indo-Pakistani War over the Jammu and Kashmir border region, conflict here has flickered on and off, unhindered by a 1949 UN-negotiated ceasefire that vaguely described the border as running 'north to the glaciers' (Ramachandran, 2016). In 1984, the glacier itself was weaponized, with specially trained mountain troops stationed at 6,000 metres. Most deaths are caused by the glacier, not enemy fire (ibid.). The Siachen, like many Karakoram glaciers, is fed primarily by avalanches from above; this makes for a highly dynamic and dangerous glacier, moving unpredictably, riddled with crevasses and icefalls (Hewitt, 2011: 189). In the largest avalanche to date, in 2012 an entire Pakistani battalion headquarters was swept away killing 129 soldiers and 11 civilians (ibid.). Soldiers have also succumbed to crevasse falls, –40°C blizzards, and pulmonary and cerebral oedema (inflammation and fluid in the lungs or brain brought on by stays at high altitude). The glacier shapes warfare: soldiers' bodies and minds, supply lines, equipment, and tactics must all conform to the unique challenges of glacial warfare (Smith, 2018).

And, as Kristen Smith has demonstrated, warfare shapes the glacier (2018). Chemicals, microplastics, heavy metals, and antibiotics have been traced from the upper waters of the Indus to the encampments on the Siachen (Kemkar, 2006: 16).The biodiversity of megafauna in the region has decreased; microbial communities living within the glacier and its meltwater appear to have developed heavy metal and antibiotic resistances. The glacier itself is littered with downed helicopters and decrepit military vehicles, telephone wire, gun barrels, burned or crushed shelters, the remains of gun shelling, and the bodies of unrecovered soldiers and civilians, in addition to the 2,000 pounds of human waste deposited into crevasses daily (ibid.). 'Years of military pollution have turned even the glacier itself into a casualty', Smith observes. 'I can't drink water', confesses an anonymous Kashmiri writer, 'it is mixed with the blood of young men who die up in the mountains' (quoted in Bose, 2003: 5).

Warfare is the most extreme example of a truth forcefully demonstrated by the liminality of the mountain cryosphere: the geopolitical is also always the environmental. It is also the domestic, something that Karine Gagné perceptively explores in her ethnography of Ladakhi relations with their home glaciers (2018). The Siachen isn't the only part of the India–Pakistan border to be militarized: the entire Jammu and Kashmir region is occupied. Tucked between the Karakoram and the Himalaya, it is home to Buddhists and Muslims who herd and farm in the high valleys fed by glacial run off. Since 1947, the region has been remade from a peripheral concern for Delhi to a zone of perpetual surveillance and patrol. As Gagné shows, this history shapes relations with glaciers. Military and bureaucratic work in centres like Leh pull young people from the mountainsides, leaving older Ladakhis to recall traditional obligations to the glaciers that go unfulfilled. Cryospheric change is experienced and understood as a matter of care and affect. Within this nexus of intrusion, imbalance, and memory, elders dismiss global warming as the reason for glacial retreat, preferring explanations grounded in unfulfilled obligations. They recall a time when young people cared for the glaciers. 'Look at what people are doing today, always running after money and not taking care of things', avers one older woman, 'to care for the glacier, you have to see the glacier, you have to know the glacier, like you know a friend. People have become careless, unmindful, and that is why there is no more snow' (ibid.: 140).

A home to be cared for, a natural national border to be claimed and secured, and a site of drastic environmental alteration, the liminality of the mountain cryosphere contains stories of overlapping geographies and histories.

Uncanny temporalities

Mountain glaciers straddle multiple timescales. Already we have seen how they are repositories of the past; they are also fellow mortals of the present, and, as commonly recognized, icons of alarming futures. These timescales coexist uneasily. Mountains have long been symbols of eternity, removed from the flow of time, their ponderous movements too slow to detect from the brief perspective of a human life. Not of perishable stuff, they are the dwelling places of deities. The four mountains circumscribing Diné territory in the American southwest are the abodes of gods (Powell, 2015: 59); Kailas in Tibet and Annapurna in Nepal house Buddhist deities; the Hebrew god met Moses on Mount Sinai; and Gabriel first appeared to Muhammad atop Jabal el-Nur (Mountain of Light). Mountains have been the abodes of holy hermits (Hollis, 2017), and dens of diabolical beasts (Bernbaum, 1998). In either register, they are symbols of eternity, regarded as unchanging, cosmic spaces. So too are mountain glaciers: the everlasting snows. And with good reason. Laid down over thousands of years through slowly accruing precipitation or by the gradual withdrawal of continental ice sheets, mountain glaciers are products of deep time.

Yet like the rest of the cryosphere today, mountain glaciers inhabit more than just geological timescales. The Great Acceleration rushes ice to its liquid form at unsettling rates. In the mountains, the disconnect between timescales is emphatic. Mountain glaciers are especially vulnerable to global warming (Zemp *et al.*, 2019). Between 2000 and 2014, remote sensing imagery showed rapid retreat and substantial downwasting of Alpine glaciers (Sommer *et al.*, 2020). The Monte Perdido Glacier in northern Spain survived the Roman period and the Medieval Climatic Anomaly only to recede dramatically in the past thirty years; none of the ice accumulated during the Little Ice Age is left (Moreno *et al.*, 2020). The glaciers of the Canadian Rockies are projected to lose roughly 70 per cent of their 2015 volume by 2100 (Clarke *et al.*, 2015: 372); the Himalayan ice streams, though they present a more complicated story, show an overall loss of mass (Das and Chakraborty, 2019: 1). Such realities give rise to narratives that figure the mountain cryosphere as an icon of global warming.

Mountain dwellers will outlive their local glaciers. Mountaineers return to routes rendered unrecognizable or impassable a decade later. Receding glaciers alter the courses of rivers in a matter of days. Our temporal expectations of mountain glaciers, rooted in ideas of eternity and deep geological time, are upended. Glaciers, products of invisibly slow geological accrual, are being unmade at rates observable in a single summer. This play of invisible and visible time places them in an uncanny position. It is as if the mountains themselves were disappearing before our eyes. Yet if glaciers are icons of global warming, they are also icons of solastalgia: psychological distress caused by rapid environmental change, a longing for 'home' induced not by travel or dispossession but by radical, externally imposed transformation (Albrecht *et al.*, 2015: S96). While many in the urban, industrialized West regard the retreat of mountain glaciers as a sign and symptom of a future climate apocalypse, for others their retreat is much more immediate and personal. This intimacy heightens the uncanniness of conflicting timescales. Some of us will outlive the glaciers of our youth that then seemed so immutable.

In the taxonomy of the frozen Earth the mountain cryosphere is an odd beast. Defined not by geographic continuity but by verticality and topographical features, it uncannily occupies multiple timescales as one of the hotspots of global warming as well as being the most accessible and directly intervened upon aspect of the cryosphere. Permeated with human bodies and refuse, infused with ideas about gender and race, abodes of dark fears and sanctuaries of high ideals, suffused with ethical feelings of home and care, and perforated with technical apparatuses of science- and state-making, mountain glaciers are thoroughly humanized. Intimate and immediate, the mountain cryosphere, and our relations with it, are as varied as its dispersed and heterogeneous manifestations making it a rich topic for the ice humanities, telling much more than simply stories of cryospheric collapse.

References

Albrecht, G., G.-M. Sartore, L. Connor *et al.* (2015) 'Solastalgia: The Distress Caused by Environmental Change', *Australian Psychiatry* 15(Suppl): S95–S98.

Bernbaum, E. (1998) *Sacred Mountains of the World*. Berkeley, CA: University of California Press.

Bose, S. (2003) *Kashmir: Roots of Conflict, Paths to Peace*. Cambridge, MA: Harvard University Press.

Carey, M. (2007) 'The History of Ice: How Glaciers Became an Endangered Species', *Environmental History* 12(3): 497–527.

Clarke, G. K. C., A. H. Jarosch, F. S. Anslow, V. Radić, and B. Menounos (2015) 'Projected Deglaciation of Western Canada in the Twenty-first Century', *Nature Geoscience* 8(6 April): 372–377.

Cruikshank, J. (2005) *Do Glaciers Listen?: Local Knowledge, Colonial Encounters, and Social Imagination*. Vancouver: University of British Columbia Press.

Das, S. and M. Chakraborty (2019) 'Mass Balance of the Himalayan Glaciers and their Regional Variations', *Journal of Geography, Environment and Earth Science International* 19(3): 1–11.

Debarbieux, B. and G. Rudaz (2015) *The Mountain: A Political History from the Enlightenment to the Present*. Chicago, IL: University of Chicago Press.

Della Dora, V. (2016) *Mountain: Nature and Culture*. London: Reaktion Books.

Dodds, K. (2018) *Ice: Nature and Culture*. London: Reaktion Books.

Elden, S. (2013) 'Secure the Volume: Vertical Geopolitics and the Depth of Power', *Political Geography* 34: 35–51.

Faraz, S. (2020) 'The Glacier "Marriages" in Pakistan's High Himalaya', *thethirdpole.net* June 3. Available at: www.thethirdpole.net/2020/06/03/the-glacier-marriages-in-pakistans-high-himalayas/ (accessed 20 August 2020).

Ferrari, M., E. Pasqual, and A. Bagnato (2019) *A Moving Border: Alpine Cartographies of Climate Change*. New York: Columbia Books on Architecture and the City, ZKM/Center for Art and Media Karlsruhe.

Gagné, K. (2018) *Caring for Glaciers: Land, Animals, and Humanity in the Himalayas*. Seattle, WA: University of Washington Press.

Goodwin, K., M. G. Loso, and M. Braun (2012) 'Glacial Transport of Human Waste and Survival of Fecal Bacteria on Mt McKinley's Kahiltna Glacier, Denali National Park, Alaska', *Arctic, Antarctic, and Alpine Research* 44(4): 432–445.

Haeberli, W., J. Oerlemans, and M. Zemp (2019) 'The Future of Alpine Glaciers and Beyond', in *Oxford Research Encyclopedia of Climate Science*. Oxford: Oxford University Press, 1–36.

Hansen, P. H. (2013) *The summits of modern man: Mountaineering after Enlightenment*. Cambridge, MA: Harvard University Press.

Heggie, V. (2019) *Higher and Colder: A History of Extreme Physiology and Exploration*. Chicago, IL: University of Chicago Press.

Hevly, B. (1996) 'The Heroic Science of Glacier Motion', *Osiris* 11(1): 66–86.

Hewitt, K. (2011) 'Glacier Change, Concentration, and Elevation Effects in the Karakoram Himalaya, Upper Indus Basin', *Mountain Research and Development* 31(3): 188–200.

Hock, R., G. Rasul, C. Adler *et al.* (2019) 'High Mountain Areas', in H.-O. Pörtner, D. C. Roberts, V. Masson-Delmotte *et al.*, eds, *IPCC Special Report on the Ocean and Cryosphere in a Changing Climate*. Cambridge: Cambridge University Press, 131–202.

Hollis, D. L. (2017) 'Rethinking Mountain Gloom', *Alpinist* 57(spring): 105–108.

Inkpen, D. K. (2018) 'Frozen Icons: The Science and Politics of Repeat Glacier Photography, 1887–2010'. PhD dissertation, Harvard University.

Kemkar, N. A. (2006) 'An Environmental Peacemaking: Ending Conflict Between India and Pakistan on the Siachen Glacier Through the Creation of a Transboundary Peace Park', *Stanford Environmental Law Journal* 25(1): 1–56.

Khadka, N. S. (2019) 'Melting Glaciers Reveal Everest Bodies', *BBC News* 21 March. Available at: www.bbc.com/news/science-environment-47638436 (accessed 30 July 2020).

Lane, K. M. D. (2009) 'Astronomers at Altitude: Mountain Geography and the Cultivation of Scientific Legitimacy', in D. Cosgrove and V. Della Dora, eds, *High Places: Cultural Geographies of Mountains, Ice and Science*. London: IB Taurus, 126–144.

Longstaff, T. G. (1908) 'A Mountaineering Expedition to the Himalaya of Gharwal', *Geographical Journal* 31(4): 361–388.

Macfarlane, R. (2003) *Mountains of the Mind: A History of Fascination*. London: Granta Books.

Moreno, A., M. Bartolomé, J. I. López-Moreno *et al.* (2021) 'The case of a southern European glacier which survived Roman and medieval warm periods but is disappearing under recent warming', *The Cryosphere* 15(2): 1157–1172.

Mort, H. (2016) *No Map Could Show Them*. London: Chatto and Windus.

Nicolson, M. H. (1997) *Mountain Gloom and MOUNTAIN GLORY: The Development of the Aesthetics of the Infinite*. Seattle, WA: University of Washington Press.

Oreskes, N. (1996) 'Objectivity or Heroism?', *Osiris* 11(1): 87–113.

Ortner, S. (1999) *Life and Death on Mount Everest*. Princeton, NJ: Princeton University Press.

Powell, D. (2015) 'The Rainbow is our Sovereignty: Rethinking the Politics of Energy on the Navajo Nation', *Journal of Political Ecology* 22(1): 53–78.

Price, M. (2015) *Mountains: A Very Short Introduction*. Oxford: Oxford University Press.

Ramachandran, S. (2016) 'Does India Really Need to Defend the Siachen Glacier?', *Diplomat* 23 February. Available at: https://thediplomat.com/2016/02/does-india-really-need-to-defend-the-siachen-glacier/ (accessed 20 August 2020).

Reidy, M. S. (2015) 'Mountaineering, Masculinity, and the Male Body in Mid-Victorian Britain', *Osiris* 30(1): 158–181.

Roche, C. (2013) 'Women Climbers 1850–1900: A Challenge to Male Hegemony?', *Sport in History* 33(3): 236–259.

Schama, S. (1995) *Landscape and Memory*. New York: Alfred A. Knopf.

Scott, J. C. (2009) *The Art of Not Being Governed: An Anarchist History of Upland Southeast Asia*. New Haven, CT: Yale University Press.

Shaheen, F. A. (2016) *The Art of Glacier Grafting: Innovative Water Harvesting Techniques in Ladakh*. Water Policy Research Highlight 8. Colombo, Sri Lanka: International Water Management Institute.

Sharma, G. (2019) 'Nepal Picks Up Four Bodies, 11 Tonnes of Garbage in Everest Clean-up', *De.Reuters.com* 5 June. Available at: www.reuters.com/article/nepal-everest-idINKCN1T61RC (accessed 21 March 2022).

Shugar. D., J. J. Clague, J. L. Best *et al.* (2017) 'River Piracy and Drainage Basin Reorganization Led by Climate-driven Glacier Retreat', *Nature Geoscience* 10(17 April): 370–375.

Simpson, J. (1997) *Touching the Void*. London: Vintage Books.

Sircar, J. (1984) 'Oropolitics: A Dissertation on the Political Overtones of Mountaineering in the East-Central Karakoram, 1975–82', *Alpine Journal*: 74–79.

Smith, K. (2018) 'Mass Balance, Accumulation Dynamics and High-Altitude Warfare: The Siachen Glacier as a Battlefield', paper given 3 October at Thinking Mountains conference, 2–5 October 2018, Banff, Canada.

Sommer, C., P. Malz, T. C. Seehaus, S. Lippi, M. Kemp, and M. H. Braun. (2020) 'Rapid Glacier Retreat and Down Wasting Throughout the European Alps in the Early 21st Century', *Nature Communications* 11(1): article 3209. https://doi.org/10.1038/s41467-020-16818-0.

Steinberg, P. and K. Peters (2015) 'Wet Ontologies, Fluid Spaces: Giving Depth to Volume Through Oceanic Thinking', *Environment and Planning D: Society and Space*, 33(2): 247–264.

Underhill, M. (1971) *Give Me the Hills*. New York: Viking Press.

Weizman, E. (2002) 'Introduction to the Politics of Verticality', *Open Democracy*. Available at: www.opendemocracy.net/ecology-politicsverticality/article_801.jsp (accessed 20 July 2020).

Whyte, K. P. (2020) 'Too Late for Indigenous Climate Justice: Ecological and Relational Tipping Points', *WIREs Climate Change* 11(e603): 1–7.

Zemp, M., M. Huss, E. Thibert *et al.* (2019) 'Global Glacier Mass Changes and their Contributions to Sea-level Rise from 1961 to 2016', *Nature* 568(7752): 382–386.

4

Glacier protection campaigns: what do they really save?

Mark Carey, Jordan Barton, and Sam Flanzer[1]

For many people, lamenting glacier retreat and ice loss is simply not enough. Over the last two decades, researchers, residents, activists, and engineers have increasingly resorted to concrete work to preserve parts of the cryosphere. Environmentalists once proposed covering Kilimanjaro with blankets to save the mountain's ice cap (Mason, 2003). The World Bank funded a project to paint rocks adjacent to Andean glaciers white, hoping this would raise the albedo and slow snow and ice melting (Collyns, 2010). Conservationists in Greenland successfully worked to list the Ilulissat Icefjord as a World Heritage site in 2004, to protect both the Sermeq Kujalleq Glacier and the ice-covered fjord (Mikkelsen and Ingerslev, 2002). There are even some dreamlike fantasies to save the world's ice, like building seafloor walls with robots to prevent warm ocean water from reaching glacier fronts, pumping cold water into glacier beds to help thicken the ice, spraying artificial snow across Antarctic ice sheets, or sending a submarine to the Arctic that makes artificial icebergs from desalinated seawater (Cox, 2018; Snowden, 2019).

These and the other glacier protection efforts discussed in this chapter demonstrate not just the utility of ice for surrounding communities, but also how glaciers have long captured the global imagination as a quintessential icon – and casualty – of global climate change (Carey, 2007; Orlove *et al.*, 2008; Bjørst, 2010; O'Reilly *et al.*, 2012; Sörlin, 2016; Mark and Fernández, 2017; Leane and Maddison, 2018). They also signal a shift from earlier efforts primarily to memorialize or lament lost ice to direct-action strategies (what we call 'glacier protection campaigns') to protect and preserve shrinking glaciers – at least that's the idea. But what does it mean to save glaciers? Who gets to speak for the ice in these glacier-saving campaigns, and who do the campaigns ultimately help? What other storylines, narratives, and agendas are embedded in glacier-protection campaigns? And how are glacier-saving efforts transforming access to and control over icy spaces?

This chapter helps answer these questions through an analysis not necessarily of the technical aspects of glacier-saving campaigns themselves, but rather of the underlying narratives and agendas embedded in media stories, news articles, lawsuits, policies, and reports about them. We thus follow this book's general approach to the 'ice humanities' by focusing on glacier representations (see the editors' introduction). Ice, after all, is more than a material object, even if existing research is dominated by studies on physical cryospheric changes and direct human impacts, rather than research on the ideas, aesthetics, emotions, discourse, and cultures of ice. Even fewer studies critically evaluate the way individuals and groups deploy ice as an object and icon, which is usually in rather one-dimensional ways that sound the climate alarm and lament lost ice without tackling root drivers of the climate crisis or addressing how the making of glacier icons can itself alter the politics and control of icy spaces, as some have noted (e.g. Bravo, 2009, 2017; Bjørst, 2010; Kaijser, 2013; Sörlin, 2015; Carey *et al.*, 2016; Dodds, 2021). Examining glacier-saving discussions and discourse through the ice humanities, however, helps expose not only how people make sense of melting ice but also how they use glaciers, which kind of solutions they propose to counter ice loss, who gets to act (and who is left out), and who or which entities have the authority to access and control ice-covered regions. Essentially, stories about glacier protection celebrate certain climate solutions while diminishing other roles and narratives. The stories also privilege a cast of main characters as the protagonists – the social actors and entities given agency, authority, and legibility to resolve the problem of ice loss and climate change, while others remain invisible, passive, or static. An ice humanities approach helps understand these actors, their authority, and the implications for mountain landscapes and residents.

Glaciers are instructive to analyse because – like polar bears, whales, and the Amazon rainforest (Slater, 2003; Born, 2019; Brito *et al.*, 2019; LeCain, 2019) – they have become high-profile imperilled icons that now inspire campaigns to save them through science, technology, laws, and policies. Analysing ways in which people celebrate or scrutinize glacier-saving campaigns helps illustrate how environmental features become icons or commodified wild nature – and what is at stake when certain environmental narratives spread (Fairhead and Leach, 1996; Davis, 2007; LeCain, 2019). These narratives also exist within discussions of climate change solutions. A climate-action imperative, sometimes inspired by scientific evidence documenting ice loss, can stimulate international development agendas that in some cases have facilitated neoliberal policies, top-down engineering and infrastructure, the hardening of social hierarchies and inequalities, and the perpetuation of colonialism (Bravo, 2009; Swyngedouw, 2010; Cameron, 2012). What's more, laws and policies implemented to protect landscapes, wildlife, and natural resources can often dispossess and harm Indigenous people, making it more urgent to understand how glacier-saving campaigns affect local residents (Guha, 1989; Nuttall, 2016; Flora *et al.*, 2018).

This chapter analyses the narratives and agendas related to five distinct glacier protection projects: (1) the use of insulating blankets to save Switzerland's Rhône Glacier (2) glacier protection at European ski resorts (3) the construction of artificial glaciers and ice stupas in India (4) Argentina's glacier protection law, and (5) the granting of legal personhood status for two glaciers in India. Our ice humanities approach to these glacier protection campaigns focuses more on the representations of ice than on the projects themselves. In fact, our point is not to criticize the projects or laws themselves, which are innovative and interesting. Rather, our goal is to expose implications of how those glacier protection campaigns are discussed and represented. We analysed hundreds of media stories, news articles, legal documents, policies, and published scholarship to uncover underlying narratives and agendas embedded in the glacier representations. Ultimately, as our chapter concludes, the narratives and agendas offer key insights about glacier *uses*, climate *solutions*, the *actors* involved, and *authority* over the ice-covered regions.

Blanketing an icon at the Rhône Glacier

Saving Switzerland's Rhône Glacier has become one of Europe's most iconic examples of glacier protection, but the approach to protecting it has involved blanketing the ice with tarps, thereby preserving some dying ice from damaging solar energy while blocking it from the very people trying to protect it. The Rhône Glacier has been a famous tourist destination since at least the nineteenth century. By 1882, Josef Seiler had opened the famous Belvedere Hotel alongside the glacier and carved a tunnel into it so visitors could literally go inside the ice and enjoy the stunning blue grotto (Agence France-Presse, 2015; Norfolk *et al.*, 2020). The glacier and the grotto have, ever since, been major tourist attractions and iconic symbols of the Swiss Alps. Even James Bond showed up at the Rhône Glacier in 1964, driving past the Belvedere as he tailed Goldfinger in the film. The glacier was much bigger then, visible from the hotel guest rooms. Now the glacier has retreated so far up the mountain that it is invisible from the Belvedere, which has closed. But still the tourists come to see the ice, to buy trinkets at the gift shop, and to venture inside the deep-blue grotto in the glacier. Yet the Rhône Glacier is more than a tourist destination. The melted ice – along with snow, rain, and groundwater – nourishes the Rhône River that flows out of the glacier and into Lake Geneva. Residents rely on the river for farming, pastures, wine production, and hydropower.

Given the importance of the Rhône Glacier for water and tourism, a group of residents from the nearby area of Obergoms started covering a section of the shrinking glacier with white blankets about a decade ago, hoping this layer of insulation would help save the ice (Treviño, 2018). David Volken, a Swiss scientist, says these blankets reduce ice melting by as much as 50–70 per cent (Hester, 2019). The Carlen family, who bought the Belvedere and gift shop from the Seilers, supported these blankets because, as the glacier shrank, they 'saw their asset simply melting away' (Norfolk *et al.*, 2020). The blankets cover their glacier grotto, helping protect the tunnel that attracts so many visitors even as the blankets block the ice, making the site look more like a construction zone than mountain scenery. The impact of the ice loss extends beyond the Carlens because the glacier helps 'boost the region's economy' as people congregate to glimpse the glacier (Bachmann, 2018). Another journalist explains that 'In the Rhône region, that shrinking [glacier] represents an economic emergency as well as an environmental one' (Letzter, 2018). Rhône representations thus underscore the glacier's economic uses and the priority of tourism. But the blankets cover a tiny portion of the Rhône Glacier: only about 2 hectares of this nearly 8-kilometres-long glacier (Bachmann, 2018 and Figure 4.1).

What's interesting in all these news stories, magazine articles, and tourism advertisements about the ice-saving blankets on the Rhône Glacier is what they fail to cover. Most news coverage and magazine articles focus solely on the act of installing the blankets, not their results. The act of offering a solution to ice loss seems more important than whether that solution works.

Figure 4.1 Rhône Glacier, Switzerland, showing the white blankets covering the ice in the lower right section.

Commentators hardly mention the influence of glacier protection on tourism, on downstream water supplies, on farming, on jobs, or on the local economy. These factors – the economics of ice – justify the use of the blankets in the first place, but stories do not address whether the practice makes any difference. The campaign is framed as a project for local people and businesses, but the main actors are tourists and scientists, not residents. Further, it is the symbolic act of draping the ice with white blankets that attracts the attention. The discussion celebrates heroic engineering solutions but not the impacts; it offers a strong justification for intervention in mountain landscapes and societies, without addressing how, why, or if that technical solution helps. The narrative ends up promoting more anthropogenic intervention, so that the blankets covering the ice are both a physical layer on the ice as well as a symbolic layering of more human influence, without addressing drivers of the climate crisis or local societal impacts.

Another key storyline with the Rhône blankets is the backward-looking lament rather than forward-looking solutions: writers and publishers essentially sell death and despair in their Rhône coverage. In doing so, they tend to celebrate the distant past – the glory days of the Ice Age when glaciers were gigantic, or the late Little Ice Age in the nineteenth century, when tourists followed the Grand Tour through Europe. The future? Apocalypse. All the glaciers will probably be gone by 2100 anyway, most commentators conclude. Here the storyline is remarkably consistent: the past was plentiful, the future condemned to destruction. The actors become those placing the tarps and those expressing climate anxiety, while local hotel workers and downstream vineyard owners vanish from the story like the melting ice. An article about a recent photography series on the Rhône's glacier-saving blankets illustrates the point clearly, noting that their photographs could be called 'an elegy, or a requiem, or a lament'. After all, 'it's a deeply disturbing scene up there [on the Rhône Glacier]. We're looking at a wreck of a glacier, beautifully swaddled for death' (Norfolk *et al.*, 2020). What do the Rhône blankets save, then? Memories, it seems. This is not a forward-looking path but backward-looking nostalgia. The Rhône tarps may preserve a small portion of the glacier and can be important for generating public awareness about climate change impacts, but the blanketing discourse creates a barrier from the local population, conceals causes of the climate crisis, and blocks bigger solutions.

The business of blankets: saving ice at ski resorts

Europe's ski industry is big business entrenched in regional cultures and identities. With 35 per cent of the world's ski areas spread across eight countries, the European ski resorts serve 120 million tourists a year. They employ millions and bring in billions of Euros to local and national economies. Yet more than two hundred ski resorts in Europe have closed down in recent years, due in part to climate change and snow loss (Parkin, 2019). Glaciers allow resorts to guarantee better snow and longer ski seasons, making them less dependent on the weather. Resorts with glaciers can in some cases even stay open year-round – at least they used to. One skiing website (www.skiresort.info/glacier-ski-resorts/) identifies forty commercial ski resorts in the world that have glacier terrain for skiing, with the vast majority in Europe. But recent rising temperatures have led most to shorten their seasons compared with previous year-round skiing (Snowhunter, 2019).

Consequently, ski resorts with glaciers have turned to glacier-saving campaigns: covering glaciers with tarps, snow farming to collect and redistribute natural snow, artificial snow making, and slope grooming with snowcats. The technology to preserve ice by covering it with blankets or insulating foams dates back to the 1960s (Shibel *et al.*, 1962; Herrmann and Stehle, 1966). Ski resort managers in Austria started testing a variety of glacier blankets in 2003, particularly at the Stubai Glacier near Innsbruck where they have used hemp, wool, polypropylene, and white fleece (Eckardt, 2005). Now engineers are experimenting with an even larger variety of materials to cover glaciers at ski resorts, such as polyester, dextrose extracted from corn, and many other geotextiles. Yet glacier-saving campaigns, like blankets on the Rhône Glacier, can only protect a small portion of the overall ski area. The white fleece blankets used initially at Stubai Glacier covered just 10 of the 1,450 total hectares of skiable terrain. An ice expert on that project, Andrea Fischer, said in 2005 that 'Since only 0.1 per cent of the overall glacier area in Austria is being covered, the application will have no effect that we can measure' (Eckardt, 2005). A recent study also suggests ski resort glacier tarps might be futile, saying 'the monetary and environmental costs to protect significant glacier areas are prohibitive for likely all cases' (Senese *et al.*, 2020).

Ultimately, the ski resort storyline embeds a singular agenda for ice: ski resorts must save their glacier slopes or go extinct. Media and news stories tend to have a similar emphasis as at the Rhône Glacier: they claim to present a solution but only discuss the ice-saving projects themselves, not their consequences. So, the narrative foregrounds scientists and ski resort managers as the main actors while offering technoscientific interventions as the principal solution, even though the stories initially justify glacier blankets by referencing skiers and local economies. The stories usually quantify the amount of ice lost or saved under the tarps, and they detail the process of covering ice with blankets, thereby making the engineers not the hotel owners or skiers the principal protagonists. The glacier representations chronicle ice protection but not job protection. So, to answer our question about what glacier-saving campaigns save, we can only answer that, at ski resorts, they save some ice. We find no details in the discourse about the economic benefits of the ice-saving projects, about the number of jobs lost or preserved, about impacts on hotels or restaurants nearby, about the number of skiers visiting the resorts before and after the implementation of ice-saving schemes, or about the cultural impacts on memory, identity, family, or recreation trends. We find little discussion within the glacier-saving stories about adaptation to ice loss, even though studies on ski resort and winter recreation adaptation under climate change exists (e.g. Scott and McBoyle, 2007; Rivera and Clement, 2019). Instead, the glacier stories offer solutions that focus on geotextiles and snow cannons, not on people or livelihoods.

Fabricating glaciers in India

Schemes to protect the planet's ice can go even beyond the saving of existing glaciers. In Ladakh, India, for example, people are making their own ice. Motivated by the increasing water needs downstream and by the reduction of water supplies upstream, local engineers have been building artificial glaciers and ice stupas to augment water supplies for farming, irrigation, and drinking earlier in the spring, when villagers most need it to plant their peas, wheat, and barley. This is not a new practice. Local communities in the Himalayan region have since the twelfth century tried what they call 'glacier mating', which entails the grafting of two glaciers to create a new, more stable glacier (Nüsser and Baghel, 2016). But the new do-it-yourself glaciers – and the two men who first engineered them – have attracted far-reaching attention in international news, as well as among researchers (Pandey, 2014; Pareek, 2014; Vince, 2014; Vince, 2009; Clouse, 2017; Nüsser *et al.*, 2019). Journalists usually celebrate the manufactured glaciers as heroic and ingenious strategies for adapting to the climate crisis.

Chewang Norphel started making artificial glaciers in the late 1980s, with the first completed around the time he retired in 1995. 'Water is the most precious commodity here', Norphel worried. 'People are fighting each other for it; in the irrigation season, even brother and sister or father and son are fighting over water. ... Peace depends on water' (Vince, 2014). Now Norphel, known globally as the Ice Man, has built ten artificial glaciers in Ladakh. The concept is straightforward. During October and November, he diverts water from river channels into empty reservoirs contained by stone walls. The artificial glaciers are typically 15–60 metres wide and 1–2 metres deep (Pareek, 2014). The main ingredient for the reservoirs, besides water, is shade, which allows the water to freeze during the subsequent winter. The ice then begins melting in April, which provides water much earlier in the season and with higher volume than the non-artificial Himalayan glaciers. Norphel's artificial glaciers offer a low-cost engineering intervention, relying primarily on temperature and gravity once residents build stone walls reinforced with steel mesh (Clouse, 2017; Nüsser *et al.*, 2019).

Journalists and activists worldwide have turned Norphel the Ice Man into a celebrity who solves climate change. And almost always, they highlight his technical work rather than the results of his artificial glaciers (exceptions include Clouse, 2016; Nüsser *et al.*, 2019). Pankaj Chandon, a regional programme coordinator for the World Wildlife Fund, calls them 'a fantastic adaptation technology for the climate changes that we are experiencing in this region' (Vince, 2014). Vince (2009) moves towards an assessment of the results by quoting one villager who reported income three times higher than before the artificial glacier, along with the ability to grow new crops. But otherwise, commentators rarely report on whether artificial glaciers provide solutions to water shortages. Have brothers and sisters, fathers and sons, stopped fighting for scarce water? Have the artificial glaciers helped create the peace that Norphel sought when he started back in the 1990s? And how is the diverted water being distributed among Ladakh communities? These questions are not in media stories and glacier representations. The ice narrative instead revolves around a single local engineer who can save society. It foregrounds his engineering solutions without probing deeper questions about who needs the water in the first place, why shortages have emerged, and how societal structures and political economy influence access to and allocation of water.

There is one study that tries to answer questions like these, but it is an unpublished NGO report, not a media story or a popular article circulating in the Ice Man stories. This study on the artificial glaciers' effectiveness was conducted in 2012 by the French NGO Renewable Energy, Environment, and Solidarity (GERES). They surveyed fifty four households (out of about five hundred in total) in three villages with artificial glaciers (Hasnain, 2012). They found that 83 per cent of the households reported increased cereal crop yields, 24 per cent identified growth in cultivable land, and more than 20 per cent reported an ability to grow new cash crop vegetables such as peas and potatoes, after the artificial glaciers were built. They also discovered indirect community benefits, including the replenishment of groundwater springs and aquifers. And this study did in fact confirm Norphel's hope for peace, at least in these three villages: they found reduced conflicts and quarrelling over water in the community. But the study does point to some problems. For one, benefits are unevenly spread across and within the various projects. The effectiveness of the *Churpon*, the community water manager who tries to distribute water equally, is also waning in some communities, thereby altering the local politics of water management. Artificial glaciers may thus be shifting water politics and access to water, but the popular media portrays the fabricated glaciers as singularly positive.

Another local engineer from Ladakh, Sonam Wangchuk, invented a different form of manufactured ice reservoirs known as ice stupas. In 2013, Wangchuk started building these ice towers by laying a pipe far off the ground that brings water from upstream to the stupa site, where it drips out of the pipe, freezes, and then builds up into a cone-shaped block of ice from 5 to 20 metres high (Clouse, 2016; Parvaiz, 2018; Stewart, 2018; Kolbert, 2019). Like Norphel's artificial glaciers, Wangchuk's ice stupas melt earlier in spring than natural glaciers, thereby providing downstream water in higher quantities and earlier in the season (Griffin, 2014). The ice stupas store significantly less water than Norphel's artificial glaciers, but they can be constructed easier and in more sites since much less land is required. The stupa shape also has an advantage in that it does not require shade to remain frozen. Beyond the water supply, ice stupas also emulate the shape of a Buddhist shrine. Wangchuk selected the name 'stupa' precisely because of this spiritual and cultural significance. He hoped the name would help build wider support and make the cones more than just an ice block serving hydrologic needs. As the stupas have proliferated, they now draw tourists as well. The Ice Cafe inside a stupa in Gya Meeru village, for example, has attracted thousands of tourists, who can have a coffee and then venture inside the ice sculpture. Wangchuk has attracted international attention, with funds flowing in to build more ice stupas through such funding as the Rolex Award for Enterprise. This money allowed for the construction of dozens of stupas to supply nearby villages with 25 million litres of water (Bhirani, 2019; Maqbool, 2019).

Like the artificial glaciers, ice stupas have attracted worldwide acclaim as adaptations to the climate crisis. When artificial glaciers and ice stupas appear in the media and news, almost every story offers singular praise for Wangchuk (and Norphel). However, the media narratives generally emphasize the individual engineer and his technological solution to climate change. Coverage focuses mostly on the engineering itself rather than on impacts or solutions. The underlying agenda usually promotes entrepreneurship and economic development through farming and tourism, but without clear indications of who benefits. Are the stupas helping? Who gets the water, and who doesn't? Some media stories do at least hint at the varied results and complexities of ice stupas. In the village of Phey, residents have complained that the ice stupas 'monopolize the water supply'. These villagers say they lose water for their farms, and their groundwater springs are not replenished any more. A professor at the University of Kashmir explains that the stupas are innovative, but they 'cannot solve water scarcity in the Leh area of

Ladakh' (Parvaiz, 2018; Maqbool, 2019). The water problem is not due solely to glacier shrinkage; rather it is a problem that the Ladakhi villagers have faced since early settlement in this cold desert climate. Thus, when the media stories prioritize only an engineering solution, they neglect to expose the roots of social inequalities, economic marginalization, climate change vulnerability, and the factors that influence access to and control of water in the first place. Ice circulates in the global media, then, as an object and icon, becoming divorced from the very social setting in which Wangchuk and Norphel built their ice structures. They make ice to provide water for people. But when the artificial glaciers and ice stupas appear in the media and news, ice turns into a climate object and an engineering icon. Objectification of ice in turn generates a narrative that fails to question the political and socio-economic drivers of water insecurity for Ladakhi residents, even though water distribution is deeply political.

Legal creep and Argentina's glacier protection law

On 28 October 2010, Argentina passed the world's first national glacier protection law. The 'Minimum Standards Regime for the Preservation of Glaciers and the Periglacial Environment' proclaimed that glaciers constitute 'goods of public character' and legally identified them as 'strategic reserves of hydrologic resources for human consumption; for agriculture and as water sources for the recharge of watersheds; for the protection of biodiversity; as a source of scientific information and as a tourist attraction' (Center for Human Rights and Environment, 2010; Ministerio de Justicia y Derechos Humanos de Argentina, 2010). To protect the ice, the law prohibited certain activities in and around glaciers, including mining and hydrocarbon exploration, industrial activities, infrastructure building, and environmental contamination. It called for a national glacier inventory to document where exactly glaciers existed. The law had far-reaching implications and set off a decade of controversy about natural resource extraction, the place of multinational mining companies in Argentina, the role of science in policy, and the question of jurisdiction (local, national, or international) over Andean ice. The law obviously focused on glaciers but ultimately had many other agendas embedded in it, primarily related to longstanding conflicts between mining and environmental protection. The glacier law inspired deep debates about who has access to and control over Argentina's ice-covered regions, thereby turning glaciers into contested national and international spaces.

Argentina's glacier law emerged from a long, contested process dating back to 2005 when Canada's Barrick Gold Corporation proposed to relocate three glaciers at its Pascua Lama mine on the Argentina–Chile border. Barrick initially proposed to bulldoze the Toro 1, Toro 2, and Esperanza glaciers so it could reach the precious metals under the ice. The company said it would simply relocate the ice, reattaching the small glaciers to a larger glacier nearby. But residents downstream in Chile's Huasco Valley and activists globally protested Barrick's proposal, claiming that the glacier destruction and mining activity would reduce and contaminate water supplies. The ensuing conflict turned the glaciers into highly contested global icons for anti-mining campaigns and global warming impacts. It also unleashed movements within both Chile and Argentina to create new glacier protection laws (Kronenberg, 2013; French *et al.*, 2015; Taillant, 2015; Li, 2016; Herrera Pérez and Segovia, 2019; Barandiarán, 2020). Social protest led to the creation of organizations like Protest Barrick and the Anti-Pascua Lama Front that sought to save the glaciers and protect nearby Indigenous communities from the consequences of ice loss and water contamination (Estrada, 2005; Wilmarth, 2012; Li, 2016; Saunders, 2019).

This Pascua Lama context helped propel Argentina's glacier protection law forward. In fact, the Argentine congress nearly unanimously passed a 2008 glacier protection law. But President de Kirchner vetoed it, remarking that 'the prohibition is excessive' because it gave 'pre-eminence to environmental aspects over activities that could be developed in perfect harmony with the environment', particularly mining (Valente, 2008). The deeper storyline was about how to define glaciers and how to manage ice-covered regions: were those spaces for economic activity and resource extraction, or for environmental protection? Ostensibly the glacier law was about ice, but underneath lay competing agendas about land and water management, and the place of multinational companies in the country's economy. Argentine activist Jorge Daniel Taillant explains that mining interests compelled the president to terminate the proposed law. He even refers to it as the 'Barrick veto' (Taillant, 2015). Despite the setback, activists and policymakers drafted a second glacier protection law, which eventually passed and became law in 2010.

But the new 2010 law met immediate resistance. Barrick and the pro-mining province of San Juan filed a lawsuit proclaiming the law unconstitutional because it unjustly transferred environmental sovereignty from local provinces to the national government. With their lawsuit tied up in courts for nearly a decade, Barrick and other companies continued their mining operations unfazed (Barrick, 2012; Taillant, 2018). The Supreme Court did eventually uphold the glacier protection law in 2019, but by then mining contamination had already occurred, particularly with three accidents releasing cyanide into waterways at Veladero between 2015 and 2017 (Healey and Martin, 2017; Corte Supreme de Justicia de la Nación, 2019). Moreover, as judges failed to find Barrick negligent for the Veladero accidents in 2017, the court and environmental activists blamed scientists who produced the national glacier inventory for failing to identify nearby glaciers appropriately. Specifically, they went after Argentine Institute of Snow, Ice and Environmental Research (IANIGLA) director Ricardo Villalba, an internationally acclaimed glaciologist, with legal charges for the supposed inventory deficiencies. Yet Villalba had followed strict scientific protocols for identifying minimum glacier size (1 hectare). The accusations led to an international outcry from the scientific community that rallied behind Villalba; they excoriated the blatant scapegoating of him by the Argentine court and environmental activists (e.g. Fraser, 2017; Tollefson and Rodríguez Mega, 2017). What's more, holding Villalba accountable because of his institute's inventory failed to acknowledge that it was mining, not glacier scientists, that caused the water contamination. Through all the discourse and news about the glacier law, the main actors have always been policymakers, mining companies, judges, scientists, and activists rather than the residents the law is supposedly designed to protect.

Argentina's glacier law remains highly contested, and discussions during the last fifteen years exemplify several key trends in the stories told about glacier-saving campaigns. Depending on the stakeholder, the glacier law can serve its stated purpose of environmental protection, as an obstacle to economic development, or even as a tool to exert control over mining companies. The law's lack of enforcement power and its focus on contamination in glacierized watersheds – rather than rapid shrinking of glaciers due to climate change – renders it significantly limited. The glacier protection law may thus appear to save glaciers, but ultimately treats the symptoms (ice loss) of a problem rather than its roots (anthropogenic warming). The glacier law seems less about ice and more about regulating the mining industry to conserve clean water, and debates about the law have rearranged authority over Andean landscapes and resources. Argentina's glaciers remain highly contested sites of protest where local, regional, national, and international entities fight over the ice, its value, its management, its narratives, and the knowledge considered as 'legitimate' in all these debates. The last fifteen years of Argentina's glacier protection laws show that glaciers are as much sites of power and authority as they are about ice, minerals, or water.

Glaciers and legal personhood in India

Another type of glacier law to protect ice recently emerged in India. In March 2017, the Uttarakhand High Court granted two glaciers the status of legal personhood. The decision aligns with several other 'rights of nature' legal decisions in Ecuador, New Zealand, and elsewhere that have granted rivers and other nonhuman features personhood status by law (Clark *et al.*, 2018; Knaub, 2018). Within India, the decision fits into a longer trajectory of legal and religious discussions: colonial Indian law granted legal personhood to temples, and this has been a complex and longstanding issue within India about national politics and religion (Mehta, 2015; Studley and Bleisch, 2018). This complex history is not explored here, where we focus strictly on the 2017 glacier ruling. India's state court in Uttarakhand declared that the Gangotri and Yamunotri glaciers are essential for both water and religion, but have been threatened by climate change and pollution (O'Donnell, 2018). Consequently, the state court decided that harming the two glaciers would be 'legally equivalent to harming, hurting and causing injury to people'. The court justified the decision saying that water from these two glaciers, which flows into the sacred Ganges (Drew, 2012; Alley, 2019), 'is [the] elixir of life and we must conserve and preserve every drop', while also noting that 'we are morally bound to hand over the same Mother Earth to the next generation' that past generations gave to us (Uttaranchal High Court, 2017: 65).

As with Argentina's glacier law that shifted authority over Andean ice, India's legal personhood ruling transformed the glaciers into global symbols of 'rights for nature' as well as contested national spaces. Four months after the initial ruling, in July 2017, India's Supreme Court issued a stay for the Uttarakhand cases about the personhood status for glaciers and rivers. The national court said that the decision failed to define harm and delineate responsibility, thereby making enforcement difficult or impossible. Other analysts say the Supreme Court's ruling was due to complex national politics and the central government's unwillingness to be liable for harm to the glaciers or rivers (Alley, 2019). The Uttarakhand decision had in fact identified several state government officials, including the Chief Secretary of State, as well as legal advisers, academics, and judges, who would act essentially as legal guardians for the glaciers (O'Donnell, 2018). Regardless, the July 2017 Supreme Court decision ended up stalling this legal and religious glacier-saving campaign. Media stories and news about the glacier ruling not only reported on the case, but were also indicating who had authority to intervene and act on these glaciers; residents vanished from the scene as discourse identified lawyers, judges, and policymakers as the principal agents of authority with jurisdiction over the ice.

As the original Uttarakhand High Court glacier decision turned ice into a platform for national debates about religion, politics, and water, it also turned the glaciers into global icons for legal battles about climate change (e.g. Clark *et al.*, 2018; Knaub, 2018; O'Donnell, 2018). Most of the media and research coverage has put this legal personhood for glaciers into a larger context of 'rights for nature' laws passed in other countries, especially New

Zealand and Ecuador but also Bolivia, Colombia, and some legal proceedings back to the 1970s United States (e.g. Knaub, 2018; O'Donnell, 2018; Alley, 2019). Attention, then, has focused largely on the legal rationale, precedence, and implications of the decision, not on the glaciers. This is an important contrast to the glacier-saving campaigns at ski resorts and artificial glaciers, for example, which focus almost entirely on the ice, not people or politics. Despite the differences, continuity across glacier-saving campaigns does exist in the lack of local voices or local actors represented, as well as the related absence of discussion about social inequities, unequal vulnerabilities, or the distribution of impacts across a society divided by race, class, gender, age, religion, geographical location, education levels, and many other factors.

Conclusions

News and media stories about glacier-saving activities do much more than explain the new projects, laws, or engineering of ice. The reporting produces specific narratives, circulates storylines, and promotes certain agendas for the use of ice, the management of mountains, the best solutions for climate change, and the actors who should supposedly be involved in these processes. It is crucial to understand this invisible work that glacier representations do. They create environmental icons, change ideas, produce perceptions, and alter the geopolitics of ice and icy spaces. While the schemes vary considerably, the larger discussions in media accounts and other glacier stories tend to focus on the campaigns themselves, usually celebrating them as heroic efforts to save glaciers. Few representations analyse underlying causes of ice loss or impacts on society. Further, the media coverage about glacier-saving campaigns can transform glacier landscapes into national or global commons, where outsiders can (and must, as the storyline goes) act to save the vanishing glaciers. Proponents often use the glacier projects to pursue agendas that transcend the ice, such as expanding tourism, boosting businesses, conserving mountain landscapes, advancing political positions, or solidifying control over regions and resources. Glacier-saving campaigns can still serve local interests, such as when artificial glaciers in Ladakh help secure downstream water supplies or when Argentina's glacier law prevents water contamination from mining. On the other hand, these glacial interventions are often led by outsiders, not local community members, and they thus contribute to the nationalization and globalization of glaciers.

This chapter has revealed ways that glacier-saving campaigns have portrayed glaciers through the actions of blanketing, creeping, and fabricating. Both ski resorts and the Rhône Glacier protectors cover ice with blankets to preserve it. But the blankets also act as a barrier: they shield the ice from view. What's more, the corresponding glacier narrative about blanketing glaciers emphasizes death and nostalgia for the lost past, rendering the present-day ice and its future invisible. The glacier-saving stories also expose a focus on glaciers themselves, which creep back as they melt while laws and policies creep into glacier spaces. Glacier protection and legal personhood laws increasingly bring the nation state and international actors into icy mountain regions, while the science of ice ushers in knowledge creep, displacing local knowledge. And finally, the anthropogenic warming that melts ice on a global scale triggers new technological and engineering interventions on local glaciers. There are manufactured blankets to cover glaciers, the construction of artificial glaciers, and the storyline that celebrates heroic engineers and their technoscientific solutions to ice loss – all of which exemplify the human fabrication of ice and underscore a larger narrative about how humans are framed as both sinners causing the climate crisis and saviours resurrecting new campaigns to save ice (Paerregaard, 2020). The solutions themselves thus layer more anthropogenic activity onto the climate crisis.

This ice humanities lens on glacier-saving efforts and the iconography of ice exposes storylines and narratives embedded in conversations about ice. Within these glacier narratives are key messages and perspectives about glacier uses, solutions for ice loss, the actors involved (and ignored), and the entities or individuals given authority over icy places. The overarching themes related to uses and solutions – heroic engineers, faith in technoscientific solutions, glaciers as economic resources, invisibility of local residents, backward-looking nostalgia versus forward-looking optimism, and authority of the nation state – are more than just glacier representations and discourse. They can influence the politics of ice, including access to and control over glaciers. This examination of glacier-saving campaigns helps better understand ice loss impacts in mountain regions. Perhaps more importantly, the analysis shows how responses to the climate crisis are themselves transformative because they alter landscapes, shift governance and environmental politics, prioritize technoscientific interventions, commodify environments, exacerbate social inequalities, and change meanings and values of nonhuman nature.

Moving forward, it is important not to take critiques of glacier-saving campaigns through this ice humanities approach as a sign to do nothing during the climate crisis. Glacier protection schemes, nevertheless, should not simply perpetuate outsider agendas, erode local autonomy, or fail to address the roots of climate change or social inequalities. Discussions of glacier-saving campaigns could instead address local issues, wishes, and contexts. Have India's artificial glaciers addressed water insecurity in local communities? If so, how? If not, what are the constraints – are they climate- and ice-related, or driven by local politics or social divisions? In short, media and news stories about glacier protection could shift the conversation to results rather than merely identifying a climate icon and mourning its death or celebrating its conservation. Stories and accounts about climate adaptation projects should do less venerating of heroic engineers or activist rescuers, a narrative that tends to identify the saviours as outsiders, usually men. This only perpetuates narratives of individualism, domination, and control. The goal should be not just to put a Band-Aid over the melting glaciers but rather to pursue projects that ultimately break down entrenched systems of inequality, disparate vulnerabilities, and systemic practices that restrict some groups' access to water, land, and other vital natural resources and human rights. A more grounded understanding of ice characterized not only by its dynamic physical properties but also by the myriad ways different social groups value and interact with glaciers could serve as an excellent starting point to help achieve greater social and environmental justice around glaciers.

Note

1 This article is based on work supported in part by the US National Science Foundation under grant #1253779.

References

Agence France-Presse (2015) 'Blankets Cover Swiss Glacier In Vain Effort To Halt Icemelt', *NDTV*, 12 December. Available at: www.ndtv.com/world-news/blankets-cover-swiss-glacier-in-vain-effort-to-halt-icemelt-1254051 (accessed 10 August 2020).

Alley, K. D. (2019) 'River Goddesses, Personhood and Rights of Nature: Implications for Spiritual Ecology', *Religions* 10(9): 502–519.

Bachmann, H. (2018) 'The Swiss Found a Way to Save this Glacier from Melting: Wrap it in a Giant Blanket', *USA Today*. Available at: www.usatoday.com/story/news/world/2018/03/13/swiss-found-way-save-their-glacier-melting-wrap-blanket/421142002/ (accessed 9 August 2020).

Barandiarán, J. (2020) 'Documenting Rubble to Shift Baselines: Environmental Assessments and Damaged Glaciers in Chile', *Environment and Planning E: Nature and Space* 3(1): 58–75.

Barrick (2012) *Pascua-Lama and Veladero Not Affected by Argentina Supreme Court Ruling*. Available at: www.barrick.com/news/news-details/2012/Pascua-Lama-and-Veladero-Not-Affected-by-Argentina-Supreme-Court-Ruling/default.aspx (accessed 20 June 2020).

Bhirani, R. (2019) 'Unique "Ice-stupa Café" Attracts Tourists in Leh's Village', *Economic Times*. Available at: https://economictimes.indiatimes.com/industry/

miscellaneous/unique-ice-stupa-cafe-attracts-tourists-in-lehs-village/who-built-this-stupa/slideshow/68477921.cms (accessed 29 July 2019).

Bjørst, L. R. (2010) 'The Tip of the Iceberg: Ice as a Non-human Actor in the Climate Change Debate', *Études/Inuit/Studies* 34(1): 133–150.

Born, D. (2019) 'Bearing Witness? Polar Bears as Icons for Climate Change Communication', *National Geographic: Environmental Communication* 13(5): 649–663.

Bravo, M. (2017) 'A Cryopolitics to Reclaim Our Frozen Material States', in J. Radin, and E. Kowal, eds, *Cryopolitics: Frozen Life in a Melting World*. Cambridge, MA: MIT Press, 27–58.

Bravo, M. T. (2009) 'Voices from the Sea Ice: The Reception of Climate Impact Narratives', *Journal of Historical Geography* 35(2): 256–278.

Brito, C., N. Vieira, and J. G. Freitas (2019) 'The Wonder Whale: A Commodity, a Monster, a Show and an Icon', *Anthropozoologica* 54(1): 13–27.

Cameron, E. S. (2012) 'Securing Indigenous Politics: A Critique of the Vulnerability and Adaptation Approach to the Human Dimensions of Climate Change in the Canadian Arctic', *Global Environmental Change* 22(1): 103–114.

Carey, M. (2007) 'The History of Ice: How Glaciers Became an Endangered Species', *Environmental History* 12(3): 497–527.

Carey, M., M. Jackson, A. Antonello, and J. Rushing (2016) 'Glaciers, Gender, and Science: A Feminist Glaciology Framework for Global Environmental Change Research', *Progress in Human Geography* 40(6): 770–793.

Center for Human Rights and Environment (2010) Argentine National Glacier Act. Center for Human Rights and Environment, Florida, 28 October.

Clark, C., N. Emmanouil, J. Page, and A. Pelizzon (2018) 'Can You Hear the Rivers Sing: Legal Personhood, Ontology, and the Nitty-Gritty of Governance', *Ecology Law Quarterly* 45: 787–844.

Clouse, C. (2016) 'Frozen Landscapes: Climate-adaptive Design Interventions in Ladakh and Zanskar', *Landscape Research* 41(8): 821–837.

Clouse, C. (2017) 'The Himalayan Ice Stupa: Ladakh's Climate Adaptive Water Cache', *Journal of Architectural Education* 71(2): 247–251.

Collyns, D. (2010) 'Can Painting a Mountain Restore a Glacier?', *BBC News*, 17 June. Available at: www.bbc.com/news/10333304 (accessed 8 July 2020).

Corte Supreme de Justicia de la Nación (2019) La Corte Suprema convalidó la constitucionalidad de la ley de preservación de los glaciares rechazando el pedido de Barrick Gold, Minera Argentina Gold y provincia de San Juan. 4 June, Centro de Información Judicial.

Cox, D. (2018) 'Two Audacious Plans for Saving the World's Ice Sheets', *NBC News*, 28 May. Available at: www.nbcnews.com/mach/science/can-these-bold-plans-keep-world-s-ice-sheets-melting-ncna877616 (accessed 9 July 2020).

Davis, D. K. (2007) *Resurrecting the Granary of Rome: Environmental History and French Colonial Expansion in North Africa*. Athens, OH: Ohio University Press.

Dodds, K. (2021) 'Geopolitics and Ice Humanities: Elemental, Metaphorical and Volumetric Reverberations', *Geopolitics* 26(4): 1121–1149.

Drew, G. (2012) 'A Retreating Goddess? Conflicting Perceptions of Ecological Change near the Gangotri-Gaumukh Glacier', *Journal for the Study of Religion, Nature and Culture* 6(3): 344–362.

Eckardt, A. (2005) '"Wrapping" Alpine Glaciers: Revolutionary Methods to Preserve Snow, the "Austrian Gold"', *NBC News*, 15 July. Available at: www.nbcnews.com/id/wbna8432120 (accessed 24 August 2020).

Estrada, D. (2005) 'Conflict Over Andean Glaciers Heats Up', *Inter Press Service News Agency*, 11 Nov.

Fairhead, J. and M. Leach (1996) *Misreading the African Landscape: Society and Ecology in a Forest-Savanna Mosaic*. New York: Cambridge University Press.

Flora, J., K. L. Johansen, B. Grønnow, A. O. Andersen, and A. Mosbech (2018) 'Present and Past Dynamics of Inughuit Resource Spaces', *Ambio* 47(Suppl 2): S244–S264.

Fraser, B. (2017) 'Argentine Scientist Indicted over Design of Glacier Inventory', *Science*, 5 December. Available at: www.sciencemag.org/news/2017/12/argentine-scientist-indicted-over-design-glacier-inventory (accessed 6 September 2020).

French, A., J. Barandiarán, and C. Rampini (2015) 'Contextualizing Conflict: Vital Waters and Competing Values in Glaciated Environments', in C. Huggel, M. Carey, J. J. Clague, and A. Kääb, eds, *The High-Mountain Cryosphere: Environmental Changes and Human Risks*. Cambridge: Cambridge University Press, 315–336.

Griffin, P. (2014) 'Ice Stupas: Conserving Water the 3 Idiots Way', *Forbes India*, 29 December.

Guha, R. (1989) 'Radical American Environmentalism and Wilderness Preservation: A Third World Critique', *Environmental Ethics* 11(1): 71–83.

Hasnain, M. (2012) 'Artificial Glaciers in Ladakh: A Socio-economic Analysis', *GERES India*, December. Available at: www.geres.eu/en/news/guides-and-studies/articficial-glaciers-ladakh-socio-economic-analysis/ (accessed 29 March 2022).

Healey, M. and F. Martin (2017) 'A Troubling Turn for Glacier Science in Argentina', *GlacierHub*, 12 December.

Herrera Pérez, J. and A. Segovia (2019) 'Ley de Protección de Glaciares: El devenir de un conflicto socioambiental', *Revista Investigaciones Geográficas* 58: 118–134.

Herrmann, M. R. and N. S. Stehle (1966) 'Protective Coverings for Ice and Snow', *Hokkaido University*. Available at: https://eprints.lib.hokudai.ac.jp/dspace/bitstream/2115/20343/1/2_p797-806.pdf (Accessed 2 August, 2019).

Hester, J. L. (2019) 'Swaddling Glaciers in Blankets Isn't the Oddest Idea for Extending Their Lives', *Atlas Obscura*. Available at: www.atlasobscura.com/articles/technology-slows-glaciers-melting (accessed 10 August 2020).

Kaijser, A. (2013) 'White Ponchos Dripping Away? Glacier Narratives in Bolivian Climate Change Discourse', in C. Methmann, D. Rothe, and B. Stephan, eds, *Deconstructing the Greenhouse: Interpretive Approaches to Global Climate Governance*. New York: Routledge, 183–197.

Knaub, S. (2018) 'Conceptualizing Human Stewardship in the Anthropocene: The Rights of Nature in Ecuador, New Zealand and India', *Journal of Agricultural and Environmental Ethics* 31(6): 703–722.

Kolbert, E. (2019) 'The Ice Stupas: Artificial Glaciers at the Edge of the Himalayas', *New Yorker*, 13 May.

Kronenberg, J. (2013) 'Linking Ecological Economics and Political Ecology to Study Mining, Glaciers and Global Warming', *Environmental Policy and Governance* 23(2): 75–90.

Leane, E. and B. Maddison (2018) 'A Biography of Iceberg B09B', *Australian Humanities Review* 63(November): 99–115.

LeCain, T. J. (2019) 'The Consumption of Humans: How the Danger of Grizzly Bear Attacks in American National Parks Became a Commodity', *Environmental History* 24(4): 680–687.

Letzter, R. (2018) 'Every Year, the Swiss Cover their Melting Glaciers in White Blankets', *LiveScience*. Available at: www.livescience.com/61951-swiss-glacier-blanket.html (accessed 9 August 2020).

Li, F. (2016) 'The Defeat of Pascua Lama', *Observatorio Latinoamericano de Conflictos Ambientales*, 9 March.

Maqbool, R. (2019) 'In Kashmir, an Ancient Solution Solves a New Problem', *Global Press Journal*, 13 January.

Mark, B. G. and A. Fernández (2017) 'The Significance of Mountain Glaciers as Sentinels of Climate and Environmental Change', *Geography Compass* 11(6): e12318, 1–16.

Mason, B. (2003) 'African Ice Under Wraps', *Nature*, 24 November. doi: 10.1038/news031117-8.

Mehta, D. (2015) 'The Ayodhya Dispute: The Absent Mosque, State of Emergency and the Jural Deity', *Journal of Material Culture* 20(4): 397–414.

Mikkelsen, N. and T. Ingerslev, eds (2002) *Nomination of the Ilulissat Icefjord for Inclusion in the World Heritage List*. Copenhagen: Geological Survey of Denmark and Greenland.

Ministerio de Justicia y Derechos Humanos de Argentina (2010) *Ley 26.639: Régimen de Presupuestos Mínimos para la Preservación de los Glaciares y del Ambiente Periglacial*. Buenos Aires: Presidencia de la Nación, Argentina. Available at: http://servicios.infoleg.gob.ar/infolegInternet/anexos/170000-174999/174117/norma.htm.

Norfolk, S., K. Thymann, and F. Hodgson (2020) 'The Shrouds of the Glacier du Rhône', *Granta*. Available at: https://granta.com/the-shrouds-on-the-glacier-du-rhone/ (accessed 11 August 2020).

Nüsser, M. and R. Baghel (2016) 'Local Knowledge and Global Concerns: Artificial Glaciers as a Focus of Environmental Knowledge and Development Interventions', in P. Neusburger, T. Freytag, and L. Suarsana, eds, *Ethnic and Cultural Dimensions of Knowledge: Knowledge and Space*. Switzerland: Springer, 191–209.

Nüsser, M., J. Dame, B. Kraus, R. Baghel, and S. Schmidt (2019) Socio-hydrology of 'Artificial Glaciers' in Ladakh, India: Assessing Adaptive Strategies in a Changing Cryosphere, *Regional Environmental Change* 19: 1327–1337.

Nuttall, M. (2016) 'Narwhal Hunters, Seismic Surveys, and the Middle Ice: Monitoring Environmental Change in Greenland's Melville Bay', in S. A. Crate and M. Nuttall, eds, *Anthropology and Climate Change: From Actions to Transformations*, 2nd ed. New York: Routledge, 354–372.

O'Donnell, E. L. (2018) 'At the Intersection of the Sacred and the Legal: Rights for Nature in Uttarakhand, India', *Journal of Environmental Law* 30(1): 135–144.

O'Reilly, J., N. Oreskes, and M. Oppenheimer (2012) 'The Rapid Disintegration of Projections: The West Antarctic Ice Sheet and the Intergovernmental Panel on Climate Change', *Social Studies of Science* 42(5): 709–731.

Orlove, B., E. Wiegandt, and B. H. Luckman (2008) 'The Place of Glaciers in Natural and Cultural Landscapes', in B. Orlove, E. Wiegandt and B. H. Luckman, eds, *Darkening Peaks: Glacial Retreat, Science, and Society*. Berkeley, CA: University of California Press, 3–19.

Paerregaard, K. (2020) 'Searching for the New Human: Glacier Melt, Anthropogenic Change, and Self-reflection in Andean Pilgrimage', *HAU: Journal of Ethnographic Theory* 10(3): 844–859.

Pandey, S. (2014) 'Crowdfunded "Ice Stupas" Help Ladakh Adapt to Climate Change', *Climate Change News*. Available at: www.climatechangenews.com/2014/12/29/crowdfunded-ice-stupas-help-ladakh-adapt-to-climate-change/ (accessed 5 August 2019).

Pareek, S. (2014) 'The Man who Creates Artificial Glaciers to Meet the Water Needs of Ladakh', *Better India*, 6 November.

Parkin, S. (2019) 'Snow Machines and Fleece Blankets: Inside the Ski Industry's Battle with Climate Change', *Guardian*, 19 December.

Parvaiz, A. (2018) 'Ice Stupas Not Cutting Ice With Some Farmers in Ladakh', *India Climate Dialogue*, 27 June.

Rivera, J. and V. Clement (2019) 'Business Adaptation to Climate Change: American Ski Resorts and Warmer Temperatures', *Business Strategy and the Environment* 28(7): 1285–1301.

Saunders, S. (2019) 'About Us'. *Protest Barrick*. Available at: http://protestbarrick.net/section.php@id=3.html (accessed 21 March 2022).

Scott, D. and G. McBoyle (2007) 'Climate Change Adaptation in the Ski Industry', *Mitigation and Adaptation Strategies for Global Change* 12(8): 1411–1431.

Senese, A., R. S. Azzoni, D. Maragno *et al.* (2020) 'The Non-woven Geotextiles as Strategies for Mitigating the Impacts of Climate Change on Glaciers', *Cold Regions Science and Technology* 173: 103007, 1–12.

Shibel, F. C., C. S. Grove Jr, and A. R. Aidun (1962) *Development of Insulating Aqueous Foams for Protection of Ice Surfaces*. Port Hueneme, CA: US Naval Civil Engineering Laboratory Polar Division.

Slater, C., ed. (2003) *In Search of the Rain Forest*. Durham, NC: Duke University Press.

Snowden, S. (2019) 'Ice-making Mini-submarines is the Latest Idea to Refreeze the Arctic', *Forbes*, 19 August. Available at: www.forbes.com/sites/scottsnowden/2019/08/19/ice-making-mini-submarines-is-the-latest-idea-to-re-freeze-the-arctic/#6d3bc24a673e (accessed 9 July 2020).

Snowhunter (2019) 'Molltal Glacier Latest Summer Ski Area to Close as Snow Melts from Glacier', *Snow-forecast.com*. Available at: www.snow-forecast.com/whiteroom/molltal-glacier-latest-summer-ski-area-to-close-as-snow-melts-from-glacier/ (accessed 6 September 2020).

Sörlin, S. (2015) 'Cryo-History: Narratives of Ice and the Emerging Arctic Humanities', in B. Evengård, J. Nymand Larsen and Ø. Paasche, eds, *The New Arctic*. New York: Springer, 327–339.

Sörlin, S. (2016) 'Do Glaciers Speak? The Political Aesthetics of Vo/ice', in J. Thorpe, S. Rutherford, and L. A. Sandberg, eds, *Methodological Challenges in Nature-Culture and Environmental History Research*. New York: Routledge, 13–30.

Stewart, J. (2018) 'Artificial Glaciers are Helping Provide Critical Water Supply in the Himalayas', *My Modern Met*, 4 June.

Studley, J. and W. V. Bleisch (2018) 'Juristic Personhood for Scared Natural Sites: A Potential Means for Protecting Nature', *PARKS (The International Journal of Protected Areas and Conservation)* 24(1): 81–96.

Swyngedouw, E. (2010) 'Apocalypse Forever? Post-political Populism and the Spectre of Climate Change', *Theory, Culture & Society* 27(2–3): 213–232.

Taillant, J. D. (2015) *Glaciers: The Politics of Ice*. New York: Oxford University Press.

Taillant, J. D. (2018) *Controversial Pascua Lama Project Closes for Good in Chile*. Center for Human Rights and Environment, 18 January. Available at: https://center-hre.org/controversial-pascua-lama-project-closes-for-good-in-chile/#:~:text=January%2018%2C%202018.,border%20between%20Argentina%20and%20Chile (accessed 29 March 2022).

Tollefson, J. and E. Rodríguez Mega (2017) 'Geoscientist Faces Criminal Charges over Glacier Survey', *Nature* 552(14 December): 159–160.

Treviño, J. (2018) 'This Swiss Town is Protecting its Glacier with a Blanket', *Smithsonian*, 12 March. Available at: www.smithsonianmag.com/smart-news/swiss-town-glacier-blanket-180968451/ (Accessed 12 July 2019).

Uttaranchal High Court (2017) *Lalit Miglani vs State of Uttarakhand and Others on 30 March, 2017*. Nainital, High Court of Uttarakhand. Available at: https://indiankanoon.org/doc/92201770/ (accessed 21 March 2022).

Valente, M. (2008) 'Argentina: Frustration Over Veto of Glacier Protection Law', *Tierramerica*. Available at: www.ipsnews.net/2008/11/argentina-frustration-over-veto-of-glacier-protection-law/ (accessed 14 July 2019).

Vince, G. (2009) 'Glacier Man', *Science* 326(5953): 659–661.

Vince, G. (2014) 'DIY Glaciers', *Conservation*. Available at: www.conservationmagazine.org/2014/10/diy-glaciers/ (accessed 28 August 2020).

Wilmarth, C. M. (2012) 'Mining Megaliths in the Argentine Andes: Where Will Victims of Environmental Degradation Find Justice?', *William and Mary Environmental Law and Policy Review* 36(3): 959–990. Available at: https://scholarship.law.wm.edu/wmelpr/vol36/iss3/10 (accessed 15 July 2019).

5

Ice futures: the extension of jurisdiction in the Anthropocene north

Bruce Erickson, Liam Kennedy-Slaney, and James Wilt

In the Arctic, where settlement is not the primary strategy for colonial states, the future has played an important role in projecting dreams of control/sovereignty. Perhaps the most unbelievable (given its unbridled imperial conceit) of these was the Royal Charter of the Hudson's Bay Company granted by Charles II on 2 May 1670, which included not only trading rights, but also governing duties to 40 per cent of what is now Canada despite relatively little knowledge of the territory and those who already occupied it. It was given on the basis of the promise of future profits, but also future settlements which only came to fruition in fits and spurts under the company. This claim to jurisdiction (still held ceremonially at the Hudson's Bay headquarters in Toronto) was challenged on many fronts – rival trading companies, internal political enemies, the French military, and certainly Inuit, Innu, Dene, and Cree hunters, traders, and other residents of those lands – and was eventually bought out by the British state and transferred to Canada in 1870 (Cavanagh, 2011).

As Arctic ice changes, we see a renewed interest in the future of the Arctic, especially in the concerns around climate change but also as a realm of resource extraction and transportation. The Arctic is now in a geopolitical resurgence that is built around the loss of sea and glacial ice, the defining feature of the Arctic in the Anthropocene. Indeed, ice has become the metonym that speaks for the drastic changes happening in the Arctic, and the general sentiment is that the future of ice is foreclosed, resulting in popular titles like *A Farewell to Ice: A Report from the Arctic* (Wadhams, 2017) and *The Arctic Melt: Images of a Disappearing Landscape* (Tuft, 2017). This chapter argues that among this newfound concern for the material state of ice we can also see a renewed effort to assert jurisdiction in the north against the efforts of local control. This jurisdictional creep is justified by the loss of ice in the future, making an iceless future Arctic the foundation through which governance happens. We focus here on the Canadian Arctic, however, the

Anthropocene as a particular rhetorical strategy has provided a new set of jurisdictional futures for an extractive-administrative colonialism. Two cases anchor our observations: the scientific approach held within the Arctic Pilot Project that attempted to produce ice as an object with a predictable and actionable future, and the securitization of polar bears through monitoring and deterrence programmes in northern towns. Before turning to those cases, we set the scene by considering the interplay between jurisdiction claims in the north, changing sea ice, and the rise of northern environmentalism.

This is, as we point out below, a story fundamentally intertwined not only with settler colonialism, but also the Anthropocene. As a proposed geological epoch, the Anthropocene justifies the jurisdictional aspirations of the Canadian state, as it projects the impact of the modern age forward, suggesting that our ecological impact will be the single most important historical feature of the modern world (Erickson, 2020). Ice, in the Anthropocene, is a widely recognized marker of humanity's impact on the geological. Yet the dynamics of ice, both material and social, are far from universally agreed upon. In the state discourse, ice is a fundamentally knowable object (even if not all is yet known about it), and because of this, ice is an object that can be disciplined – made to be productive. As such, the vision of ice that is privileged by the settler state, in sovereignty practices, extraction licenses, and wildlife management plans, does away with the political dimensions of ice in favour of a universalist approach (Linton, 2010).

Our chapter considers, in the first place, the pace and place of settler-colonial jurisdiction in the Canadian North. The two cases (the Arctic pilot project and the conservation of polar bears) we deal with below highlight how this universalist approach not only limits potential relationships and understandings of ice, but also consolidates authority over ice within settler institutions. Finally, we conclude by suggesting that contemporary engagements with the Anthropocene have been drafted by the Canadian state to not only protect a burning, melting, and thawing environment, but also to reassert state jurisdiction and ensure the continuity of the settler state in the north and elsewhere.

Jurisdiction and northern environments

Our argument is built on the renewed interest in settler-colonial jurisdiction in recent years (e.g. Simpson, 2014; Pasternak, 2017; McCreary and Turner, 2018; Stanley, 2019). As Shiri Pasternak (2017) argues, jurisdiction 'is the apparatus through which sovereignty is rendered meaningful, because it is through jurisdiction that settler sovereignty organizes and manages authority' (3). When challenged, settler states often hang their sovereign claims on the ability to survey and police the territory in question. In the north, we are reminded that jurisdiction is both a strategy of internal governance and an assertion of sovereignty. Attempts to solidify sovereignty have tremendous ramifications on northern residents, a point disastrously illustrated by the High Arctic Relocation project of the Canadian government (Qikiqtani Truth Commission, 2010; Wakeham, 2014). Sovereignty then becomes not just an assertion, but a continual struggle for achievement. Indeed, Pasternak argues that insomuch as Canada is trying to fuse sovereignty to jurisdictional claims over Indigenous land, its project is, and will continue to be, incomplete. In the Arctic, jurisdiction, whether around upholding colonial laws, game restrictions and patterns, providing social services or restricting access to certain parts of the territory, was mostly accomplished poorly and for the most part with the help of Inuit. The fact that the territory is both land and ice has caused some significant barriers to the operation of colonial jurisdiction, as the history of the Northwest Passage illustrates. However, as the colonial state becomes more confident in its knowledge over ice, ice is transformed into opportunities, or, at the least, a manageable obstacle.

This pattern continues into the age of devolution and the resolution of land claims in Northern Canada as the Canadian state downloaded jurisdictional practices onto the territories and Indigenous governments. These changes were put into motion with the challenge to federal jurisdiction in the 1960s and 1970s from both external (e.g. the Cold War and the SS *Manhattan* transit of the Northwest Passage) and internal (e.g. the Mackenzie Valley Pipeline inquiry) forces. Joint management initiatives, self-governing agreements, and other modern jurisdictional compromises initiated a new period in settler jurisdiction whereby the state officially cedes some control, but maintains what Pasternak (2017) describes as 'the authority to have authority' (2).

Emilie Cameron and Tyler Levitan's (2014) work on Impact Benefit Agreements in Nunavut and Warren Bernauer's (2019) examination of the power imbalances of modern land claims suggest that jurisdiction still favours the settler and capitalist orientation of the federal government. They also highlight the importance of environmental concerns in conflicts over jurisdiction. In many cases the desire for jurisdiction in the north focused around the resources of the territory (on land and under ice). The transfer of the Arctic islands from British to Canadian control in 1880 started with a request to purchase land for a mine on Baffin Island (Grant, 2002). At the start of the twentieth century, southern interests in the nature of the north manifested into a set of anxieties about the future of northern resources, including regulations and control of large mammals in Northern Canada (Sandlos, 2011). These early game preservation (and production, in the case of the attempt to introduce reindeer herding into the western Arctic) practices were extensions of southern jurisdiction into the far north (Sandlos, 2011; Stuhl, 2016).

The future north

The important discursive switch from the future imagined by the Hudson's Bay Charter and that which started to take place in the latter decades of the twentieth century is from a future of abundance (furs, profits, settlements) to a future of less (ice, polar bears, control). The transits of the Northwest Passage by the *Manhattan* in 1969/70 make this clear. At issue for Canada was the status of the Northwest Passage: Canada believes it to be sovereign waters, while the US sees it as an international strait connecting two expanses of high seas. Ice plays a key role here: during the transit, the Canadian Coast Guard accompanied the *Manhattan*, eventually freeing the tanker when it was stuck in the ice. The Coast Guard's success emboldened the Canadian state's claim to the Northwest Passage. In the face of increasing marine traffic through the Northwest Passage, the Canadian Government (Pierre Trudeau at the time) passed the Arctic Waters Pollution Prevention Act (AWPPA), which enabled the policing of ships throughout the Northwest Passage in the name of protecting the Arctic environment (Utton, 1972). Ice again was central to this jurisdictional claim: ice creates the conditions of possibility to label the Arctic water more dangerous, thereby requiring greater state control over who is on the water. The ability to search and limit all ships in the Northwest Passage meant that the AWPPA worked as a de facto claim to sovereign waters. The danger provided by ice-covered waters and the concern for the pristine waters of the Arctic were thus drafted by what David Meren and Bora Plumptre (2013: 174) call an 'environmentalism of convenience' in the service of jurisdictional claims. The Arctic became mobilized here – and later in the inclusion of Article 234 of the United Nations Convention on the Laws of the Sea (UNCLOS) – as a fragile place which must continually be protected from the encroachment of (inevitable) industrial development.

In this environmentalism of convenience, jurisdiction is provided by the need to protect the environment, a justification that attaches a future claim to the jurisdiction: in the future, the AWPPA will have protected these ice-covered waters. An imagined success story is essential, even if unwritten, for without the protected environment in the future, then the contemporary efforts to protect them (and the jurisdiction that follows) are unnecessary. This ideal future is one in which not only the environment is intact, so too are the jurisdictional claims, projecting the state into an idealized environmental-legal future. If the Arctic waters are protected in the future, the story goes, it will be in part due to the AWPPA.

With the adoption of Article 234 of UNCLOS in 1982, Canada was able to get the principles established in AWPPA justified (broadly, at least) by international law (Solski, 2021). The article establishes greater responsibility (and therefore potential sovereign control) over ice-covered areas of the sea because of the 'major harm or irreversible disturbance of the ecological balance' (116) that could arise from a pollution event in those waters. This article not only drafts the increased hazards of marine travel in ice-covered water into the service of increasing jurisdiction, it also prioritizes scientific understandings of ice, as the laws and regulations covered by Article 234 need to be 'based on the best available scientific evidence' (ibid.). Ice, under this framework, becomes an object of legislative concern that is made legible by Western science, and the pathway to the future comes from that scientific knowledge.

Knowable ice

At the same time as Article 234 was being negotiated (Solski, 2021), the rush to understand ice was underway to profit on the discovery of oil and gas resources in the Canadian Arctic Archipelago. The Arctic Pilot Project (APP) of the late 1970s and early 1980s was a highly ambitious and hyper-nationalistic partnership between Petro-Canada, Alberta Gas Trunk Line, and Melville Shipping to transport liquefied natural gas (LNG) from Melville Island in the Canadian Arctic Archipelago to the country's east coast using large icebreaking tankers (*Star-Phoenix*, 1977; Campbell, 1977). The APP emerged in the wake of the Polar Gas pipeline proposal that would have transported gas from Melville Island to Northern Ontario using unprecedented pipeline installation techniques across five deep-water Arctic channels. Polar Gas eventually failed for many reasons, including the discovery of lower-cost gas reserves in Alberta, Inuit opposition and resistance, and the development of competing proposals including the APP.

Like Polar Gas, the APP sought to produce reliable scientific knowledge to apply technical solutions to the problems of ice, a material whose slippery processes of 'formation, movement, and melting hide or complicate ideal notions of permanence so valued by states' (Bravo, 2017: 48). In contrast to Inuit knowledge and practices with ice developed over thousands of years rooted in networks of kinship and relationality (see: Watt-Cloutier, 2015; Karetak and Tester, 2017; Pfeifer, 2018; Smith, 2021), the imperatives of fossil capital necessitated that ice be rendered a stable 'condition of production' to transport ancient fossil fuels that would run factories and buildings in the south. While Polar Gas mostly sought to strengthen and increase the thickness of ice for operation of heavy trenching and installation equipment on, the APP approached ice as a superfluous nature that could be destroyed and reconstituted at will, but one that posed potential material and social threats if not adequately studied and managed. Rosemary-Claire Collard and Jessica Dempsey's (2017) analytic of five orientations of capitalist natures helps make sense of this: Polar Gas oriented ice as 'the underground' (nature that was 'useful, but unvalued') while the APP oriented it as both 'outcast surplus' ('of no interest to capital') and 'threat' ('that which is deemed to endanger capitalist production') (79). These orientations are 'dynamic, contested and in some cases overlapping' (Collard and Dempsey, 2017: 84), with ice both shaping and shaped by capital in a dialectical relation. Scientific knowledge production was foundational to this reorienting of ice for capitalist production. Ice science also expanded the sovereignty of the colonial state over the threatening, but potentially useful, Arctic through discourses of expertise and environmental management that promised certainty and authority (see: Stuhl, 2016; Powell, 2017).

One of the key sites of this scientific knowledge production was at Bridport Inlet on the southern shore of Melville Island, where a 160-kilometre pipeline from the Drake Point gas field would end and the liquefaction and storage facilities would be built on top of barges. Specifically, the APP sought to develop an 'ice management system' in the inlet itself to cultivate ideal environmental conditions for shipping. This approach oriented ice as an 'outcast surplus' to be physically melted. In order to keep the inlet ice-free throughout the year, enabling carefully scheduled shipping, the consortium proposed the managed release into the inlet of warm salt-water effluent generated from the liquefaction process itself. This specific process was estimated to increase the average temperature in the inlet by 3 degrees Celsius and limit ice thickness to 20 centimetres between arrivals of icebreakers (Arctic Pilot Project, 1978: 5–6; Sidjak, 1978: 5). Put another way, the waste from condensing Arctic gas to ship for use in southern factories and buildings was to be used to create the very conditions in which shipping could take place and socially necessary turnover time could be achieved. However, the APP acknowledged there were many unknowns about the impacts of the proposed intervention, including the possibility of the warm effluent undermining soil strength in the inlet, the possible melting of frozen effluent in the spring 'rather suddenly,' and elimination of fifty different species of ice algae altogether (Arctic Pilot Project, 1978; Arctic Pilot Project, 1979). The consortium reported that 'these changes could have a variety of biological effects, none of which are well understood' (Arctic Pilot Project, 1978: 6–45). Despite continued uncertainty about possible impacts of the ice management system in Bridport Inlet, the environmental assessment and review process (EARP) report gave it the go-ahead, writing that 'the thermal regimes of the water and seabed are being studied in a competent manner and that with proper engineering, *it does not appear that any exceptional problems will arise* [emphasis added]' (APP Environmental Assessment Panel, 1980: 53). The review panel's recommended conditions included ongoing monitoring of the subsoil, weather, and biology of Bridport Inlet (APP Environmental Assessment Panel, 1980: 98). The potential orientation of ice as 'threat' due to a novel engineering scheme was managed through the deployment of 'proper engineering' to be proven at some point in the future.

However, the APP's general orientation to ice as 'outcast surplus' was quickly shifted to potential 'threat' due to organized militancy by Inuit to the project – including the formation of the Inuit Circumpolar Conference in 1980 – that politicized and opposed the most visible harms of their proposed activities in the region (*Asbury Park Press*, 1980). APP executives had recently witnessed what could happen if such opposition was not effectively addressed in the form of the Berger Inquiry that had shut down short-term prospects of a Mackenzie Valley pipeline. Many concerns by Inuit about the APP were articulated and formalized through the EARP in 1980 and National Energy Board (NEB) hearings in 1982. The APP readily admitted throughout these processes that it did not possess adequate knowledge of many of the potential social and environmental impacts of its operations in the Arctic Islands. In response, it proposed that monitoring and assessment activities could mitigate nearly all impacts, with knowledge production of ice and impacts on Inuit harvesting and mobility occurring simultaneously. This relation relies on a projected colonial future of industrial extraction *and* environmental protection in which the knowledge gathered in the proposal process ensures success on both fronts. John Amagoalik, the then president of Inuit Tapirisat of Canada (now Inuit Tapiriit Kanatami), criticized this aspect of the supposed pilot project, describing the APP as 'only the cutting edge to facilitate oil tanker traffic' (Lewington, 1982). The fact that environmental impacts could only be measured once environmental impacts occurred guaranteed reoriented ice towards the imperatives of capital, greatly undermining Inuit futures that relied on the presence of ice.

One of the most significant problems identified by Inuit was the potentially fatal hazards the project posed to Inuit hunters who travelled over the large channels that lay in the wake of icebreaking tankers. George Eckalook of Resolute told the federal environmental review panel that ships 'will be breaking a big distance of ice and it's going to be very wide. The hunters in Resolute are afraid about that' (APP Environmental Assessment Panel, 1980, p. 60). Previous studies had indicated that ice broken by a ship would likely not refreeze during the warm summer months, making hunting far more difficult and dangerous. The APP responded to this by commissioning several scientific investigations to produce recommendations that would ostensibly reduce potential harm. One study, conducted by Acres Consulting Services (1982), attempted to devise a variety of technical solutions to problems posed by large channels including far-fetched ideas like spraying actual liquefied natural gas onto the channel to accelerate refreezing of the ice, creating an ice bridge by using the tanker to 'bulldoze' pieces of broken ice together, and establishing bridges that were inflatable or made of imagined biodegradable foam (Acres Consulting Services, 1982: 42). The firm also speculated about imposing radical spatio-temporal changes onto Inuit hunters, including altering hunting seasons and establishing caribou farming on Cornwallis Island to prevent hunters from having to cross the channel. Another study saw APP partner with Dome Petroleum, Gulf Canada, and Esso Resources on a field investigation conducted on ship tracks left in McKinley Bay near Tuktoyaktuk. The investigation's methodology was highly improvised, involving participants walking out onto the freshly consolidated ice to evaluate stability. Inuit hunters and trappers from nearby communities were used by the researchers to provide recommendations about ice safety. If deemed safe, a snowmobile weighted down with ice blocks was driven across the ice several times. Based on the number of intact blocks in the channel, the report concluded: 'Their behaviour suggested that a person stepping directly into the slush would probably not sink into it above his knees. Because waterproof footwear had not been taken along, this test was not attempted' (Danielewicz *et al.*, 1983: 5). The report suggested that channels would not pose serious obstacles to travel for Inuit hunters but noted that similar trials would have to be conducted in new locations under different conditions. However, Arctic communities continued stressing the hazardous variability of conditions at any given site, including during bad visibility and under-ice currents. Recommendations devised by the company in response included sounding a foghorn in bad weather to alert hunters, performing helicopter reconnaissance of the channel, conducting further tests in areas of traditional use, and issuing community radio messages to inform local residents of changing conditions. The APP's recognition that failing to adequately address Inuit concerns about impacts on ice could lead to ice itself becoming a threat to capitalist development resulted in pledges of future monitoring, weather reports, and surveillance.

The APP was officially declared dead by the media after the NEB closed the project's application in August 1984. A post-mortem written by Jennifer Lewington (1987) argued the project was defeated by an absence of 'clear and compelling purpose,' along with a combination of falling commodity prices and a prolonged recession (163). The complicated nature of ice and determined Inuit resistance to the project also contributed to its downfall.

But as with many large-scale industrial projects, there were lingering effects despite its failure (Peyton, 2017). Elements of the APP proposal live on in the imaginations of various fossil fuel proponents. A 2005 report prepared by the Canadian Energy Research Institute – a research organization funded by Natural Resources Canada, Alberta Energy, and the Canadian Association of Petroleum Producers – examined the economic case for producing and transporting natural gas from the Arctic Islands by tanker (Chan *et al.*, 2005). In 2008, Petro-Canada officially formed a 'small team' to again examine its feasibility, with the company's CEO describing it as 'the project to die for' (Lee, 2007). In 2014, Michael Bell – previously of Melville Shipping – made an appeal to 'Arctic sovereignty via LNG', calling on a resuscitation of it as a means of 'putting a Canadian sovereignty stamp on the Canadian High Arctic' (Bell, 2014). Each of these built on the APP's orientation to ice as a material that could be managed through iterative study and adaptation, in order to minimize its development into a potential threat.

Beyond the technical specifics of the project, the legacy of the APP lives on to advance claims by capital and state about ice – and increasingly ice-free channels – through promises of future knowledge and expertise. Both the APP and environmental review process assumed the supposed adaptability of the pilot project to new information, allowing it to reduce total impacts throughout the lifetime of the project. More recently, Lackenbauer and Lajeunesse (2014) emphasized the need for development of 'safe shipping lanes' in the Canadian Arctic through improved surveying, navigational charts, marine infrastructure, and oil spill detection and mitigation (5). A 2016 report by Pew Charitable Trusts proposed the creation of an 'integrated Arctic corridors framework' in Canada, heavily dependent on the ability to 'systematically monitor all variables' using an 'iterative and adaptive approach' (24–27). These claims to future perfection of shipping impacts on Arctic ice can be traced back to at least partial origin in the APP and demonstrate the ways that colonial sovereignty projects itself forward through the discourses of environmental management and mitigation. The capitalist and colonial impulse to imbue an increasingly unpredictable materiality with permanence, or at least predictability, continues through appeals to future knowledge, subordinating Inuit usage and desires for what remains of the ice.

The future perfect of polar bears

The projection of the Canadian state onto northern environments and futures has depended on the reworking of relations into resources, be they mineral or animal. While the Arctic Pilot Project insisted that the hazards of ice could be known, mitigated, and appropriated through state-funded science, a similar logic of expertise had long been at work optimizing wildlife. The long process that has made the polar bears of the Canadian Arctic into a state-managed natural resource helps illustrate how past dependence on ice, as an environment and resource for wildlife, has underpinned state jurisdiction over lives and livelihoods. Polar bears and ice are intimately entwined in both narrative and nature. The history of polar bear management in Canada has depended on sea ice as the regenerative substrate for the animal and taken it mostly for granted. As the sea ice retreats under a changing climate, Canadian wildlife managers have faced a conflicting set of challenges to both preserve this species and respect Inuit rights to harvest the animal and protect themselves from it. The newest innovations in polar bear management promise to reduce or eliminate the number of bears killed while ensuring the safety of an increasingly populated human Arctic. However, modifications to hunting quotas and attempts to deter polar bears away from humans still fail to address the thawing ice on which the bears depend. While the drivers of ice loss are beyond the control of any polar bear management policy, most desires to preserve the polar bear harbour a nostalgia for an irretrievably icy Arctic. Such icon-first preservation extends and entrenches Canadian colonial jurisdiction over polar bears and the people who share a landscape with them.

In the early histories of Canadian polar bear management, the ice was understood as a surface on which a polar bear resource reproduced. Early conservation strategies focused on limiting polar bear harvest and scarcely imagined preserving habitat. In 1956, the Canadian Wildlife Service listed the polar bear as a species of special concern and began drafting conservation strategies (Kulchyski and Tester, 2007). In 1968, Canada imposed the first polar bear hunting quotas in response to a harvest level described as 'unprecedented' (Schweinsburg, 1981: 1) despite a lack of reliable long-term data to make that claim (Prestrud and Stirling, 1994). The quota system implemented a new way of understanding the polar bear: as a calculable resource that could be developed, conserved, optimized, and potentially exhausted (Larsen and Stirling, 2009; Schreiber, 2013). It also introduced disciplinary mechanisms by which hunters were rewarded or punished, depending on whether or not they cooperated with management (Kulchyski and Tester, 2007). The quota was intended to supplant Inuit understandings of the polar bear with the values held by the wildlife technocrats of the Canadian state. Drafts of polar bear management plans from the 1980s consigned Inuit wildlife expertise to an archaic and irrelevant past, asserting that 'many of the ancestral beliefs and traditions in polar bear hunting have been abandoned along with stone weapons' (Department of Renewable Resources, 1988: 1). The future of polar bear management, and the future of polar bears, would be made through state science that metaphorically and empirically mapped this species to ice.

In the narratives that have coalesced into the scientific knowledge of the animal, polar bears have come to be understood as ice beasts. Some of the earliest names for the polar bear, such as the Norwegian *isbjørn* (ice bear), connect the animal to a glaciated landscape (Engelhard, 2017). As late as the mid-twentieth century, some naturalists hypothesized that all polar bears comprised a single, migrating pan-Arctic community that was coterminous with the sea ice (DeMaster and Stirling, 1981). In some places, biologists have described witnessing polar bears literally emerging from the ice. In the Beaufort Sea north of Alaska, polar bears were observed constructing maternity dens on drifting pack ice (Armstrup and Gardner, 1994). Polar bears require the sea ice to hunt seals and chase the feeding glut that allows them to subsist for the summer months on land with little or no food (Molnár *et al.*, 2020). Their dependence on ice for feeding is the most important connection for narratives that leverage the polar bear as a portable and compelling icon to demonstrate the impacts of Anthropocene-era climate change (Slocum, 2004; Smaill, 2015; Born, 2018). It is now hardly possible to discuss polar bears without also discussing climate change given the well-tracked and forebodingly predicted Anthropocene 'death spiral' for Arctic sea ice (Wadhams, 2014: 9; Gibbens, 2017).

While polar bears are certainly affected by climate change, the specifics of how this species responds to a changing climate cannot be easily generalized across time or space. This is partially the result of the geographic variability of sea-ice patterns. Within the central Canadian Arctic Archipelago, polar bear populations appear largely stable at present (Lunn *et al.*, 2018). However, at the peripheries, even in the high-Arctic Kane Basin, there is evidence that the populations are in decline (York *et al.*, 2016). Many scholars have warned that the declines in sea ice will result in broad extirpation or complete extinction by the end of the century (Castro de la Guardia *et al.*, 2013; Hamilton *et al.*, 2014; and Molnár *et al.*, 2020). The scale of their concerns is matched by a similar scale of recommendations that draw a direct connection between the threats to ice and polar bears and the excesses of capitalist economies: they recommend the reduction of greenhouse gas emissions. However, navigating the local contexts of the thaw has no obvious course for remedy.

The environmental concern about a future that threatens imminent extinction has intensified management of the interactions between humans and bears in the interest of preserving the threatened species. Alongside the restrictions on polar bear hunting and harvesting that are typical of wildlife conservation, there has been a rise of urban safety protocols that focus on protecting Arctic communities against polar bears through non-lethal force (Zerehi, 2016; Wilder *et al.*, 2017; Dickie, 2018). These polar bear monitoring and deterrence programmes attempt to create secure spaces for people around

Arctic settlements without threatening the lives of polar bears, as was normal in the past. While some programmes have been operational since the 1980s, they are by no means widely adopted. Deterrence and monitoring programmes are a proposed technological fix to the growing tension between the goals of conservation and the demand for safety. In reworking the relationships between people and polar bears, they draft the future loss of an ice-covered Arctic into the changing jurisdictional claims over northern urban spaces.

The first such programme was put in place in Churchill, Manitoba. The town was sited along a wedge of land that protrudes into Hudson Bay and coincidentally serves as a gathering point for polar bears who wait for the sea ice to form in autumn. In response to increasing conflict between people and bears, the Polar Bear Alert Program was started as a volunteer Halloween night-patrol and later formalized into a state-directed array of policies and technologies (Kearney, 1989). In the 1960s and early 1970s, polar bears in or near Churchill were shot and killed by game officers of the Manitoba Department of Mines and Resources. While many residents were concerned about human safety, there was a sustained distaste for lethal means of dealing with the bears (Archibald, 2017). This has amplified today, as the town's economy is largely driven by polar bear tourism (D'Souza *et al.*, 2021). This economic imperative to avoid killing bears is coupled with an ethical tension, since many tourists could not imagine why anyone would intentionally kill a polar bear. Beginning in 1971, representatives from the International Fund for Animal Welfare (IFAW) organized funding and logistics for the first relocation programmes for nuisance polar bears in the town of Churchill. Members of IFAW also encouraged the residents to form the Churchill Wildlife Association to focus efforts contesting what many considered to be the unnecessary killing of bears. The IFAW-led relocation programme existed uneasily alongside the official provincial policy of killing bears until the creation of the Polar Bear Alert Program in 1980. In the following year, the Department of Mines and Resources issued the first policy directive that considered the safety of people, the protection of property, and ensuring that 'polar bears are not unduly harassed or killed' (Durnin, 1983: 27). In 1982, the Province renovated a former military compound to detain problem bears until they could be relocated to land-fast ice when it formed along the coast of Hudson Bay. While the location of the town in relation to sea ice gave the impetus for Alert, the ice was a necessary part of the programme's design.

Reports on the Polar Bear Alert Program emphasize its rationality and assume that sea ice will dependably act as a sequestering infrastructure. Alert is considered by many to be a cutting-edge, world-class conservation strategy targeting human–polar bear interactions (Schmidt and Clark, 2018). Churchill has been referred to as the world's first 'polar bear safe community' (Dickie, 2018) because of Alert. The programme organizes the space of

Churchill into categories of decreasing bear tolerance towards the centre of the town (Kearney, 1989). It is an arrangement that equates the urban with safety and the wilderness with hazard. Polar bears found within the town limit are held within a facility known colloquially as the 'polar bear jail'. When there is sufficient ice on Hudson Bay, the bears detained in the facility are relocated by airlift to the frozen water. Before their release, the bears are marked with a green dye to help conservation officers in Churchill identify repeat offenders. The sea ice mediates the Alert programme's capacity to provide safety for bears and people through distance. However, like elsewhere in the Arctic, the ice on Hudson Bay has a shortened winter duration and this abbreviating trend is expected to continue (Hamilton and Derocher, 2019). To resolve the problem of this disappearing platform, the town of Churchill and the environmental advocacy group Polar Bears International have proposed the nearby Wapusk National Park as a functional, but less desirable, alternative release site (Brace, 2020). The Alert programme was devised for the sea ice of the 1980s, and is having to reconcile with a thawing Hudson Bay. In recent years, green-spotted polar bears released by the Alert programme have been seen returning to the lands of the coast in Arviat, Nunavut (Zerehi, 2016).

Global interest in polar bear conservation has pushed the expansion of non-lethal deterrence practices beyond Churchill. The town of Arviat is the other site with a comprehensive polar bear monitoring and deterrence programme. Partly inspired by Alert, the town has partnered with the World Wildlife Fund (WWF) since 2010 to implement the Human–Polar Bear Conflict Reduction Project (WWF, 2013). The project is funded by the WWF with the hope of demonstrating alternatives to lethal bear deterrence. Under the programme, garbage storage has changed, and polar bears have been actively deterred from the townsite by a patrol. But these adaptations are not always enough to address the increasing presence of polar bears in the area. Sea ice loss may increasingly make clear the shared vulnerability of polar bears and humans in a shared landscape. In July of 2018, Aaron Gibbons of Arviat was killed by a polar bear outside of the town while collecting goose eggs with his children (Rogers, 2018). In a report on the incident, the Kivalliq Wildlife Board (KWB) tentatively suggested that the habituation of bears to human presence in Churchill may have been a factor in the attack (KWB, 2018).

Also in 2018, Darryl Kaunak of Naujaat was killed by a polar bear on a hunting trip outside of town (Hutchins, 2019). Months afterwards, elders from Naujaat implored the hunters in their community to kill an approaching polar bear. An Inuk named Laurent Kringayark took the situation and the advice of the elders seriously. Kringayark was charged with the killing of a polar bear without a hunting tag and defended his actions in court in

March 2019 where he was let off with a warning. Despite the fact that he was relieved of prosecution, the legal implications of polar bear safety sided against him, producing dangerous ambiguities for those who have to share landscapes with polar bears. Kringayark expressed a sentiment felt by many in Naujaat and in other communities across Nunavut: 'we're captives in our own town now – because of the bears' (Hutchins, 2019). This viscerally felt threat is an immediate expression of the destabilized sea ice, but the sense of captivity is exacerbated by a legal and political framework that foregrounds the mournability of polar bear deaths. These frameworks advance a somewhat contradictory set of conservation values that police direct interactions between polar bears and humans while failing to address the structural causes of stress on the species. Under Canadian state jurisdiction, polar bear mortality may be consistently tragic, but it is only made criminal when there is someone who pulls the trigger.

In contrast to deterrence programmes, communities in Nunavut have proposed adjustments to the quota system so that bears killed in the interest of human safety are counted separately from those that are hunted (Brown, 2018). Nunavik Inuit have proposed increases in the polar bear hunting quota, partially in response to the increasing risk of polar bear–human conflict (Forrest, 2019). While not increasing the quota, the 2019 polar bear management plan in Nunavut marks the first official policy on monitoring and deterrence in the Territory. It pledges to 'continue to support communities in the development and implementation of polar bear monitoring and safety plans; [and] hire, train, and equip more community polar bear monitors' (Government of Nunavut, 2019: 30). The 'people and bears' objectives of the plan are centred on minimizing the number of bears killed in defence of life and property. The quota has been the primary means of managing bears since the 1960s and communities have frequently demanded quota increases to deal with problem bears. Previous drafts of the plan had considered increasing quotas to accommodate defence kills (Canadian Press, 2018). Inuit hunters from the coast of Nunavut along Hudson Bay have expressed strong dissatisfaction with recent polar bear management, but have variously supported increases in the quota, or an accounting of defensively killed bears outside of the quota altogether (Lokken *et al.*, 2019). Ice haunts these management discussions, both as a known driver of the ecological conditions that strain the relationships between people and polar bears, and as an unwinding fabric beyond the control of those who live within it.

The push for monitoring and non-lethal deterrence comes at the expense of supporting Inuit communities who see lethal deterrence as a reasonable response to the threat of polar bears. Already, Inuit risk criminalization for defending themselves and their communities from polar bears. Monitoring and deterrence programmes place the burden of adaptation on those most directly affected by climate change. They propose solutions that compromise Inuit capacity to change with their landscape in order to forestall the anticipated declines of polar bears driven by forces outside the Arctic. Kringayark's frustration highlights the limited protection that monitoring and deterrence affords Inuit lives and livelihoods. Gathering goose eggs, knowing the ice, hunting country foods and having embodied encounters with the landscape are all necessary for the ongoing production of Inuit cultures and economies (Watt-Cloutier, 2015; Arnaquq-Baril and Thompson, 2016). Both climate change and many proposed adaptations to it add barriers to these practices. Over time, the official policy stance of Nunavut has shifted from one of climate change mitigation to one of adaptation, while the necessity of Inuit cooperation in this project has been normalized (Cameron *et al.*, 2015). Adaptation implies a dynamic response to change. However, adaptation and resilience projects have more often been imposed on Inuit than they have been led by them (Cameron, 2012). As the experiences in Naujaat and Arviat illustrate, individual communities require support for their own safety strategies, even if those include killing polar bears. This does not fit in the logic that insists on exporting elaborate technical systems from Churchill to keep polar bears deterred while their habitat disintegrates.

Polar bear monitoring and deterrence programmes may help contend with a present and a future of thawing ice in which an Arctic icon is both a risk and at risk. However, there is a disconnect between the situated capacities of these programmes and the global scale of climate change and species extinction. Alert, and similar monitoring programmes, rely on methods of deferral and displacement not to forestall the hazard posed by bears, while circumventing the political authority of the people who share a landscape with polar bears in the present. In this, the Anthropocene future of polar bears is used as a rationale to impose romanticized and tenuous policies of harmony on those who have to live with them. Through their production of different material spheres of safety and security in the northern landscape, monitoring and deterrence programmes indefinitely extend a set of colonial conservation sensibilities into both bear and human lives. These sensibilities are built on a nostalgia for a landscape of ice, and sublimate an entire space and species into an object of salvage. This sort of polar bear conservation has reinforced the colonial Canadian state's authority over safety, wildlife, and the future.

Anthropocene futures

In both of these cases, the future of ice – a knowable, scientific ice and the lost bear habitat – is folded into the jurisdictional claims to Arctic space. In them, the future that will have been becomes a fundamental part of the authority on the ground, whether to securitize Arctic space or to experiment with fossil fuel extractions. There is an interesting temporality happening here, in which the discussion of ice in the future is focused on universal ice – ice as known through the scientific method. Ice as a plural object may be allowed into the discussion, but usually only as it pertains to ice in the past, through traditional knowledge, oral history, or descriptions of activities that have been or are proposed to be disrupted. One significant reason for this is that the ice of the scientific method is a disciplined ice, an object that is meant to be isolated, explained, and managed. Such a material object is infinitely more productive to the modern state.

In her discussion of the temporality of Indigenous peoples in late liberalism, Elizabeth Povinelli (2011) describes the 'governance of the prior', in which Indigenous peoples are relegated to the past of the nation. Contemporary settler colonialism, she argues, does not eliminate Indigenous peoples from their stories, symbols, or myths. Instead, they are separated within those discourses to a different temporality. 'When we look at these different narrative structures we find that although all people [governed by the nation] may belong to nationalism, not all people occupy the same tense of nationalism' (37). The settler state reserves the future perfect, which justifies its jurisdiction, for those discourses that reproduce settler-oriented capitalist growth.

Looking at the securitization of bear–human interactions and the integration of scientific experimentation into the environmental assessment process, it is possible to see the future perfect of the Anthropocene asserting a continued settler state. These discourses, importantly, are being contested by Indigenous actors, who have increasingly contested colonial control in the Arctic. One way to support this contestation is to illustrate the ways in which the temporal narratives of ice attempt to restrict the influence of Indigenous jurisdiction and modes of governance on our vision of the future.

References

Acres Consulting Services Limited (1982) *Study of Ship Track Crossing Methods*. Calgary, Canada: Petro-Canada and Arctic Pilot Project.

Amstrup, S. C. and C. Gardner (1994) 'Polar bear Maternity Denning in the Beaufort Sea', *Journal of Wildlife Management* 58(1): 1–10.

Archibald, K. (2017) 'Animal Capital: Managing Polar Bears in Churchill, Manitoba', in J. Dean, D. Ingram, and C. Sethna, eds, *Animal Metropolis: Histories of Human–Animal Relations in Urban Canada*. Calgary, Canada: University of Calgary Press, 255–283.

Arctic Pilot Project (1978) *Draft Environmental Statement – Melville Island Components Volume v, Part b, Book 2*. Calgary, Canada: Arctic Pilot Project.

Arctic Pilot Project (1979) *Arctic Pilot Project Environment Statement Supplementary Information*. Calgary, Canada: Arctic Pilot Project.

Arctic Pilot Project Environmental Assessment Panel (1980) *Arctic Pilot Project (Northern Component): Report of the Environmental Assessment Panel*. Ottawa, Canada: Federal Environmental Assessment Review Office.

Arnaquq-Baril, A. (producer and director) and B. Thompson (producer) (2016) *Angry Inuk* [film]. National Film Board of Canada.

Asbury Park Press (1980) 'Greenland Eskimos Fight to Keep Canadian Gas Tankers from Shore', *Asbury Park Press*, 14 October, B3.

Bell, M. H. (2014) 'Rather than Lay Claim to the North Pole, the Federal Government Should Tap Gas in our Arctic Islands', *Financial Post*, 29 January. Available at: https://financialpost.com/uncategorized/rather-than-lay-claim-to-the-north-pole-the-federal-government-should-tap-gas-in-our-Arctic-islands (accessed 21 March 2022).

Bernauer, W. (2019) 'The Limits to Extraction: Mining and Colonialism in Nunavut', *Canadian Journal of Development Studies/Revue canadienne d'études du développement*, 40(3): 404–422.

Born, D. (2018) 'Bearing Witness? Polar Bears as Icons for Climate Change Communication in National Geographic', *Environmental Communication* 13(5): 649–663.

Brace, M. (2020) 'Safer Co-existence with polar bears', *Polar Bears International*. Available at: https://polarbearsinternational.org/news/article-polar-bears/safer-co-existence-with-polar-bears (accessed 21 March 2022).

Bravo, M. (2017) 'A cryopolitics to reclaim our frozen material states', in J. Radin and E. Kowal, eds, *Cryopolitics: Frozen Life in a Melting World*. Cambridge, MA: MIT Press, 27–57. https://doi.org/10.7551/mitpress/10456.003.0005

Brown, B. (2018) 'Don't deduct polar bear defence kills from quotas, Inuit say', *Nunatsiaq News*, 15 November. Available at: https://nunatsiaq.com/stories/article/65674dont_deduct_polar_bear_defence_kills_from_quotas_inuit_say/ (accessed 21 March 2022).

Cameron, E. (2012) 'Securing Indigenous Politics: A Critique of the Vulnerability and Adaptation Approach to the Human Dimensions of Climate Change in the Canadian Arctic', *Global Environmental Change* 22(1): 103–114.

Cameron, E. and T. Levitan (2014) 'Impact and Benefit Agreements and the Neoliberalization of Resource Governance and Indigenous–State Relations in Northern Canada', *Studies in Political Economy* 93(1): 25–52.

Cameron, E., R. Mearns, and J. T. McGrath (2015) 'Translating Climate Change: Adaptation, Resilience, and Climate Politics in Nunavut, Canada', *Annals of the Association of American Geographers* 105(2): 274–283. https://doi.org/10.1080/00045608.2014.973006

Campbell, T. (1977) 'Two Tankers Proposed to Haul Gas from Arctic', *Edmonton Journal*, 1 November, 104.

Canadian Press (2018) '"So Many Bears": Draft Plan Says Nunavut Polar Bear Numbers Unsafe', *CBC News*, November 12. Available at: www.cbc.ca/news/canada/north/too-many-polar-bears-1.4901910 (accessed 21 March 2022).

Castro de la Guardia, L., A. E. Derocher, P. G. Myers, A. D. Terwisscha van Scheltinga, and N. J. Lunn (2013) 'Future Sea Ice Conditions in Western Hudson Bay and Consequences for Polar Bears in the 21st Century', *Global Change Biology* 19(9): 2675–2687.

Cavanagh, E. (2011) 'A Company with Sovereignty and Subjects of its Own? The Case of the Hudson's Bay Company, 1670–1763', *Canadian Journal of Law & Society/La Revue Canadienne Droit et Société* 26(1): 25–50. https://doi.org/10.3138/cjls.26.1.025.

Chan, L., G. Eynon, and D. McColl (2005) *The Economics of High Arctic Gas Development: Expanded Sensitivity Analysis*. Calgary, Canada: Canadian Energy Research Institute. Available at: http://library.Arcticportal.org/1562/1/HighArcticGasReport.pdf (accessed 21 March 2022).

Collard, R.-C. and J. Dempsey (2017) 'Capitalist Natures in Five Orientations', *Capitalism Nature Socialism* 28(1): 78–97. http://dx.doi.org/10.1080/10455752.2016.1202294.

D'Souza, J., J. Dawson, and M. Groulx (2021) 'Last Chance Tourism: A Decade Review of a Case Study on Churchill, Manitoba's Polar Bear Viewing Industry', *Journal of Sustainable Tourism*, 1–19 [online first view].

Danielewicz, B., E. Pessah, and S. Cornett (1983) *Field Investigation of Tracks Left by Ice Breaking Vessels*. Calgary, Canada: Dome Petroleum/Canmar.

DeMaster, D. P. and I. Stirling (1981) '*Ursus Maritimus*', *Mammalian Species* 145: 1–7.

Department of Renewable Resources (1988) *Northwest Territories Polar Bear Management Plan*. (G-2013–012 1–11), NWT Archives, Prince of Wales Northern Heritage Centre, Yellowknife, NWT.

Dickie, G. (2018) 'As Polar Bear Attacks Increase in Warming Arctic, a Search for Solutions', Yale Environment 360, December. Available at: https://e360.yale.edu/features/as-polar-bear-attacks-increase-in-warming-Arctic-a-search-for-solutions (accessed 21 March 2022).

Durnin, J. (1983) *Polar Bear Alert Program, 1983*. Manitoba Department of Natural Resources, Wildlife Branch (Manitoba NatRes WB Annual 1983). Manitoba Legislative Library, Winnipeg, MB.

Engelhard, M. (2017) *Ice Bear: The Cultural History of an Arctic Icon*. Seattle, WA: University of Washington Press.

Erickson, B. (2020) 'Anthropocene Futures: Linking Colonialism and Environmentalism in an Age of Crisis', *Environment and Planning D: Society and Space* 38(1): 111–128.

Forrest, M. (2019) 'Federal Court Sides with Ottawa in Decision to Reduce Nunavik Polar Bear Hunt', *National Post*, 19 November. Available at: https://nationalpost.com/news/politics/federal-court-sides-with-ottawa-in-decision-to-reduce-nunavik-polar-bear-hunt (accessed 21 March 2022).

Gibbens, S. (2017) 'Your Questions about Our Starving Polar Bear Video Answered', *National Geographic*, 11 December. Available at: https://news.nationalgeographic.com/2017/12/starving-polar-bear-video-climate-change-spd/ (accessed 21 March 2022).

Government of Nunavut (2019) *Nunavut Polar Bear Co-Management Plan*. Iqualuit, Nunavut: Government of Nunavut. Available at: www.gov.nu.ca/sites/default/files/nwmb_approved_polar_bear_comanagement_plan_sept_2019_eng.pdf (accessed 21 March 2022).

Grant, S. D. (2002) *Arctic Justice: On trial for Murder, Pond Inlet, 1923*. Montreal: McGill-Queen's University Press.

Hamilton, S. G., L. C. De La Guardia, A. E. Derocher, V. Sahanatien, B. Tremblay, and D. Huard (2014) 'Projected Polar Bear Sea Ice Habitat in the Canadian Arctic Archipelago', *PLOS one* 9(11): e113746.

Hamilton, S. G. and A. E. Derocher (2019) 'Assessment of Global Polar Bear Abundance and Vulnerability', *Animal Conservation* 22(1): 83–95.

Hutchins, A. (2019) 'To Kill a Polar Bear', *Maclean's Magazine*, 5 April. Available at: www.macleans.ca/to-kill-polar-bear/ (accessed 21 March 2022).

Karetak, J. and F. Tester (2017) 'Introduction: Inuit Qaujimajatuqangit, Truth and Reconciliation', in J. Karetak, F. Tester, and S. Tagalik, eds, *Inuit Qaujimajatuqangit: What Inuit Have Always Known to be True*. Black Point, Nova Scotia: Fernwood Books, 1–19.

Kearney, S. R. (1989) 'The Polar Bear Alert Program at Churchill, Manitoba', in M. Bromley, ed., *Bear–People Conflicts: Proceedings of a Symposium on Management Strategies*. Yellowknife, Northwest Territories: Northwest Territories Department of Renewable Resources, 83–92.

Kulchyski, P. and F. Tester (2007) *Kiumajut (Talking Back): Game Management and Inuit Rights, 1950–1970*. Vancouver: University of British Columbia Press.

KWB (Kivalliq Wildlife Board) (2018) *Written Submission to the Nunavut Wildlife Management Board Public Hearing to Consider the Government of Nunavut Proposal on the Revised Polar Bear Co-Management Plan*. Rankin Inlet, Nunavut: Kivalliq Wildlife Board.

Lackenbauer, W. and A. Lajeunesse (2014) *On Uncertain Ice: The Future of Arctic Shipping and the Northwest Passage*. Calgary, Canada: Canadian Defence and Foreign Affairs Institute. https://doi.org/10.11575/sppp.v7i0.42493

Larsen, T. S. and I. Stirling (2009) *The Agreement on the Conservation of Polar Bears – its History and Future*. Rapportserie nr. 127. Tromso: Norsk Polarinstitute.

Lee, H. Y. (2007) 'Petro-Canada CEO: Arctic LNG would be a project "to die for"', *Dow Jones Newswires*, 29 November. Available at: www.rigzone.com/news/oil_gas/a/53475/petrocanada_ceo_Arctic_lng_would_be_a_project_to_die_for/ (accessed 21 March 2022).

Lewington, J. (1982) 'Inuit Want Claims Settled Before Projects Started', *Globe and Mail*, 9 March, B9.

Lewington, J. (1987) 'Lessons of the Arctic Pilot Project', in F. Griffiths, ed., *Politics of the Northwest Passage*. Montreal: McGill-Queen's University Press, 174–191.

Linton, J. (2010) *What is Water? History of a Modern Abstraction*. Vancouver: University of British Columbia Press.

Lokken, N. A. A., D. A. Clarck, E. G. Broderstad, and V. H. Hausner (2019) 'Inuit Attitudes Towards Co-managing Wildlife in Three Communities in the Kivalliq Region of Nunavut, Canada', *Arctic* 72(1): 58–70.

Lunn, N. J., M. Branigan, K. Breton-Honeyman *et al.* (2018) 'Management on polar bears in Canada 2009–2016', in G. M. Durner, K. L. Laidre, and G. S. York, eds, *Polar Bears: Proceedings of the 18th Working Meeting of the IUCN/SSC Polar Bear Specialist Group, 7–11 June 2016, Anchorage, Alaska*. Gland, Switzerland and Cambridge: IUCN, 33–67.

McCreary, T. and J. Turner (2018) 'The Contested Scales of Indigenous and Settler Jurisdiction: Unist'ot'en Struggles with Canadian Pipeline Governance', *Studies in Political Economy* 99(3): 223–245.

Meren, D. and B. Plumptre (2013) 'Rights of Passage: The Intersecting of Environmentalism, Arctic Sovereignty, and the Law of the Sea, 1968–82', *Journal of Canadian Studies/Revue d'études Canadiennes* 47(1): 167–196.

Molnár, P. K., C. M. Bitz, M. M. Holland, J. E. Kay, S. R. Penk, and S. C. Amstrup (2020) 'Fasting Season Length Sets Temporal Limits for Global Polar Bear Persistence', *Nature Climate Change* 10(8): 732–738.

Pasternak, S. (2017) *Grounded Authority: The Algonquins of Barriere Lake Against the State*. Minneapolis, MN: University of Minnesota Press.

Pew Charitable Trusts (2016) *The Integrated Arctic Corridors Framework: Planning for Responsible Shipping in Canada's Arctic Waters*. Available at: www.pewtrusts.org/~/media/Assets/2016/04/The-Integrated-Arctic-Corridors-Framework.pdf (accessed 21 March 2022).

Peyton, J. (2017) *Unbuilt Environments: Tracing Postwar Development in Northwest British Columbia*. Vancouver: University of British Columbia Press.

Pfeifer, P. (2018) 'From the Credibility Gap to Capacity Building: An Inuit Critique of Canadian Arctic Research', *Northern Public Affairs* 6(1): 29–34.

Povinelli, E. A. (2011) *Economies of Abandonment: Social Belonging and Endurance in Late Liberalism*. Durham, NC: Duke University Press.

Powell, R. C. (2017) *Studying Arctic Fields: Cultures, Practices, and Environmental Sciences*. Montreal: McGill-Queen's University Press.

Prestrud, P. and I. Stirling (1994) 'The International Polar Bear Agreement and the Current Status of Polar Bear Conservation', *Aquatic Mammals* 20(3): 113–124.

Qikiqtani Truth Commission (2010) *QTC Final Report: Achieving Saimaqatigiingniq*. Iqaluit, Nunavut: Inhabit Media.

Rogers, S. (2018) 'Nunavut Man Dies in Kivalliq Polar Bear Attack', *Nunatsiaq News*, 4 July. Available at: http://nunatsiaq.com/stories/article/65674nunavut_man_dies_in_polar_bear_attack/ (accessed 21 March 2022).

Sandlos, J. (2011) *Hunters at the Margin: Native People and Wildlife Conservation in the Northwest Territories*. Vancouver: University of British Columbia Press.

Schmidt, A. L. and D. A. Clark (2018) '"It's Just A Matter of Time": Lessons from Agency and Community Responses to Polar Bear-inflicted Human Injury', *Conservation and Society* 16(1): 64–75.

Schreiber, D. (2013) 'Immobilizing Polar Bears/Inuit: Productivity and Interspecies Wildlife Management in the Canadian Arctic', *Anthropologica* 55(1): 157–176.

Schweinsburg, R. E. (1981) *A Brief History of Polar Bear Management in the N.W.T.* NWT wildlife notes, no. 2. Yellowknife, Northwest Territories: Government of Northwest Territories, Department of Renewable Resources.

Sidjak, W. (1978) *The Arctic Pilot Project*. Chemical Engineering Conference. Calgary, Canada: Arctic Pilot Project.

Simpson, A. (2014) *Mohawk Interruptus: Political Life Across the Borders of Settler States*. Durham, NC: Duke University Press.

Slocum, R. (2004) 'Polar Bears and Energy-efficient Lightbulbs: Strategies to Bring Climate Change Home', *Environment and Planning D: Society and Space* 22(3): 413–438.

Smaill, B. (2015) 'Tasmanian Tigers and Polar Bears: The Documentary Moving Image and (Species) Loss', *NECSUS: European Journal of Media Studies* 4(1): 145–162.

Smith, J. R. (2021) '"Exceeding Beringia": Upending Universal Human Events and Wayward Transits in Arctic Spaces', *Environment and Planning D: Society and Space* 39(1): 158–175. https://doi.org/10.1177/0263775820950745.

Solski, J. J. (2021) 'The Genesis of Article 234 of the UNCLOS', *Ocean Development & International Law* 52(1): 1–19.

Stanley, A. (2019) 'Aligning Against Indigenous Jurisdiction: Worker Savings, Colonial Capital, and the Canada Infrastructure Bank', *Environment and Planning D: Society and Space* 37(6): 1138–1156.

Star-Phoenix (1977) 'Size of Tankers Poses Problem', *Star-Phoenix*, 17 August, 43.

Stuhl, A. (2016) *Unfreezing the Arctic: Science, Colonialism, and the Transformation of Inuit Lands*. Chicago, IL: University of Chicago Press.

Tuft, D. (2017) *The Arctic Melt*. New York: Assouline Books.

Utton, A. E. (1972) 'The Arctic Waters Pollution Prevention Act, and the Right of Self-Protection', *University of British Columbia Law Review* 7(2): 221–234.

Wadhams, P. (2014) 'The Polar regions', in P. S. Dasgupta, V. Ramanathan, and M. S. Sorondo, eds, *Sustainable Humanity, Sustainable Nature: Our Responsibility*. Vatican City: The Pontifical Academy of Sciences, 225–238.

Wadhams, P. (2017) *A Farewell to Ice: A Report from the Arctic*. Oxford: Oxford University Press.

Wakeham, P. (2014) 'At the Intersection of Apology and Sovereignty: The Arctic Exile Monument Project', *Cultural Critique* 87: 84–143.

Watt-Cloutier, S. (2015) *The Right to be Cold: One Woman's Story of Protecting her Culture, the Arctic and the Whole Planet*. Toronto: Penguin Canada.

Wilder, J. M., D. Vongraven, T. Atwood *et al.* (2017) 'Polar Bear Attacks on Humans: Implications of a Changing Climate', *Wildlife Society Bulletin* 41(3): 537–547.

WWF (2013) 'Hamlet of Arviat and WWF-Canada Celebrate Success of Human Polar Bear Conflict Reduction Project', World Wildlife Fund Canada, 31 January. Available at: https://Arcticwwf.org/newsroom/news/hamlet-of-arviat-and-wwf-canada-celebrate-success-of-human-polar-bear-conflict-reduction-project/ (accessed 21 March 2022).

York, J., M. Dowsley, A. Cornwell, M. Kuc, and M. Taylor (2016) 'Demographic and Traditional Knowledge Perspectives on the Current Status of Canadian Polar Bear Subpopulations', *Ecology and Evolution* 6(9): 2897–2924.

Zerehi, S. S. (2016) 'Nunavut's Polar Bear Problem Growing in Hudson Bay Communities', *CBC News*, 6 April. Available at: www.cbc.ca/news/canada/north/problem-polar-bears-hudson-bay-nunavut-1.3522375 (accessed 21 March 2022).

Part II

Working with ice

6

White spots on rivers of gold: imperial glaciers in Russian Central Asia

Christine Bichsel

Central Asia was an important study area for Soviet glaciology due to the high mountains and numerous glaciers in the region.[1] Scientific insights from Central Asian glaciers significantly shaped Soviet glaciology, alongside similar studies in the Altai, the Caucasus, the Arctic, and the Antarctic. But Central Asian glaciers became scientific objects even before the Soviet period (1917–91). Fundamental knowledge in Soviet glaciology was formed from Russian scientific research of glaciers in the nineteenth and early twentieth century in Central Asia (or 'Russian Turkestan', as the region was then called). Take, for example, the Abramov Glacier (Figure 6.1) in Kyrgyzstan: a key site of Soviet glaciology in Central Asia with an onsite research station hosting up to seventy scientists and support staff. The glaciological and meteorological data collected continuously from the Abramov Glacier since 1967 informs ongoing research on mountain glaciers and climate change today (see for example Denzinger *et al.*, 2021). Yet the very name of the Abramov Glacier points to a history of glaciology that predates the Soviet Union: the glacier was likely named after Alexander Konstantinovich Abramov (1836–86), a Russian major general and member of the Imperial Russian Geographical Society (Glazyrin *et al.*, 1993). Abramov played a key role in the military conquest of Central Asia, but was also actively involved in scientific exploration of the region.

Early Russian glaciology in Central Asia is thus integral to the foundations for Soviet and post-Soviet glaciology in the region. In this chapter, I examine the scientific biographies of glaciers to establish why and how these glaciers became objects of and for imperial Russian science. The investigation focuses on two glaciers: the Abramov Glacier and the Zeravshan Glacier. There is no shortage of glaciers to study in the region, as Central Asia has an estimated ice cover of 16,000 km^2 in the Tian Shan (Sorg *et al.*, 2012: 727) and 12,000 km^2 in the Pamir Mountains (Lambrecht *et al.*, 2014: 233). I selected the Abramov and Zeravshan glaciers because their scientific biographies, which

Figure 6.1 Overview map of glaciation in Central Asia. The sites of Zeravshan and Abramov glaciers are marked with red boxes.

date back to the nineteenth century, are particularly instructive for revealing the history and relevance of Russian glacier research. Scientific observation of both glaciers continued throughout the twentieth century and was relevant to the development of Soviet glaciology. Additionally, Russian scientific observations of these two glaciers are richly documented and thus accessible. Scholars published their insights in Russian and, occasionally, international outlets.[2] Lastly, the two glaciers' scientific biographies are closely connected in terms of intersecting geopolitical context, forms of knowledge, and leading protagonists.

The Abramov and Zeravshan glaciers are located in the Pamir Mountains of Central Asia. The Abramov Glacier is located in the Alai mountain range. It has a total area of 24 km^2 and ranges from 3,600 to nearly 5,000 metres altitude. The glacier feeds the river Kök Suu, eventually draining into Central Asia's biggest river, the Amu Darya. It was intensely researched after being selected as a representative glacier for scientific measurements during the UNESCO programme of the International Hydrological Decade (1965–74) (Grosval'd and Kotlyakov, 1969: 20). The second glacier is the Zeravshan Glacier, nested between the Zeravshan and Turkestan ranges, and among the larger glaciers in Central Asia. At an altitude between 2,800 and 5,200 metres, the glacier is about 26 km long and has a total area of 39 km^2. It consists of a main lobe with about twenty smaller confluences originating from the lateral mountain ridges. It feeds the third biggest river of Central

Asia, the River Zeravshan. Today, the glacier lies within the boundaries of Tajikistan. It was also the first glacier in Central Asia visited, in 1880, by an imperial Russian scientific expedition.

Scientific exploration of glaciers in Central Asia began during the nineteenth century in conjunction with the establishment of Russian colonial rule in the region. Russian conquest of Central Asia is usually connected with the fall of Tashkent in 1865, although it was a gradual process during the eighteenth and nineteenth centuries (Morisson, 2014). Military conquest and colonial administration went hand in hand with establishing scientific knowledge for political control. Military campaigns were almost always accompanied by a broad range of scholars including cartographers and surveyors, but also zoologists, botanists, geologists, and anthropologists (Middleton, 2019: 8). Moreover, most of these scholars were themselves officers of the General Staff of the Russian army (Schimmelpenninck van der Oye, 2019: 151). In turn, scientific research for ethnographic, geographic, and natural-scientific knowledge also served the purpose of 'reconnoitring', thereby laying the logistical and intellectual groundwork for later territorial expansion in and beyond Central Asia (Bailey, 2008: 14). The Imperial Russian Geographical Society played a key role in funding and organizing scientific expeditions to Central Asia. The famous expedition led by its member Petr Petrov Semenov to the Tian Shan Mountains in 1855 marked the beginning of a series of expeditions to explore and map Central Asia (Maslova, 1956, 1962, 1971). Early Russian research of the Abramov and Zeravshan glaciers was part of this exploratory framework.

Approaching the history of Russian glaciology in Central Asia from a political environmental history perspective (Kim and Pianciola, 2019), the term 'glaciology' addresses the scientific research of glaciers, even if Soviet glaciology as a scientific discipline only emerged during the 1960s. Although the scientific interest in ice is older – as many contributions to this edited collection demonstrate – glaciology as a formally acknowledged discipline is comparatively young and institutionalized only during the mid-twentieth century (Sörlin, 2015: 329). My political environmental history centres on earlier entanglements between politics and the production of scientific knowledge on glaciers. I seek to reveal the imperial political configuration, or 'imperial formation', as Ann Laura Stoler (2008) has it, within which the scientific biographies of the Abramov and the Zeravshan glaciers took form. As Stoler noted, 'Imperial formations are relations of force. They harbour political forms that endure beyond the formal exclusions legislating against equal opportunity, commensurate dignities and equal rights' (Stoler, 2008: 193). Stoler stresses the durability of these relations of force to shape objects, knowledge, but also sentiments beyond their formal existence. 'Imperial glaciers', therefore, refer to the Abramov and Zeravshan glaciers as scientific objects constituted by seemingly neutral and apolitical knowledge that nonetheless harbours relations of political violence, economic exploitation, and ethnic discrimination.

Central Asian glaciers as the 'white spots' of imperial Russian science highlight two different aspects that are germane. Literally speaking, 'white spots' refer to the perception and representation of glaciers as white and highly reflective surfaces. Glaciers, like many ice bodies, can disorientate the human observer and traveller. Metaphorically, however, they denote blind spots – areas in the field of vision that remain either invisible or lack sufficient information. And where dominant powers sent their mappers and surveyors to 'fill in the gaps' on prevailing imperial maps.

First and foremost, Central Asian glaciers became 'white spots' as they were seen through what Mary Louise Pratt (1992: 201) called 'imperial eyes': an imperial vision bestowing the scientific observer with the power to be the 'monarch of all I survey'. In well-known colonial fashion, imperial Russian science conceived of Central Asia as unknown and uncharted territory. Glaciers became 'white spots' alongside other geographical features in Central Asia. 'Filling the white spots' [Russ. *zakryt' belye pyatna*][3] with scientific knowledge was the foremost duty and, if achieved, granted the highest praise to Russian explorers of Central Asia. The figure of the heroic-explorer [Russ. *puteshestvennik*] was a consequence of this idea. Explorers 'discovered' features of the landscapes, with discovery [Russ. *otkrytie*] framed as a – mostly male – heroic deed, achieved by overcoming the hardship and danger of a scientific expedition.

Russian scientists laid claim to the intellectual property of glaciers and excluded Central Asians as bearers of legitimate knowledge. Insights from these expeditions were shared and discussed in scientific circles but also made available to the wider Russian public. Imperial Russian science of Central Asia was thus directed back at Russia. Filling the 'white spots' had the important function of producing 'self-knowledge' [Russ. *samopoznanie*] (Bailey, 2008: 1) about the lands and people in the empire. More specifically, glaciers became 'white spots' as science was enlisted in Russia's military political economy strategy of occupying Central Asia. After the defeat of Central Asia's main cities and lowlands, populations in the mountainous areas continued to elude or resist Russian military control. Early Russian glaciology took shape in the context of military campaigns to reconnoitre these little-known areas and subdue their populations.

Second, glaciers became 'white spots' through their cartographic representations resulting from military surveys. Russian military maps were utilitarian as they showed information of strategic and economic interest to Russia as an imperial power – settlements, routes of communication, and rivers. While irrigated agriculture ranked high on Russia's priority for Central Asia, glaciers were not yet understood to be relevant for this purpose. Seen from this perspective, glaciers represented obstacles rather than opportunities and did not justify a detailed survey in difficult terrain. Therefore, maps hardly indicated glaciers; they appeared on them literally as poorly defined 'white spots'.

Third, glaciers became 'white spots' through the scientific priorities of Russian colonial administration. The first Russian scientists to study Central Asian glaciers were trained as mining engineers and tasked to conduct geological surveys to assess the region's mineral resources for industrial development and colonial exploitation. Studying glaciers was neither part of their scientific background nor their research priority. All three points need to be borne in mind as the Zeravshan and Abramov glaciers emerged as scientific objects for imperial Russian science.

The Zeravshan: ice feeding the golden river

In the spring of 1870, General Alexander Konstantinovich Abramov led the military-scientific Iskanderkul expedition to explore the upper reaches of the Zeravshan River. Abramov had made his military career during the Russian conquest of Central Asia, participating in the siege of Samarkand and Tashkent (Ostrovskii, 2002: 39). Although Samarkand had fallen to the Russian army in 1868, the nearby mountains were still ruled by independent principalities with shifting allegiances (Yastrebova, 2003). While the lowlands were relatively well researched, Russian military and scientific expeditions had only just begun to explore the more remote mountains areas of Central Asia (Maslova, 1956: 4). Departing from Samarkand, Abramov's expedition was declared peaceful, but certainly had a military outlook. It consisted of two hundred Cossack soldiers, an infantry company, and mountain artillery. It also included a mining engineer, an 'orientalist', two naturalists, and three military topographers. Alongside its political and scientific aims, the expedition was motivated by 'rumours of natural resources' (Aminov, 1873: 50) about the presence of gold-bearing sands in the upper reaches of the Zeravshan River. The name 'Zeravshan' is translated from Persian and Tajik as 'gold-spreading' or 'golden'. Indeed, Alexander Leman had observed local gold prospecting in the upper reaches of the river during an earlier Russian expedition in 1841 (Solov'ev, 1936: 146). Mining engineers also expected to find stone coal deposits. During its search for mineral riches, the expedition reached the Zeravshan Glacier.

The expedition's mining engineer, Dimitrii Konstantinovich Myshenkov, was the first to describe the Zeravshan Glacier in Russian science. However, his report presented only a few observations of the glacier and was mainly concerned with geological mapping and barometric measurements. For the expedition, the glacier represented 'a natural barrier impeding further travel' at an altitude of 2,700 metres. Myshenkov referred to it tentatively as 'Zeravshanian' [Russ. *zeravshanskii*] after the river that, as he established, indeed originated 'from underneath' the glacier (Myshenkov 1871: 271). This name later became the full toponym of the glacier, today still called *Zeravshanskii lednik* in Russian. Myshenkov added a few reflections on Central Asian glaciers in his report. He stated that according to information by the locals, all glaciers in the area were advancing. He saw this as confirmed by his observation that the Zeravshan Glacier's end moraine was directly connected to its frontal lobe. Myshenkov struggled to reconcile his observations with the general opinion held by Russian scientists at that time, that precipitation in Central Asia was decreasing rather than increasing, and that the overall climate in the region was becoming drier. Previous expeditions had reported that levels of lakes had dropped, river flow had decreased, and rivers had become shorter. Unable to find a climatic explanation, Myshenkov (1871: 286–287) suggested a continued but unnoticed uplift of Central Asia as the cause for advancing glaciers.

Myshenkov's report piqued the curiosity of Ivan Vasil'evich Mushketov, a geographer and geologist of the Imperial Russian Geographical Society. In the summer of 1880, he led a geological expedition to the Zeravshan Glacier: the first official Russian scientific expedition to study a glacier in Central Asia. Trained at the Mining Institute in St Petersburg, Mushketov had been sent to Tashkent in 1874 to carry out a geological survey for stone coal deposits in Central Asia. It was believed that stone coal would enable the building of a planned railway between Russia and Tashkent and boost regional industrial development (Baskov, 1986). Affiliated with the General Staff of the Russian Army, Mushketov carried out several mineralogical and petrographic surveys during excursions to the Tian Shan and Pamir Mountains during the 1870s. This last expedition to the Zeravshan Glacier, in his words, was to fill some remaining gaps in his reflections on Central Asia. It was expected to provide evidence for the scientific hypothesis of Central Asia becoming drier. Myshenkov's report of an advancing Zeravshan Glacier contradicted Mushketov's and other scientists' observations of retreating glaciers in Central Asia's mountains, and the Zeravshan expedition was meant to review Myshenkov's claim by providing further glaciological data (Mushketov, 1881: 4 and Figure 6.2).

The political conditions for Mushketov's expedition were certainly more favourable than they had been ten years prior. With the fall of the Khanate of Kokand in 1875, Russia's control over Central Asia had expanded and consolidated: unlike Abramov's heavily armed expedition, Mushketov's was accompanied by only seventeen Cossack soldiers. It also included a mining

Figure 6.2 Native building and agriculture close to the Zeravshan Glacier. Original legend: 'Dwelling and field of the Matchintsy in the Matcha Valley'.

engineer, a topographer, an officer 'on special duty', two translators, and an unspecified number of native porters. As the expedition headed upstream along the Zeravshan River through the Matcha district, Mushketov devoted a part of his report to notes on local life. He perceived the inhabitants of Matcha [Russ. *Matchintsy*] as stubbornly clinging to primitive life rather than descending from the mountains to meet civilization in the lowlands of Central Asia. Noting the increasing use of rough stone for native building and work at higher altitudes (Figure 6.2), he wrote that 'The closer to the glacier, the clearer and clearer the peculiar features of this, albeit remote, stone age relative of the human manifest themselves, not being ashamed to live in sight of the glacier on almost bare rocks' (Mushketov, 1881: 12). By living in the high-altitude environment of stone and ice, according to Mushketov, natives remained caught in their pre-modern forms of life and failed evolution towards modern civilization. Mushketov's ambivalent feelings about ice and this high-altitude environment became even more pronounced when he compared the debris-covered lower part of the glacier and the absence of vegetation (Figure 6.3) to the Kyzyl-Kum desert, 'with the only difference that instead of sand dunes here were boulder cones' (Mushketov, 1881: 15 and Figure 6.3).

Mushketov's relations with native communities appear conflicted. All too aware of the limited local resources, he sheepishly admitted that the expedition heavily affected the valley's economy, and that feeding its forty horses probably

Figure 6.3 Mushketov's expedition approaches the Zeravshan Glacier. Original legend: 'Crevasses turning into ice shafts among the dense detritus cover of the lower Zeravshan Glacier'.

used up half a year's stock of fodder. However, he hastened to assert that in any case, the natives preferred the silver money that the expedition offered in exchange for the stock (Mushketov, 1881: 11). He marvelled at the natives, formidable walkers with suitable footwear for the mountainous terrain. 'On the very first day [on the glacier], we had to learn by experience that our European boots with heels and thick soles were unsuited, we exchanged them for the natives' leather stockings without heels' (Mushketov, 1881: 15). He found it all the more difficult to understand native porters' refusal to walk on the glacier – in his perspective an irrational stance. They had to be persuaded repeatedly to continue the strenuous glacier crossing from its frontal lobe to its upper part at the Matcha Pass over four days (Figure 6.3). The conflict escalated on the third day, when the expedition had to set up camp on bare ice, burning wooden sledges and instrument handles to keep warm. The porters walked away from the expedition in protest. Mushketov, worried by the expedition's dependence on the porters and yet certain that they would not be able to go very far, scornfully remarked on their return that 'these people are too cowardly, they believe too much

Figure 6.4 Mushketov's expedition on the Zeravshan Glacier close to the Matcha Pass. Original legend: 'Needle-shaped mountains in the firn area of Zeravshan Glacier'.

in the strength of the Russians to dare attacking us'. In Mushketov's representation, the porters not only lacked courage but also were simple-minded, 'completely reconciled after the handing out of *nasvai*' [Russ. for: snuff tobacco] (Mushketov, 1881: 21 and Figure 6.4).

The expedition mapped the Zeravshan Glacier, establishing its length at around thirty *verst* [thirty-two kilometres]. They also mapped its confluents from the adjacent ranges, at times retaining existing local names, but more often renaming them after expedition members. One of the confluents was named after Akhun, the head of the native porters. Two further confluents received the names of the Cossack soldiers Tolstov and Skatchkov who almost perished in a crevasse (Mushketov, 1881: 16). Mining engineer Dimitrii L'vovich Ivanov visually recorded the expedition in seventeen drawings (Figures 6.2, 6.3, and 6.4), and topographic measurements resulted in the first map of the Zeravshan Glacier. Mushketov (1881: 12) clearly rejected Myshenkov's claim of an advancing Zeravshan Glacier, calling the latter's observations a 'grave error'. He argued that all of the observed glaciers in the upper reaches of the Zeravshan River, including the main one, were retreating rather than advancing. Mushketov supported his argument with evidence of huge end moraines and information provided by local people. He stated that the Zeravshan Glacier had been by far longer – and thicker – in the 'recent past', possibly extending about fifty *verst* further down the Zeravshan Valley than it did in 1880 (Mushketov, 1881: 24). Mushketov did not, however, propose any explanation of what might have led to such a massive glacier surge in the past.

Mushketov's report provoked another member of the Imperial Russian Geographical Society, Alexander Ivanovich Voeikov, to write a critical intervention. Voeikov, a trained meteorologist, was later to become one of Russia's most influential climate scientists (Coen, 2011). Clearly frustrated with the quality of Russian science, he wrote that 'There would be far less erroneous theories and hypotheses in geology and physical geography, if [scholars] bothered to establish the precise facts beforehand, and did not content themselves with too little, and, what is more, inaccurate [knowledge]' (Voeikov, 1882: 4). He then outlined the state of glaciology in Switzerland, discussing possible explanations for observed phases of glacial retreat and advancement. He was critical not so much of the missing evidence for glacier change in Central Asia, but rather of Russian scientists jumping to erroneous conclusions: 'It is very well possible that some of the glaciers in Central Asia indeed advance, and this would by no means prove an uplift [of the mountain range] or be in contradiction with a decrease in precipitation or lake levels in neighbouring countries, because the advancing of glaciers could be the result of a prolonged period, possibly of 100 or more years in the past.' Acknowledging that long-term and in-depth research was far more difficult to carry out in remote Central Asian locations than in Switzerland, he proposed to elaborate a 'short and practical scheme' which would allow the non-specialist members of expeditions to the region to gather useful, rather than speculative, data on glaciers (Voeikov, 1882: 16–18).

The Abramov: a hidden passage over ice

If data on the Zeravshan Glacier was scarce, even less was known about the Abramov Glacier during this period. Siberian plant scientist Sergei Ivanovich Korzhinskii was probably the first to describe the glacier in his 'Essay on the vegetation of Turkestan' (Russ: *Ocherki rastitel'nosti Turkestana*) (1896). Korzhinskii encountered the glacier during his expedition to Central Asia in 1895. Supported by the Imperial Russian Academy of Sciences and the Russian Ministry of Agriculture and State Property, the expedition aimed to both look for plants for botanical systematization and gather insights on the region's potential for agricultural development. Korzhinskii's expedition

Figure 6.5 Abramov Glacier with mountain pass Bok-Bashy (saddle in the centre of the image) in the background. Photograph taken by a Swiss-Kyrgyz research expedition in the summer of 2018 for drilling ice cores.

arrived at the glacier while trying to establish the upper limit for plant growth in the Alai range. According to Korzhinskii's words, they were able to identify this limit solely at one location, on the slopes of the mountain massif Bok-Bashy, but under circumstances that 'could hardly be called normal. This massif is located at the upper reaches of the rivers Katta-Karamuk and Kok-Su; its broad flat summit is covered by a giant glacier called *Abramov Glacier*' (Korzhinskii, 1896: 69, emphasis in original). The expedition traversed the glacier with horses when crossing the Bok-Bashy Pass, the shortest connection from neighbouring Karategin into the Ferghana Valley and a dangerous passage used by locals only during a very short period in summer.

Korzhinskii was a plant scientist and not a glaciologist. For him, the Abramov Glacier was primarily of interest for its particular microclimatic conditions and their effect on the local flora. Establishing the upper limit for plant growth just below 4,000 metres on the Bok-Bashy massif, he acknowledged that this limit might lie considerably lower than elsewhere in the Alai range due to the 'huge accumulation of ice and snow'. His focus was thus less on the presence of ice but rather on the absence of plants. He described the glacier as a massive ice sheet, from which 'only here and there low, bare cliffs stood out, devoid of vegetation' (Korzhinskii, 1896: 70). Nevertheless, he mentioned the spectacular view of the glacier from the Bok-Bashy Pass: 'This glacier descends along wide valleys, marking its path with two tracks of piled-up stones. One of these glaciers looks like a broad river ... On its surface, scattered rocks and detritus are being carried downhill in rows. This glacier ends with piles of rock debris, and a small stream is flowing out from underneath them. This is the beginning of the river Kok-Su' (Korzhinskii, 1896: 70). It was along this river that Korzhinskii's expedition descended from the high-altitude mountain environment back to the more familiar lush green pastures of the Alai Valley.

Korzhinskii did not explain why he referred to this glacier as 'Abramov Glacier'. It is unlikely that he himself named it. Russian cartography of Central Asia had started with the establishment of the Turkestan Military Topography Directorate in Tashkent in 1869. Under its auspices, the Tashkent astronomical observatory began to operate in 1873. The main task of the military astronomers and geodesists working at the observatory was to establish a network of exact coordinates through fieldwork, thereby providing the basis for topographic research, semi-instrumental and instrumental surveys, and cartography (Maslova, 1971: 5). The heyday of Russian military cartography of Central Asia during the 1870s and 1880s (Middleton, 2019) resulted in the first detailed map of Turkestan at the scale of ten *verst* to the inch (1: 420,000). This map continued to serve as the basis for research expeditions well into the twentieth century (Finsterwalder, 1932: 3).

The first detailed Russian survey of the Pamir Mountains began with the military campaign led by general Mikhail Dimitrievich Skobelev in 1876 (Marshall, 2006: 133). The year before, Skobelev had led the Russian army to defeat the Khanate of Kokand, which was annexed to Russia (Middleton, 2019: 12). However, Kyrgyz groups living in the Alai Valley continued their raids across the high mountain passes of the Alai range into the Ferghana Valley – a practice that the Khan apparently had been unable to prevent. In 1876, Skobelev sent a military expedition to the Alai to subdue the Kyrgyz tribes and to demonstrate the superiority of the Russian army: 'Having occupied the former Khanate of Kokand, the Russians could only rely on the peace and order of [its] population if the Kara-Kyrgyz [here: Kyrgyz] would admit complete surrender and obedience, and for this it was necessary to demonstrate that the Russian army can at all times appear *amidst their glaciers*, in the Alai, where their livestock is concentrated – their key wealth and source of subsistence. This was the aim of the Alai expedition' (Kostenko, 1877: 357–358, emphasis added). Following its military aim, the expedition consisted of eight military companies, four hundred Cossack soldiers, cavalry, three mountain guns and a missile battery (Tolbukhov,

1916). Since the expedition also meant to conduct science, it included a geographer and statistician, a naturalist, a geodesist, and eight topographers.

The expedition mapped more than 4,000 km^2 at a scale of 1: 84,000 (two *verst* to the inch) in the Alai and northeastern Pamir Mountains (Middleton, 2019: 14). On their way back from the Alai Valley, the expedition crossed the Alai range via the nearby Kara-Kazyk pass, travelling along the river Kök Suu which originates from the Abramov Glacier. Captain Lev Feofilovich Kostenko provided a description of the pass in his inventory of 'routes of communication across the Alai range' (Kostenko, 1877: 366–367). While his account is detailed and reads like a traveller's guide, there is no mention of the sizeable Abramov Glacier. In fact, Kostenko does not mention glaciers at all, which he certainly encountered while crossing the Alai range. Richard Finsterwalder (1932: 3), who used the resulting Russian military maps fifty years later in the Alai and Pamir Mountains, remarked that 'the military and economic aims of the maps implied that they basically represented inhabited districts and areas which could be developed [German: *das nutzbare Gelände*] ... The mighty glaciers are hardly indicated. Contrarily, the roads and river courses on the fringes of the map, for example in the Alai Valley, turned out to be correct.' Glaciers, unlike settlements, roads and rivers, simply had no military or economic importance, and therefore were left out of Kostenko's account, even if mapped.

A further expedition visited the area of the Abramov Glacier in 1878, this time led by the above-mentioned Mushketov. Mushketov was tasked with the survey of the geological structure and mineral resources in the newly annexed Ferghana Valley and the Alai range. For permission and support to travel to this remote and still restless border zone he turned to another already familiar figure – Alexander Konstantinovich Abramov. In the meantime, Abramov had become the military governor of Ferghana province (Baskov, 1986). He still had a keen interest in science, and Mushketov (1912: 10) spoke highly of the military governor's kind and generous efforts to support the organization of his survey. Mushketov followed Kostenko's tracks in reverse by taking the same route across the Kara-Kazyk pass. His account provided details of the area close to the Abramov Glacier: 'The river Kok-su receives its name after the confluence of [the river] Kara-Kazyk and the river Bok-Bashy, upstream along which the path separates, [and] two paths branch out: the western one across the Godai-Zhol Pass, leading to Kara-Tegin, and the southern one across the Bok-Bashy Pass to the Alai and Kara-Muk. ... The muddy water of the river [Kok-su] is of bluish colour, after which the river received its name' (Mushketov, 1912: 30–31). Judging from his description, Mushketov must have stood in front of the Abramov Glacier – but makes no mention of it. It is difficult to imagine that the same scholar who three years later led the first Russian glaciological expedition in Central Asia actually missed a glacier of this size. As with Kostenko, his attention was focused on other phenomena, this time on rock formations and the presence of syenite, diabase, diorite, and limestone in the area.

There are only hints as to how the glacier that Korzhinskii visited received the name 'Abramov'. This area was certainly surveyed during the Skobelev expedition in 1876. However, it is possible that the glacier received its name sometime in the 1880s. Russian military maps of Central Asia were composites, including information provided by several surveys and older maps. This is illustrated by the map 'Upper reaches of the Amu Darya' at a scale of 1: 1,260,000 (30 *verst* to the inch) issued in 1886 by the Military Topography Directorate of the General Staff. Its main source of information was the topographic survey of the Pamir during the big expedition led by Dimitrii Vasil'evich Putyata in 1883. However, 'particularly on the fringes of the map, also other valuable geographical data, for which we are indebted to partly Russian, partly English explorers of previous years, found their place' (*Ot redaktsii*, 1886: 221). The survey of 1876 provided the basic geographical data for the area of the Abramov Glacier. As for the toponym itself, it is very likely that the glacier was named after Abramov during the time of his appointment as the military governor of the Ferghana province, perhaps in gratitude for his continued support to expeditions to the Alai and Pamir. These expeditions usually started their journey from the Ferghana Valley (Maslova, 1971). For example, Abramov also supported Mushketov's subsequent travels in 1878 (Mushketov, 1910: 379, note 1), and later the big expedition led by Putyata in 1883 (Gokov, 2005: 34). None of these expeditions returned to the area of the Abramov Glacier.

Imperial glaciology: armed conquest, military topography and mining engineering

The scientific biographies of the Abramov and Zeravshan glaciers show that early Russian glaciology in Central Asia was informed by military considerations and a product of warfare. Scientific research was mostly carried out during campaigns. This prevailing context defined the sites and brought with it research constrictions. 'To conduct research attached to a [military] detachment of several hundred men, from which we were not allowed to remove ourselves as far as desired, as required by the study, moreover in a country that is alien and almost completely unknown to us – these were circumstances not entirely conducive for successful work' (Myshenkov, 1871: 270). The very knowledge that these campaigns pursued responded to military needs, as Kostenko's account of the routes of communication suggests.

If not primarily military in nature, scientific expeditions still required permission and support from the military administration of Russian Turkestan and were accompanied by Cossack soldiers. Not only did the expeditions combine military and then scientific aims, the very figure of the explorer – termed by Marshall (2006: 51) the 'soldier-geographer' or 'soldier-ethnographer' – embodied both domains. Soldier-geographers were men with military career paths shaped by the Russian conquest of Central Asia, who also had scientific interests and aspirations in exploring its geography and geology. Russian officers were charged with producing geographical and statistical studies, and authors such as Kostenko (1880) published wide-ranging surveys, combining descriptions of the geography of Russian Turkestan with an analysis of Central Asians' warfare strategies and tactics.

The geographical location of the glaciers had military significance for the Russian army. Both the Abramov and Zeravshan glaciers were located in high-mountain areas inhabited by populations eluding or resisting Russian military control after the conquest of the main cities – Samarkand and Kokand – and their surrounding lowlands. Moreover, these areas were located at the interstices of the former khanates, often not having been entirely under control by the Khans' military troops. They presented both an obstacle and annoyance to the Russian army. In Kostenko's words, it was necessary to 'tame' the Kyrgyz:

> who, like restless ants, came and went down the valley of the Alai and over the passes to their unknown pasturing grounds beyond, paying no tribute to the fact that the new authorities looked to them for humble submission and for systematic tribute. Their retiring habits, their sulky disposition, their evasiveness, their rude and extravagant notions of inoffensive independence, were contrary to all the well-established maxims of military rule and administration (Kostenko in Mitchell 1877: 18).

The military campaigns aimed to demonstrate to the Kyrgyz, but also to the former rulers of Central Asia, Russian superiority in warfare, as well as physical mastery of the 'white spots'.

Scientific knowledge of glaciers was a product of Russian cartographic practices in Central Asia. Maps of Russian Turkestan were issued by the General Staff of the Russian army and topographers, geodesists, and astronomers had a military background. Russian military maps provided the first cartographic representations of glaciers in Central Asia. They recorded existing or new toponyms for glaciers, some of them in use until today. In Central Asian traditions, toponyms were physical and geographical, and to a lesser extent, mythical and legendary (Horseman, 2006: 283). During imperial Russian rule personal names were introduced. Glaciers were named after Russians participating in the scientific expeditions. The Abramov Glacier is a key example here, named after a 'soldier-ethnographer', a general and military governor of the Russian army, and a member of the Russian Geographical Society since 1870. By overwriting local forms of toponyms with personal names, Russian cartography created mnemonic records of Russian science in Central Asia. Thereby, Russia symbolically occupied Central Asian space by fixing references of its military conquest to its cognitive and cartographic landscape.

The first Russian scientists to study Central Asian glaciers were trained as geologists. Myshenkov, Ivanov, and Mushketov all had backgrounds in mining engineering. The context of their research on glaciers was Russian colonial prospection and extraction of mineral resources in Central Asia – a history that has still received little attention. Researching the geology of Central Asia was a priority of the military government. The Governor-General of Turkestan, Konstantin Petrovich von Kaufmann, commissioned and funded a large survey of the geological structures and mineral resources in Central Asia in 1874. This survey, carried out by Mushketov and mining engineer Gennadii Danilevich Romanovskii, led to the first geological map of Turkestan (Mushketov, 1897: 10–11). Mineral resources were required for projects such as the building of railways, thereby strengthening Russian military control, facilitating the transport of raw cotton from and Russian settlers to Central Asia, and fostering industrial modernity in the region. In this context, the presence of gold and stone coal as mineral resources in the Zeravshan Valley may well be the main reason why Russian glaciological data from Central Asia is available for the Zeravshan Glacier. In this sense, the glacier indeed represented a white spot on the river of gold. However, the context of geological surveys also led to particular blind spots: scientists were more interested in stone, strata, and structure than in ice.

Early Russian glaciology in Central Asia had a weak connection to emerging Russian climate sciences. Scholars debated whether reports of advancing or retreating glaciers could be explained with an increase or decrease in precipitation in the region. Interestingly, they did not yet link these presumed changes in precipitation patterns to the possibility of a warming or cooling Central Asian climate. In these debates, glaciers again represented a kind of 'white spot'. First, existing data was too scarce to draw conclusions on possible correlations between glacier and precipitation changes. Russian scientific expeditions lacked trained specialists and resources to conduct in-depth onsite research on Central Asian glaciers. Moreover, existing knowledge in glaciology derived its insights from glacier changes in Switzerland with different climatic conditions and was therefore not directly applicable to Central Asia. Voeikov (1882) assumed that Central Asia was not covered by a coherent ice shield during the glacial maxima in contradistinction to European countries. Second, as the connections between glaciers and surface waters were not yet well understood, receding rivers and lowering lake levels presented a much bigger concern to Russian plans in Central Asia than did glacier changes. The rivers in Central Asia were supposed to supply water to produce 'white gold', the large-scale industrial production of cotton through irrigated agriculture in the region. In this sense, the term 'rivers of gold' acquires the meaning of a key resource for industrial agriculture – a meaning that increasingly gained in importance but was connected to Central Asian glaciers only later during the twentieth century.

Conclusion

Central Asian glaciers were 'white spots' in imperial Russian science of the late nineteenth century. I do not use 'white spots' merely as a pun by referring to the perception and representation of glaciers as white. Rather, I am interested in the term's polysemy which includes the meaning of blind spots: areas in the field of vision that remain invisible or that lack sufficient information. I argue that thinking of Central Asian glaciers as 'white spots' has analytical value for understanding imperial Russian science. This value lies in the potential to not only shed light on Russian scientists' vision of glaciers but also to reveal how and why these glaciers remained invisible to imperial science despite their prominence in the mountain landscapes of Central Asia. Analysing glaciers as 'white spots' in imperial science invites us to ask how such invisibility was constructed – epistemologically, but also politically and economically. After all, the practice of 'unseeing', as China Miéville (2009) suggests, hints at a political configuration shaped by stark power differentials, regimes of prohibition, and pervasive violence. 'White spots', in other words, tell us about the condition of symbolic and material imperiality in Russian Central Asia. Thereby, insights on imperial glaciers in Central Asia contribute to understanding the decentred and multisited forms of imperial science beyond the often-held assumptions of Euro-American imperialism.

Imperial Russian glacier narratives bear some similarities to other scientific and popular narratives. For example, Mushketov's idea that as one travelled higher the landscapes and settlements became more backward and were home to 'wild' people was common to many Europeans explorers and adventurers (Carey 2007: 501–502). Moreover, Russian narratives of manly heroism, risk, and physical discomfort resonate British ideas of glaciology as a male field science (Hevly, 1996). However, these themes of heroic experience and conquest were even more pronounced in Central Asia, reinforced by expressions of Russian military masculinity. Similar to other contexts, Central Asian glaciers became the sites of colonial encounters and the negotiation of social, racial, gender, and ethnic identities (see also Carey, 2007: 506). By naming and charting Central Asian glaciers, Russian scientists laid claims to these 'white spots'. 'White' acquires a racial meaning here, implying Russian self-identification as Europeans as opposed to the natives. Yet glaciers turned out to be slippery ground for establishing Russian supremacy. The Russians' confidence in their advanced modern technology was unsettled as their equipment proved to be unsuited for traversing glaciers. Despite clearly established ethnic hierarchies and unequal terms, on the glaciers Russian scientists found themselves in precarious situations of mutual dependence with native communities. These scientists' self-proclaimed rationality and moral superiority became little more than a brittle facade, barely maintained by relegating the sensation of fear and the condition of irrationality to the natives in the face of real danger from crevasses and cold.

Russian scientific narratives lack descriptions of the very material of glaciers – the ice. This may be related to the ice's lack of visibility, as Russian reports state that Central Asian glaciers were often covered by rock debris. Clearly, Russian scientists were much more at home when describing the stratification of minerals in a rock face, or the colour and flow of water for a river. They devoted much attention to the shape and location of moraines, rather than to the ice itself. Of course, debris and moraines were indicators of glacier mechanics and past glaciations – contemporary concerns of nascent glaciology in Europe and North America during the nineteenth century (Carey, 2007: 506). Ice appears to have been an unfamiliar material with strange qualities, moved by unpredictable and uncontrollable forces, and related in yet unknown ways to the lapse of time and changes of the climate. The alienating effect of glacial materiality was reinforced by the imperial framing of Central Asia as a foreign and hostile terrain, and the political as well as epistemic impossibility that native narratives of glaciers could have actual scientific value. Imperial glaciers thus remained alien objects despite their territorial, cartographic, and cognitive conquest.

Notes

1 I would like to thank the editors of this volume, as well as Katja Doose, Horst Machguth, and Tomas Saks for their thoughtful and constructive feedback on the first draft of this chapter. I also thank Stanislas Kutuzov for valuable help in accessing selected Russian literature, and Florian Denzinger for the map. Lastly, I am most grateful to the anonymous library workers who must have spent hundreds of hours scanning the periodicals of the Russian Geographical Society to render them accessible online and make this research possible during the

COVID-19 pandemic. The research behind this chapter was funded by a Swiss National Science Foundation grant, 'Timescapes of ice: Soviet glacier science in Central Asia, 1950s–1980s'.

2 I mainly draw on scientific publications in the periodicals *Bulletin of the Russian Geographical Society (Izvestiya Russiskogo Geograficheskogo Obshestva)*, *Transactions of the Imperial Russian Geographical Society in general geography (Zapiski Imperatorskogo Russkogo Geograficheskogo Obshestva po obshei geografii)*, the *Transactions of the Military Topography Directorate of the General Staff (Zapiski voenno-topograficheskogo otdela Glavnogo shtaba)*, and the *Military Digest (Voenniy Sbornik)*, but also expedition reports and monographs available in electronic form on the Russian Geographical Society's website (www.rgo.ru). I also draw on the *Transactions of the Imperial Academy of Sciences (Zapiski Imperatorskoi Akademii Nauk)*, available from the website of the Russian Academy of Sciences (http://arran.ru). The digital library *Scientific Heritage of Russia (Nauchnoe Nasledie Rossii)* has proven very helpful for finding rare monographs (http://e-heritage.ru). Additionally, I draw on historical texts made available on the *Eastern Literature (Vostochnaya Literatura)* website (www.vostlit.info). Lastly, O. V. Maslova's *Survey of Russian Travellers and Expeditions in Central Asia (Obzor Russkikh puteshestvii i ekspeditsii v Srednyuyu Aziyu)*, available from the Faculty Digital Archive of the New York University, has been of invaluable help (www.archive.nyu.edu).

3 The transliteration of Russian words follows the Library of Congress system. Soft signs and hard signs that are particular to Russian are rendered with one or two primes, respectively. If personal, organisational, or geographical names already have a fixed transliteration in English, I use these rather than altering them. An exception is the name of the river which emerges from the Abramov Glacier: I keep the Russian transliteration 'Kok-su' in original quotations, but use the correct Kyrgyz term 'Kök Suu' in my own text. All translations are my own unless otherwise indicated.

References

Aminov, P. A. (1873) 'Iskander-Kul'skaya ekspeditsia 1870 goda', *Turkestanskie vedomosti* 13 (27 March), 49–50.

Bailey, S. C. M. (2008) 'Travel, Science, and Empire: The Russian Geographical Society's Expeditions to Central Eurasia, 1845–1905'. PhD dissertation, University of Hawai'i.

Baskov, E. A. (1986) *Ivan Vasil'evich Mushketov, 1850–1902*. Leningrad: Nauka.

Carey, M. (2007) 'The History of Ice: How Glaciers Became an Endangered Species', *Environmental History* 12(3): 497–527.

Coen, D. R (2011) 'Imperial Climatographies from Tyrol to Turkestan', *Osiris* 26(1): 45–65.

Denzinger, F., H. Machguth, M. Barandun *et al.* (2021) 'Geodetic Mass Balance of Abramov Glacier from 1975 to 2015', *Journal of Glaciology* 67(262): 331–342.

Finsterwalder, R. (1932) *Wissenschaftliche Ergebnisse der Alai-Pamir Expedition Teil 1. Geodätische, topographische und glaziologische Ergebnisse*. Berlin: Dietrich Reimers/Ernst Vohsen.

Glazyrin, G. E., G. M. Kamnyanskii, and F. I. Pertsinger (1993) *Rezhim lednika Abramova*. St Petersburg: Gidrometeoizdat.

Gokov, O. A. (2005) 'Ofitsery General'nogo Shtaba vo vneshnei politike Rossii v otnoshenii Afganistana i Indii (70-e gg. XIX v.–1914 g.)', *Russian History* 32(1): 23–46.

Grosval'd, M. G. and V. M. Kotlyakov (1969) 'Present-day Glaciers of the USSR and Some Data on their Mass Balance', *Journal of Glaciology* 8(52): 9–22.

Hevly, B. (1996) 'The Heroic Science of Glacier Motion', *Osiris* 11(1): 66–68.

Horseman, S. (2006) 'The Politics of Toponyms in the Pamir Mountains', *Area* 38(3): 279–291.

Kim, L. and N. Pianciola (2019) 'Introduction: Watering the Land-based Empires', *Journal of the Economic and Social History of the Orient* 62: 525–559.

Korzhinskii, C. I. (1896) 'Ocherki rastitel'nosti Turkestana. I–III. Zakaspiskaya oblast, Fergana i Alai', *Zapiski Imperatorskoi Akademii Nauk po fisiko-matematicheskomu otdeleniyu* 4(4): 1–112.

Kostenko, L. (1877) 'Voenno-nauchnaya ekspeditsiya na Alai i Pamir', *Voennyi Sbornik* 4(April): 357–382.

Kostenko, L. F. (1880) *Turkestanskii krai: Opyt voenno-statisticheskogo obozreniya Turkestanskogo voennogo okruga: Materialy dlya geografii i statistiki Rossii*. St Petersburg: A. Traishel.

Lambrecht, A., C. Mayer, V. Aizen, D. Floricioiu, and A. Surazakov (2014) 'The Evolution of Fedchenko Glacier in the Pamir, Tajikistan, During the Past Eight Decades', *Journal of Glaciology* 60(220): 233–244.

Marshall, A. (2006) *The Russian General Staff and Asia, 1800–1917*. London: Routledge.

Maslova, O. B. (1956) *Obzor russkikh puteshestvii i ekspeditsii v Srednyuyu Aziyu. Chast' II. 1856–186*. Tashkent: Sredneaziyatskaya gosudarstvennaya universitet im. V. I. Lenina.

Maslova, O. B. (1962) *Obzor russkikh puteshestvii i ekspeditsii v Srednyuyu Aziyu. Chast' III. 1869–1880*. Tashkent: Tashkentskii gosudarstvennyi universitet im. V. I. Lenina.

Maslova, O. B. (1971) *Obzor russkikh puteshestvii i ekspeditsii v Srednyuyu Aziyu. Chast' IV. 1881–1886*. Tashkent: Fan.

Middleton, R. (2019) *The Russians in the Great Game*. Cultural Heritage and Humanities Unit's Research Paper 2. Bishkek: University of Central Asia.

Miéville, C. (2009) *The City & The City*. London: Macmillan.

Mitchell, R. (1877) 'The Russian Expedition to the Alai and Pamir', *Journal of the Royal Geographical Society of London* 47: 17–47.

Morrison, A. (2014) 'Introduction: Killing the Cotton Canard and Getting Rid of the Great Game: Rewriting the Russian Conquest of Central Asia, 1814–1895', *Central Asian Survey* 33(2): 131–142.

Mushketov, I. V. (1881) *Geologicheskaya ekspeditsiya na Zeravshanskii lednik v 1880 godu*. St Petersburg: V. Bezobrasova i K.

Mushketov, I. V. (1897) 'Uchastie gornykh inzhenerov v poluvekovoi deyatel'nosti Imperatorskogo Russkogo geograficheskogo obshestva, 1845–1895', *Gornyi zhurnal* 73: 1–17.

Mushketov, I. V. (1906) *Turkestan: Geologicheskoe i orograficheskoe opisanie po dannym sobrannym vo vremya puteshestvii s 1874 g. po 1880 g. Tom II*. St Petersburg: M. M. Stasyulevich.

Mushketov, I. V. (1910) 'Kratkii ocherk geologicheskogo puteshestviya na Alai i ozera Chatyr-kul letom 1878 g.', in *Sobranie sochinenii Ivana Vasil'evicha Mushketova. Vypusk 1. 1872–1992 gg, Zapiski Imperatorskogo Russkogo Geograficheskogo Obshestva po obshei geographii* 39: 386–97.

Mushketov, I. V. (1912) 'Puteshestvie na Alai i Pamir v 1877 godu', in *Sobranie sochinenii Ivana Vasil'evicha Mushketova. Vypusk 2. 1877 g., Zapiski Imperatorskogo Russkogo Obshestva po obshei geografii* 29(2): 1–99.

Myshenkov, D. K. (1871) 'Geologicheskiya nablyudeniya vo vremya Zarafshanskoi ekspeditsii', *Zapiski Imperatorskogo Russkogo Geograficheskogo Obshestva po obshei geographii* 4: 269–287.

Ostrovskii, A. B. (2002) 'Boevoi general – nachal'nik Ferganskoi oblasti', *Voenno-istoricheskii zhurnal* 9(509): 1–80.

Ot redaktsii (1886) 'Po povodu karty verkhov'ev Amu Dar'i', *Izvestiya Imperatorskogo Russkogo Geograficheskogo Obshestva* 22: 221–223.

Pratt, M. L. (1992) *Imperial Eyes: Travel Writing and Transculturation*. London: Routledge.

Schimmelpenninck van der Oye, D. (2019) 'Handmaidens of Continental Colonialism? The Ambiguities of the Imperial Russian Geographical Society's Central Asian Expeditions', in D. Schorkowitz, J. R. Chávez, and I. W. Schröder, eds, *Shifting Forms of Continental Colonialism: Unfinished Struggles and Tensions*. Singapore: Palgrave Macmillan, 151–173.

Solov'ev, M. M. (1936) *Uchenaya ekspeditsiya v Bukharu v 1841–1842 gg. pri uchastii naturalista Alexandra Lemana*. Moscow: Akademiya Nauk SSSR.

Sorg, A., T. Bolch, M. Stoffel, O. Salomina, and M. Beniston (2012) 'Climate Change Impacts on Glaciers and Runoff in Tien Shan (Central Asia)', *Nature Climate Change* 2: 725–731.

Sörlin, S. (2015) 'Cryo-history: Narratives of Ice and the Emerging Arctic Humanities', in B. Evengård, J. N. Larsen, and Ø. Paasche, eds, *The New Arctic*. Cham, Switzerland: Springer, 327–339.

Stoler, A. L. (2008) 'Imperial Debris: Reflections on Ruin and Ruination', *Cultural Anthropology* 23(2): 191–219.

Tolbukhov, E. (1916) 'Skobelev v Turkestane (1969–1877)', *Istoricheskii vestnik* 12: 639–667.

Voeikov, A. I. (1882) 'Noveishiya izsledovaniya lednikov i prichin ikh izmenenii', *Zapiski Imperatorskogo Russkogo Geograficheskogo Obshestva po obshei geografii* 8(2): 1–19.

Yastrebova, O. M. (2003) 'Dokumenty 19 veka iz verkhov'ekh reki Zeravshan v Otdele rukopisei Rossiskoi natsional'noi biblioteki', *Vostochnoi sbornik* 6: 87–117.

7

The many ways that water froze: a taxonomy of ice in nineteenth- and early twentieth-century America

Jonathan Rees

> The parched inhabitants of Madras Bombay – Calcutta – Havana – Charleston & New Orleans drink at my well – While I incredulous read the vast cosmogonal philosophy of Ancient India – in modern New England … And Concord fixed air is carried in that ice to mingle with the sultry zephyrs of the Indus & the Ganges'. Henry David Thoreau, *Journals* Volume 2 (1842–48), page 371.

> 'How much ice have we got left?' George Orwell, *Burmese Days* (1931)

The first ice industry in world history began in 1806, when the American merchant Frederic Tudor sent a shipment cut from Fresh Pond, near Cambridge, Massachusetts, to the island of Martinique in the Caribbean. Ice had been harvested and stored in both ancient Rome and ancient China. However, this was the first time that ice became an elemental and dynamic object of international commerce. Just two decades later packaged ice was making its way from Boston Harbor to India. For every 180 tons of ice packed and shipped in the United States, around 100 tons would arrive intact in Calcutta (cited in Sudan, 2016: 94). While this practice would spread across the developed world, it was never quite as popular elsewhere as it was in the United States, where the ice industry lasted well into the twentieth century (for a longer survey of ice and its relationship to the United States' industrial ice and refrigeration history, see Rees, 2013, 2018).

To discuss the nature of the ice that this industry sold requires a basic understanding of how the firms that sold ice acquired their product. The first ice sold in the United States pre-dated mechanical refrigeration. To get it, ice companies harvested ice from lakes and streams like Fresh Pond, stored it for months, then packed it onto ships and sold it in much warmer climes around the world, mostly so that consumers, including those colonial administrators in India as George Orwell so memorably remarked upon, could drop the product into drinks. This period in American ice history lasted until about 1860. At that point, there was substantial interest in ice throughout the United States, even in winter, so taking the trouble to pack it into ships and dispense it around the world ceased to be a money-making proposition for ice pioneers such as the Tudor Ice Company.

From approximately 1860 through 1910, the American natural ice industry turned inwards, selling most of its product to fellow Americans who had developed a taste for ice, starting in the warm climes of the American South and gradually moving northwards. That desire for ice was served both by harvested ice and the product of the first artificial ice machines, which could produce the product no matter what the weather. While the first artificial ice machines produced flimsy ice that melted quickly, mechanical refrigeration steadily improved throughout this period. As refrigeration technology improved, the size and cost of ice machines shrank, making this ice increasingly accessible to the kinds of businesses that once had to buy their ice from ice companies rather than produce it themselves.

The development of the home refrigerator is what transformed the American ice industry. The first home refrigerators, developed between 1915 and 1925, used ice as a selling point to get wealthy consumers to try this comparatively expensive appliance. When the first, low cost, reliable, modern refrigerator appeared in 1927, the old business model of muscled delivery men delivering ice straight to your kitchen quickly became obsolete. The American ice industry of today consists of machines that manufacture and bag ice for use by consumers whenever their home freezers cannot produce it fast enough for their needs.

However, it is worth noting that the ice industry's ice served many other purposes besides keeping drinks cold. Throughout the life of this industry, ice companies harvested and produced different kinds of ice using different technologies, depending on who was using that ice and why. Natural ice, for instance, varied considerably even when cut from the same body of water. Much of it proved completely unsuitable for consumer uses.

There are certain qualities of ice that transcend its mode of manufacture or its particular use. Ice is always cold. Ice is always slippery. Under the right circumstances, ice always melt. However, the purpose of this chapter is to briefly examine the history of ice harvesting and production in the United States, then review the surprising number of variables that determined the value of that product. Clear or opaque, in blocks or in cubes, natural or artificial, dirty or clean – Americans both harvested and manufactured different kinds of ice for different markets. Some of this variety was the result of nature. Some of this variety was the result of deliberate decisions by the American ice industry.

To understand that all ice is not the same implies that it has many uses. In the opening section of this chapter, we consider how natural and artificial ice were harvested, differentiated, and used. To consumers interested in its taste (or lack thereof), clear ice was an ideal. Food distributors interested only in its coldness didn't care whether their ice was clean. To ice harvesters, the slipperiness of ice was an obstacle to do their jobs. Everyone interested in commercial ice depended on nature for its creation, but they could manipulate nature in clever ways to create a surprising range of ice shapes, sizes, and internal appearances. With the advent of artificial refrigeration during the early twentieth century, manipulating the cold to make different types of ice became even easier. Thereafter, we consider the role of artificial ice and its relationship to the development of American urban life. The invention of refrigeration transformed domestic everyday life and industrial food practices respectively. Ice was at one time a luxury but by the twentieth century a necessity for the artificial chilling and preservation of food and beverages. By the 1930s and 1940s, the ice cube tray enabled a miniaturization, accessibility, and portability of ice that previous generations would have looked upon with envy. The new 'ice age' underpinned, as this chapter concludes, the emergence of modern America.

Varieties of natural ice

The early American ice industry was built on the model of sending a cold product to hot climates, like New Orleans or any number of Caribbean islands. Perhaps the most notorious ice trade route was Frederic Tudor's successful attempt to land New England ice in India. When it first landed in Calcutta in 1833, a local newspaper, the *Calcutta Courier*, reported that 'These blocks are beautifully transparent and of great size, some of them nearly a foot thick (quoted in Weightman, 2003: 140). What this fails to recognize is that large quantities of ice melted on its way across the ocean, sometimes as much as 75 per cent of the cargo if the weather on the journey proved particularly warm. The outside ice protected the inside ice. Much of the surviving ice would be covered in sawdust, used to protect the structure of the blocks so that they could be removed more easily on reaching their destination.

Later in the nineteenth century, when the supply chain shortened and consumers were less shocked that ice could be an article of transoceanic commerce, the taste and aesthetic qualities of ice became an important issue. The middlemen who bought ice from harvesters preferred ice that was both cold and uniform because their customers preferred it. Uniform ice was also far easier to store in icehouses until summer came. As a result, harvesters didn't just cut natural ice – they marked it off and cut it multiple times in order to make sure that the product maintained its structure. '[T]he width of our plows are graded so that the deeper ones may follow the shallow ones easily', explained one manufacturer of ice equipment in their 1878 catalogue. '[I]t is very important that they should be used in their regular order of depths' or else sections of the blocks would break off into irregular shapes (Wood, 1876). Irregular shapes would limit the efficiency of the industry's storage methods and the insulating power of outside ice when it was transported.

While the commercial imperative of structural uniformity was objective, there were commercial imperatives that were entirely subjective. To customers, clarity determined quality. 'There is no doubt that the public prefer clear ice to semi-opaque ice', one expert wrote in 1903, 'and that they will be willing to pay a higher price for it' (Williams, 1903: 204). With respect to natural ice, running water or ice removed from the tops of still bodies of water tended to be clearer and therefore fetched the highest price. Yet any water in streams and lakes has other stuff in it. Leaves and other detritus on the top; sediment from whatever occupies that water on the bottom. When ice was a relatively new luxury, people willingly accepted the fact that there might be dirt when they reached the bottom of their drink, as long as getting the rare specimen of clear natural ice remained very rare.

When natural ice began to compete against the artificial product, its opacity and its dirtiness became natural ice's main disadvantage. '[T]he harvesting of dirty, snowy, broken or shattered ice to New York City to be there sold to wagon dealers, and forced upon unwilling customers … does an incalculable injury to the natural ice trade', complained one large harvester in 1905 (*Ice and Refrigeration*, 1905). Unfortunately for many dealers, they could not control nature and were often left with no other options. After all, dust or sticks or leaves were just as natural as the ice they harvested.

A different result of the natural freezing process was pockets of air inside the ice blocks harvesters cut. '[N]atural ice often presents layers or streaks of bubbles scattered through the block from top to bottom', explained T. Mitchell Prudden, a pathologist writing for *Harper's* in 1892. 'These bubbles are probably in part air which has risen from the bottom of the lake or pond or stream on which the ice was forming' (Prudden, 1892: 374). Even if there weren't pockets of air inside the blocks, ice almost never froze evenly. As a result, some of it would be thicker and harder than other parts of the ice. The more air inside, the faster the ice would melt. That meant less ice to deliver to customers and less ice in the hands of customers.

The easiest way to avoid both sediment and air pockets was to draw ice from a deep, running river. All these adjectives describe the Kennebec River in Maine, which became a hotspot for ice production around 1880 when New York's Hudson River crop failed due to a warm winter. It continued to produce ice for the next twenty years because consumers in the northeast United States developed a taste for its high-quality product and were willing to pay more for it. In 1892, the *Portland Daily Press* reported that the ice fields on the river were 'clear and smooth as a mirror', which was the kind of result which gave Kennebec ice its stellar reputation (*Portland Daily Press*, 1892). Each year, the ice was packed into icehouses, and later – when the summer came – barges, then sent as far south as Washington, DC where it competed favourably with ice cut from local, often-polluted water sources all along the coast.

As industrialization in America progressed, pollution became a growing problem with every city's ice. One ice machine manufacturing company made this explicit pitch in one of their catalogues: 'The constantly increasing contamination of water sources in the neighbourhood of large cities and health-resorts have of late given the question of ice consumption considerable attention' (De La Vergne Refrigerating Company of New York, 1890). Philadelphia's Schuylkill River, for example, became infamous for the chemicals dumped into it by industrial facilities of all kinds. Ice consumers there began to reject local ice as early as the 1880s because they had terrible health-related experiences with the water that derived from it.

Bacteria, particularly bacteria from human waste, was a similar problem. 'Difficult though it is to bring a direct charge of typhoid infection against these sources', explained a 'muckraking' reporter for the *Atlantic Monthly* in 1909, referring to the lakes and streams that supplied most of New York City's ice at that time, 'there is a perfectly reasonable probability that many cases of intestinal diseases have originated in such dirty masses' (Godfrey, 1909). While it is impossible to know exactly how many people succumbed to typhoid by consuming contaminated frozen water, there were periodic panics over natural ice across the country. After researchers demonstrated that typhoid bacteria could survive many months frozen in ice, these fears became even more commonplace and the market for natural ice suffered as a result.

Yet even the filthiest ice could still find a market since it didn't have to touch perishable food directly to keep that food cold. Ice stuffed outside of beer barrels, for instance, never touched the beer so it could still be used for keeping that product cold even if the water it was made from was unclean for any reason. Once artificial ice became a strong competitor for the natural product, ice harvesters had to do everything possible to provide consumers with a uniform product, even though the natural ice they cut was seldom uniform. Sometimes this required piling up layers of thin ice to make it look like the artificial blocks or planing layers of sediment off the irregular-shaped blocks they harvested to make it look more attractive for wholesalers.

Besides beer, another industry that separated ice from its product was meat packing. In order to concentrate the meatpacking industry on the

South Side of Chicago, meat packers had to be able to ship the remains of the cattle they killed in the slaughterhouses there in rail cars refrigerated by ice. This required the separation between the ice and the meat on those railway cars because if the ice ever touched that meat, it would turn black. Both the packers and the brewers required not so much ice itself as the cold coming off it, which, of course, was not a 'real' thing. Nevertheless, the demand for cold during the late nineteenth century proved so strong, many entrepreneurs experimented with producing enough cold to improve on the ice that nature made.

Early artificial ice and its role in shaping everyday life in American cities

In 1883, the author Mark Twain published an account of his visit to a New Orleans ice factory. While the place itself was unremarkable and Twain admitted that he could not understand the process by which ice was made, he noted that the product was 'hard, solid and crystal-clear'. In other words, it was everything that most natural ice was not. That clarity came from the common process of stirring the water while it froze. Ice makers adopted this practice for entirely aesthetic reason since consumers believed that clear ice was purer, even though ice made from impure water could still be clear as long it had no visible impurities. That's why, as Twain explained, the ice makers froze 'flowers and things' in their product, to show off its resemblance to 'plate glass' (Twain, 1883: 410).

Manufacturing ice required mechanical refrigeration, which was difficult to create because the technology to do so took over a century to develop. There were three different kinds of refrigerating machine, each using a different refrigerant, during the nineteenth century. All of them could chill water or keep perishable food fresh, but different entrepreneurs in different countries tended to favour models for very distinct reasons. By 1875, nobody was entirely certain which kind of machine would prove most effective and even what sort of chemicals as refrigerant would be favoured (Thevenot, 1979: 53; Rees, 2013). What matters for a discussion of ice making in America is that because refrigeration machinery got better over time, ice making got better over time too.

The earliest ice machines appeared in cities in the South, like New Orleans, where there was little or no competition from natural ice because of the climate. They produced little ice and it was mostly of very poor quality. Discussing the product of one such machine, the journal *Ice and Refrigeration* remembered that 'the thickness of the ice and its quality still left much to be desired. It was well known that the whiteness of the ice was due to the presence of air in the water', but nobody had yet figured out the best way to remove it to create a more aesthetically pleasing product for consumers (*Ice and Refrigeration*, 1902: 13).

It is also worth noting that these early machines were huge contraptions, large enough to take up entire warehouses. In 1890, the largest model produced by one ice machine manufacturer, the Frick Company of Waynesboro, Pennsylvania, for example, was 27 feet long and 17 feet wide. Therefore the size of the first artificial machines were more often measured by the tons of ice they produced in a day rather than their physical size. The largest of these Frick machines ('Eclipse refrigerating machines') in 1890 could produce 90 tons of ice in 24 hours. However, constant experimentation led to smaller, cheaper machines by 1890, which is when artificial ice really began to spread throughout the country.

Apart from the improvement in the technology, an important reason for the growing popularity of artificial ice was its uniformity and purity. The commercial markets for ice required that uniformity and purity, so successful commercial ice machines provided it. As T. Mitchell Prudden explained for *Harper's*, 'When water freezes out-of-doors, on still pools or on streams, the ice forming does not usually go on steadily and without interruption', but when ice is manufactured with mechanical refrigeration, 'the cooling and the freezing go steadily and relentlessly on, as regularly as the stroke of the piston in the great engines' (Prudden, 1892: 376). The result of steady freezing was a uniform product that consumers generally preferred to natural ice. To get around the problem of pollution, despite the poor quality of some municipal water systems during the late nineteenth century, ice manufacturers used distilled water. Even though the cost of doing so considerably increased the cost of the resulting ice, this is what it took to make most household customers happy (*Ice and Refrigeration*, 1902: 13).

While some industrial customers only needed ice that kept or made their products cold, things were very different for firms that bought ice to use as an ingredient in food products. One example of this would be ice cream makers, who grew rapidly even in the years before the advent of home freezers. Ice cream makers used new ice-breaking machines to turn the large blocks that artificial ice makers turned out into chips that could be mixed into their product. These blocks had to be pure since people ate the result, so they only used clean artificial ice wherever possible. That ice cream would then be kept cold by more crushed ice packed around it to prevent it from melting on the way to customers. Cocktails were another way to use the new supply of clean, clear ice that artificial refrigeration made readily available. But like ice companies or ice cream makers, that ice had to be processed. Bartenders had to be able to transform ice into many different forms depending on whichever drink they happened to be making. Shaved ice was common for many mixed drinks in the late nineteenth century, but for others the bartender needed larger cubes so that it would melt more slowly (Wondrich 2015: 47).

Ice for home consumers, whether natural or artificial, had to be delivered. Until trucks became commonplace, that meant by wagon. Ice wagons held large blocks of ice, which were then divided for individual use on the back of the wagon. Payment depended upon weight, which meant that every ice wagon required a scale, an ice-shaving tool, a bucket, and often a broom. Respectable ice dealers always cleaned the ice before delivering it to consumers. Much of the ice dealer's job, in other words, was to turn giant ice blocks into a product that individual consumers could use to meet their needs.

Since wagons couldn't enter home kitchens, that meant that ice delivery men had to carry them on their shoulders into people's kitchens. These deliverymen served even the poorest neighbourhoods as urban Americans from all races and classes developed a taste for ice over the course of the nineteenth century. The usual place for these deliveries were iceboxes, however, and before the advent of electric home refrigeration during the 1920s, those iceboxes were generally called 'refrigerators'. These boxes of ice generally had compartments, one for the ice and one for the food. This was not just because ice could be dirty. Most perishable food spoiled quicker in a damper environment, so this helped those iceboxes do their job right. In the poorer neighbourhoods of American cities, selling ice chips off the back of the truck was not unheard of either.

Icebox owners preferred clean ice, even if they didn't consume it directly, because clean ice was easier to clean after it melted. At the turn of the twentieth century, most refrigerators had pipes to help drain the melt water to make cleaning them easier. Most of this hard work was the result of mould, but the job became infinitely worse if sediment from the ice block got caught there too. Frozen water went in through the ice compartment and out in liquid form through the drain. Eventually, the advent of electric household refrigeration provided consumers with a way for ice to leave the refrigerator through the front rather than from the bottom.

Ice cubes: making ice for oneself

Whether natural or artificial, ice dealers provided an inconvenient product. If you needed ice to keep your icebox cold, you had to count on the delivery-man coming every day because otherwise most of it would have been gone. Ice melts, except when mechanical refrigeration is available to keep it cold. Electric household refrigeration provided consumers that cold. This technology debuted in the United States around 1915, but wasn't really perfected until

1927 with the release of General Electric's 'Monitor Top' refrigerator, the first reliable, low-priced electric household refrigerator in world history. By World War II, most American households had electric household refrigerators in them, far sooner than elsewhere in the world.

For the electric refrigerator to become popular, refrigerator manufacturers had to take on both ice and the iceman. 'Eliminate the Iceman and His Ice', read a 1919 advertisement for Isko, a company that sold machinery to turn your icebox into a domestic electrical refrigerator (*The Cleveland Jewish Society Book*, 1919). That was about the reliability of the ice delivery. Not only did delivery cost money, if the iceman skipped a day it could ruin all the perishable food you were keeping. But refrigerator manufacturers also attacked ice itself as a source of refrigeration. 'Your food is kept at the same temperature the year 'round', explained a pamphlet for a Kelvinator refrigerator. 'This is impossible with ice because it melts, and in its melting the amount of refrigeration, and subsequently the temperature of your food, is constantly changing' (Kelvinator, 1922). Successful mechanical refrigeration provided more reliable cold.

The ice cube tray is one of those inventions that was so obvious that nobody can really claim credit for it. In the first electric household refrigerators, the area next to the machinery was close enough to freeze water. Since there was a tremendous demand for ice, people used that space to create ice cubes. The earliest commercial ice cube tray appeared with the DOMELRE (which stood for Domestic Electrical Refrigerator, an early domestic electrical refrigerator created by Fred C. Wolf, Jr, whose father sold industrial refrigeration equipment during the late 1800s.

While that model refrigerator failed, the idea of making your own ice became an important selling point for refrigerators of all kinds. 'Many women freeze maraschino cherries, sprigs of mint or violets in these ice blocks', explained a different refrigerator manufacturer in a 1923 advertisement. 'They also make mousses, ices, sherbets, custards and salads by placing them in the Kelvinator ice trays for a few hours' (*Good Housekeeping*, 1923: 145). Kelvinator also suggested freezing gold numerals for birthday parties and even freezing place cards (Kelvinator, 1922: 5). All of this reflected the novelty of household ice production during the early years of this technology. As time passed, easy access to ice cubes became something that many Americans took for granted.

Once direct control of refrigeration became more common, later ice cube trays became more about convenience than either wonder or aesthetics. Early ice cube trays were made from nickel-plated copper. To get cubes out of the tray, consumers had to put the back of it under warm running water. The first flexible trays made from rubber were produced in the late 1920s, allowing consumers to simply twist them to get the ice cubes out. By 1940, whether by flexing the tray or pulling up a steel rod running down its centre, removing ice cubes from trays had become child's play.

Yet this raises the question of why anyone wanted ice cubes in the first place. Drop ice in your drinks and you inevitably water down whatever drink you put the ice into. While there are some mixed drinks that benefit from this process, there is little benefit to watering down a beverage like iced tea. What makes this particularly American habit particularly weird is that refrigerators are so commonplace in the modern-day United States that it would be easy to chill whatever drink you want to consume if you simply left it there for an hour or two before drinking it. While impossible to prove, it certainly seems possible that the American ice habit may have to do with two centuries of its wide availability within the United States.

The taxonomy of ice

There were different types of ice because American consumers cared about ice for different reasons. Understanding how ice was made is a prerequisite for understanding whether the needs of consumers of ice were met. Companies that harvested or manufactured different kinds of ice competed over which ones of them would best meet those needs. Those needs changed as the technology changed, which also helps explain why the commercial ice industry in the United States has shrunk in recent years, mostly to just supplying bags for parties.

An 1897, a reporter for the *Fort Worth Morning Register*, on visiting an ice factory, bragged to his readers on behalf of the entire human race: 'Jack Frost is to be driven out of the ice making business. He cannot compete with the ice machines which make artificial ice. Nature may still defy man to make an egg, but human inventive genius has excelled her in other lines of industry. Icemaking is one of them' (*Forth Worth Morning Register*, 1897: 5). While the transition from natural to artificial ice was a milestone in the history of the ice industry, it was not nearly as significant a milestone in the history of ice. After all, natural ice still had its aesthetic, commercial, and strategic uses. And despite the eventual superiority of artificial ice, natural ice persisted as a stopgap source of cold for the poorest consumers for at least another twenty years.

In many ways, the physical differences between natural and artificial ice were minimal because ice has many qualities that transcended the method by which it was made. Whether cut from a lake or pulled from a refrigerator, ice is cold. That's why it melts outside a freezing temperature. When it melts it helps to keep other things cold, whether in a drink or, if you happen to have an icebox, the leftovers you're storing there. Yet all ice was not the same: it varied in composition, structure, and the context in which it was used. By reviewing each of these factors in the context of the American ice and refrigeration industries as described above, it becomes possible to understand some timeless qualities about the nature of ice in general, as well as how its differences determined the course of the American ice industry.

The most obvious quality of any ice is its clarity (or lack thereof). American consumers preferred clear ice, although clear ice could easily be populated with microbes that would have made consuming it unhealthy. Moreover, opaque ice was not necessarily harmful. The cloudy parts in pure ice are air pockets that had no effect on whether that ice was safe to consume. Only visible sediment in natural ice was an obvious detriment. While not necessarily unhealthy, nobody wanted to find dirt or leaves at the bottom of their glass. Even then, if this ice could keep something cold that it did not touch, it still served an important purpose.

The structure of ice is another quality that helped define its usefulness. Harvesters cut ice into blocks because it was easier to store and transport. Ice machines produced ice in blocks for the same reason. Natural ice harvesters often had to manipulate their product to produce the blocks they preferred. Any covering of snow had to be removed. After the harvesters poked holes through the ice, they applied their weight to it: 'The weight of the men causes the water to rise through the holes thus made, and to overflow the surface', read a description of the harvesting process from 1884. 'The snow is by this means slowly reduced, and the water derived from the flooding and the melting of the snow freezes into ice, the lower portion of which is solid and clear' (*Longman's Magazine*, 1884: 414). The result was both more marketable and easier to turn into regular blocks. In this manner, what nature created became commoditized.

No matter how ice was made, suppliers could shave it or crush it and sell the result by the bag. The idea here was to take advantage of its shape to increase the points of contact between it and whatever it happened to be keeping cold. In the case of fish, ice is preferable to mechanical refrigeration because submerging it in ice chips can keep the product both colder and moister. This, in turn, keeps it fresher. Salmon can last a week this way. Other species of fish can last for as long as three weeks. The slipperiness of the ice is what matters since it was the water coming off the melting ice that kept the product fresh. In this kind of ice, the shape mattered far less than the amount of surface area that touched the fish.

No matter how it was made or what shape it's in, ice melts. Natural ice companies had an enormous incentive to sell as much ice as they could early in the season because later in the season – not just summer, but any season – they invariably had less ice since much had melted away. The less ice anyone had the faster it would melt because, as was the case with the ships taking ice to warm places around the world, the outside ice protected the inside ice and when there was less outside ice every part of the supply faced the inevitable effects of time. Besides the quality of their ice, the other advantage that mechanical refrigeration ice companies had over harvesters was the ability to produce on demand to prevent panic selling at the end of the season. Their shorter cold chains from the point of production to the point of consumption were also good protection against rapid melting.

For electric household refrigerator owners, the best thing about their ice was its convenience. It didn't so much matter what their cubes looked like since they knew it was made from the same water they drank every day. What mattered was that they could have it whenever they wanted without depending on any outside company to provide it for them. The best ice was all the ice they wanted at practically no cost, and if it could be removed easily from their ice cube trays then so much the better. All the earlier distinctions between natural and artificial ice became irrelevant when the cold chain for providing it collapsed entirely inside people's residences.

These days, modern craft cocktail makers are trying to do people's refrigerators one better. 'In hipster meccas from Portland to Williamsburg', explained the magazine *Mother Jones* in 2014, 'bars are serving up their drinks on extra-dense, extra-clear cubes, produced through a laborious process of freezing and carving. Cocktail connoisseurs swear the difference in flavor is worth the extra effort: In addition to being more aesthetically pleasing, the cubes' density and relatively large size mean they melt more slowly and dilute your drink less.' These cubes depend on a special freezer called a Clinebell machine. While most freezers freeze water from the top down, the Clinebell freezes from the bottom up and stirs the remixing water to eliminate unsightly air bubbles in the final product, just like the early artificial ice producers did. Since Clinebell machines are too expensive to go in most bars, modern artisanal ice makers have to deliver their product too (Cohen, 2014). What once was old is new again.

Ice has gone from the only available source of refrigeration during the early nineteenth century to a by-product of mechanical refrigeration today. The changes in the ice industry have made it cheaper and cleaner, but its purpose hasn't really changed. Ice is mobile refrigeration for perishable foods or drinks. Even today it has uses in fish markets or for picnics because it is superior to mechanical refrigeration for such purposes. Looking back is a useful exercise because it should prevent anybody who does so from taking this technological marvel for granted.

References

The Cleveland Jewish Society Book (1919) *The Cleveland Jewish Society Book* 3, March.

Cohen, R. (2014) 'Lay Off the Artisanal Ice, You Ignorant Hipsters', *Mother Jones*, 29 October. Available at: www.motherjones.com/environment/2014/10/giant-ice-cubes-hipster-craft-cocktail/ (accessed 21 March 2022).

De La Vergne Refrigerating Company of New York (1890) *Third Catalog*. New York: De La Vergne Refrigerating Company.

Fort Worth Morning Register (1897) *Fort Worth Morning Register*, 11 July.

Godfrey, H. (1909) 'The City's Ice', *Atlantic Monthly* 104(July): 119–129.

Good Housekeeping (1923) *Good Housekeeping* 76, March.

Ice and Refrigeration (1902) *Ice and Refrigeration* 22.

Ice and Refrigeration (1905) *Ice and Refrigeration* 28.

Kelvinator (*c*.1922) 'Kelvinator – and Some of Its Uses' [pamphlet].

Longman's Magazine (1884) *Longman's Magazine* 3(February): 414.

Portland Daily Press (1892) *Portland Daily Press*, 26 January.

Prudden, M. (1892) 'Ice and Ice-Making', *Harper's New Monthly Magazine* 85(August): 370–383.

Rees, J. (2013) *Refrigeration Nation: A History of Ice, Appliances and Enterprise in America*. Baltimore, MD: Johns Hopkins University Press.

Rees, J. (2018) *Before the Refrigerator: How We Used to Get Ice*. Baltimore, MD: Johns Hopkins University Press.

Sudan, R. (2016) *The Alchemy of Empire: Abject Materials and the Technologies of Colonialism*. New York: Fordham University Press.

Thévenot, R. (1979) *A History of Refrigeration Throughout the World*, trans. J. C. Fidler. Paris: International Institute of Refrigeration.

Twain, M. (1883) *Life on the Mississippi*. Boston, MA: James R. Osgood and Company.

Weightman, G. (2003) *The Frozen-Water Trade*. New York: Hyperion.

Williams, H. (1903) *Mechanical Refrigeration*. New York: Whittaker & Co.

Wondrich, D. (2015) *Imbibe!* New York: Penguin.

Wood, W. T. (1876) 'Price List of Ice Tools', Maine Maritime Museum, Bath, ME, Mark W. Hennessy Collection, MS-18, Box 13.

Bibliographic sources

American Society of Heating, Refrigeration and Air-Conditioning Engineers (n.d.) *DOMELRE First Electric Refrigerator*. Available at: www.ashrae.org/about/mission-and-vision/ashrae-industry-history/domelre-first-electric-refrigerator (accessed 21 March 2022).

Anderson, O. (1953) *Refrigeration in America: A History of a New Technology and Its Impact*. Princeton, NJ: Princeton University Press.

Cold Storage and Ice Trade Journal 40.

Freidberg, S. (2009) *Fresh: A Perishable History*. Cambridge, MA: Harvard University Press.

Frick Company (1890) '"Eclipse" Refrigerating Machines' [pamphlet].
Hiles, T. (1893) *The Ice Crop: How to Harvest, Store, Ship and Use Ice*. New York: Orange Judd Company.
Rees, J. (2008) 'The Natural Price of Natural Ice in America, 1880–1910', *Business and Economic History On-Line* 6: 5–6. Available at: https://thebhc.org/sites/default/files/rees.pdf (accessed 21 March 2022).
Rees, J. (2015) *Refrigerator*. New York: Bloomsbury.

8

Drift, capture, break, and vanish: sea ice in the Soviet Museum of the Arctic in the 1930s

Julia Lajus and Ruth Maclennan

Staging the Soviet Arctic

The 1930s was a period of transformation for both the economy and ideology of the Soviet Union. The eve of the new decade known as the Great Break was when the planned centralized economy with its five-year plans was founded to bring about the rapid industrialization of the country. It was the time of the collectivization of peasants accompanied by terrible famine, and the massive migration of people – both forced by government and voluntary, but driven by a search for survival. The late 1920s and the early 1930s are known as a time of cultural revolution that was directed first towards mass literacy. The peripheral territories of the large country were more and more involved in all these processes and the north, as well as the Arctic, were no exception.

The 1930s are also widely recognized as a period of deep changes in the Soviet Arctic both in terms of practices and discourses. It was in the 1930s that the Soviet Union finally began to develop its own distinct Arctic policy (Lajus, 2013). In this decade imagery of the Arctic began to shift from depicting the Arctic as the Soviet Union's resource backyard to becoming its 'shop window' for signposting national ambition and mastery over nature – a process later amplified by the securitization of the Arctic during the Cold War (Eklund *et al.*, 2019). The 1930s was a time of integration of the Arctic into the symbolic construction of national space (Frank, 2010). During this decade books and films contributed powerful images of a hostile environment defeated by brave Soviet people that were circulated widely in official propaganda (McCannon, 1998). Museum exhibitions along with films were mobilized during the period of cultural revolution for a fundamental reform, which aimed to transform them into political-educational institutions accessible and attractive to the broadest masses of the population (Chlenova, 2017).

In 1932, the ice-ship *Malygin* was the first ship to navigate along the Siberian coast in one season. This led to the establishment of a new central governing body for the Soviet Arctic, 'The Main Administration of the Northern Sea Route'. Sea ice along this route – from Murmansk to the Bering Strait along the entire Arctic coastline of the Soviet Union – was no longer considered an insurmountable obstacle to navigation. The success was possible in part because of a period of substantial warming of the Arctic, which had been carefully measured by Soviet scientists. Nonetheless this does not diminish the achievements of technology and the bravery of the protagonists.

In 1934, to develop scientific forecasting and predictions of ice, the Soviet government established the Interdepartmental Bureau of Ice Prognosis under the leadership of Otto Schmidt, who was unofficially named the Commissar of the Ice (McCannon, 2007). Leading specialists in oceanography and climate studies Nikolai Zubov, Vladimir Wiese, and Boris Dzerdzeevsky, became members of the bureau. Thus, sea ice officially became an object of Soviet governance. In two decades, it became internationally acknowledged that the Soviet Union had obtained more knowledge on sea ice than any other country.

In this chapter, we will not rehearse well-known aspects of the history of studies of sea ice in the Soviet Union in the 1930s (Sörlin and Lajus, 2013). Neither will we address the broad cultural context that other scholars have studied through looking at literature and film (Bugaeva, 2018; Kaganovsky *et al.*, 2019). Literary scholars Susanne Frank and Tim Young, in studying Soviet fiction and non-fiction literature on Arctic explorations of the 1930s, including accounts of major events involving sea ice such as the *Chelyuskin* disaster and North Pole-1 expedition, have come to the conclusion that the rejection of the environment from Soviet Arctic discourses of that time was due to the environment being seen as an obstacle to the integration of the northern regions into Soviet territory (Frank, 2010; Youngs, 2010). We prefer to stress the rejection of Arctic 'nature', instead of 'environment', as an ahistorical eternal domain, incompatible with human life, but which stimulates human efforts and technology and fuels the process of environing (Sörlin and Wormbs, 2018). The gradual disappearance of sea ice as an obstacle in political and cultural discourses was a part of that environing. To smooth this process, however, sea ice needed to be carefully studied to better understand and predict its movements. Whenever possible the ice should become friendly, along with the rest of the Arctic that was becoming 'friendly', as it lost its most dangerous features thanks to human-induced transformation.

The interest to history of the relations between people and sea ice has been conceptualized as a part of cryo-history, stressing its growing strategic importance even before the Cold War (Sörlin, 2018: 31). The possibility of sea ice being considered an extension of a country's territory, with legal status, has been an important element within a diverse set of the Arctic's imaginaries (Steinberg *et al.*, 2015: 45–64). In this chapter our method is to examine how sea ice was reimagined and engaged with through looking at the Museum of the Arctic and Antarctic in St Petersburg, from its inception in the 1930s to its present incarnation. We demonstrate how, on the one hand, the ways of representing sea ice and attitudes towards it that were established in the 1930s continue to exert a powerful influence today, and on the other, that the importance of images of sea ice in the museum has considerably diminished. We consider the spaces and collections, and how they represent and reflect the policies, technologies, and attitudes towards sea ice. We look closely at materials, objects, and images and their configurations in the spaces that greet visitors, and what stories they tell. We also unpack icebreakers as important objects and protagonists in the transformation of Arctic sea ice, which continue to exert power as heroic heritage as well as contemporary symbols of Russian Arctic development and dominance; some, like the icebreaker *Krasin* in St Petersburg, and the *Lenin* in Murmansk, have themselves become museums. We also pay attention to how the museum engages the visitor affectively – through admiring and identifying with Soviet heroes amidst the ice and/or touching stuffed polar bears. The groups of visiting school children seem to be performing a role too, connected to the history of the museum as descendants of earlier visitors.

Museum of the Arctic (and Antarctic) in St Petersburg

The Museum of the Arctic opened in Leningrad on 8 January 1937. It stages a unique version of the history of human encounters with Arctic environments, and in particular with Arctic sea ice. The concept of the museum was part of the transformation in 1930 of the Institute for the Study of the North into the Arctic Research Institute (Arctic and Antarctic, from 1958). The museum was organized within the institute, which soon became well-known internationally for sea-ice research. There were already enough museum objects in the institute to organize temporary exhibitions, but in 1933 the future museum received a building. The museum was housed in the neoclassical Old Believers church of Saint Nicholas which had been forcibly closed in 1932. The church setting suggests or at least primes the viewer for a heightened, possibly spiritual, experience. The ideology that replaces it and fills its spaces declares its own 'revealed' truth: the progress, through science and engineering, towards building a socialist Soviet Arctic. The museum replaced the furnishings and icons of the church with relics from Arctic

Figure 8.1 3434: Ruth Maclennan, Museum of the Arctic and Antarctica, St Petersburg.

voyages and scientific research, donated by explorers. The museum remained a department of the institute until 1998, and then became an independent organization under the authority of the Russian Federal Service for Hydrometeorology and Environmental Monitoring.

In the museum, the fictional, the imaginary, the symbolic, and the real spaces are superimposed, and at times inextricably tied up with each other. Journeys through Arctic sea ice are documented, memorialized, and imagined. The museum is a place to share them, to proselytize, to inform, and inspire – and as a visitor to be a participant, in a quasi-religious way. Recently Vlad Strukov describes the Museum of the Arctic's layout and use of specific areas of the church to 'elevate the Arctic so that early explorers appear as new saints and the Arctic as a shrine' and draws attention to what he sees as the emphatic links between the idea of a Christian notion of culture as Noah's Ark and the theme of the exploration of the Northern Sea Route (Strukov, 2021: 256). He argues that the location of the Museum of the Arctic in the neoclassical church 'signifies the projection of Russian modernity, always cultivating classical notions of beauty and "progress" such as colonization of new territories, which had been appropriated by Stalin's regime to recontextualise Soviet modernity' (Strukov, 2021: 261–262). Walking through the museum then and today the visitor is caught up in a life-sized diorama, playing a part, representing the human scale in the vast landscapes that are imagined and referred to in the exhibits. The dioramas are instrumentalizing 'haptic visuality', by showing the layers above and below the ice for instance, giving an otherwise impossible perspective to visitors, and reinforcing the message of conquest contained in the maps (Strukov, 2021: 259–261).

The ground floor area of the Museum of the Arctic is still full of objects that were designed for it in the 1930s. Although it has seen several transformations, they were not radical, and the museum today could be seen as a shrine to the 1930s museum: a museum of museums. In the late 1940s, there was major restoration when the museum's objects were returned to Leningrad, having been evacuated to Krasnoyarsk in Siberia during the war. In 1953, there was another major rehang when the exhibition on Soviet economic development in the Arctic was reduced and the historical sections of the museum were expanded. This change was connected to the weakening position of the Northern Sea Route Administration in 1953. At this point, many new art objects, especially newly commissioned paintings, were added to the museum displays. Another set of paintings was added in the 1970s (Andreev *et al.*, 2013).

The museum's Arctic exhibit is designed like a theatre or film set that conjures the Arctic as a stage for stories of heroism and rescue, of scientific discovery, feats of engineering, mastery, and conquest. The museum was one of many museums founded to display the diverse geography of the new country: from the Central Geographical Museum established in Leningrad in 1919 which survived until 1940, to the Earth Science Museum of Moscow State University opened in 1950, not to mention numerous local museums in every regional town. All these museum-types were organized according to a strict principle with a department of natural history, a department of history and culture, and a department dedicated to the building of socialism (Gavrilova, 2019: 182). And this is precisely how the Museum of the Arctic was organized when it opened. Today the museum has three departments: Natural History, History of the Northern Sea Route, and History of Antarctic Exploration. The first two were the original departments of the museum in 1937 and are what concern us here. The Antarctic exhibit is on an upper floor and only opened in 1958 after the first Soviet expedition to Antarctica. As space within the church building was limited, it was necessary to dismantle the exhibitions on the economy of the Arctic region and Indigenous people that previously occupied the second floor.

The museum uses the religious architecture and the symbolism afforded by its spaces to suggest and perform relationships to the Arctic exhibits, many of which are relics of past expeditions and scientific activities with and on Arctic sea ice. The first director of the museum was the researcher and artist Nikolai Pinegin. Leading scientists of the Arctic Institute – its

Figure 8.2 3297: Ruth Maclennan, Museum of the Arctic and Antarctica, St Petersburg.

director Rudolf Samoilowitsch, Vladimir Wiese, Yakov Gakkel – took part in the planning of the museum's exhibitions as well as leading geographers from other Leningrad institutions including Yulii Schokalsky and Veniamin Semenov-Tian'shansky. The latter founded the unique Central Geographical Museum but, in 1936, lost his directorship there and transferred his attention and expertise to the Museum of the Arctic, which had even been planned initially as a department of the Geographical Museum (Muzei Arktiki, 1931).

Opening at the height of Stalinism, everything about the original design is inflected, at least in retrospect, with the ideological priorities of that time. A newspaper article covering the opening of the museum describes the objects that form the basis of the museum collections as being the 'scientific "trophies" brought back from Arctic expeditions as well as accidentally discovered relics of previous expeditions' (Ianvarsky, 1936). There are several mentions of the 'conquerors' and the 'conquest' of the Arctic as well as 'battling' with the Arctic. The museum is praised for panoramas made by an artists' brigade, but also as the future repository of documents and objects that will show the stages of the socialist construction of the Far North. This is indeed what happened – new exhibits were added as the museum kept pace with discoveries and events happened in the Soviet Arctic.

The Arctic is conjured in this article as a place that is being transformed and will be further transformed as part of a process of Soviet 'mastery' (*osvoenie*) and transformation by means of 'advanced technologies' and 'the best people'. And finally, the article states that the museum has established links with overseas museums and institutions involved in Arctic research. The museum thus was seen at the time (or at least described here) as a means not only of representing and promoting a Soviet Arctic domestically, but also internationally by belonging to the Arctic museum community that was being developed at that time across the circumpolar north, and projecting a particular idea of the Soviet Arctic abroad (Ianvarsky, 1936). Among the polar museums serving as repositories for quasi-sacred 'relics' of Arctic expeditions are the Fram Museum in Oslo (1936), the Scott Polar Research Institute with its Polar Museum in Cambridge, England (1930), and, more recently, the Peary-MacMillan Museum in Brunswick, Maine (1967) and the Andrée Museum in Gränna, Sweden (1977). Despite how different they are from each other, they each use objects as evidence, props that contribute to stories of national self-image, or counter-narratives (Houltz, 2013).

Ice as imagined object

Imagined from afar, from the temperate zones of metropolitan life, the Arctic is about ice: frozen seas, icebergs, polar bears, and lone explorers tramping across the frozen wastes to reach the North Pole. For a museum, this is a challenge: how do you demonstrate the material qualities and seasonal movement of sea ice? How do you communicate the vastness of sea ice, the scale of the view from an icebreaker or aircraft? Or the cold? It is very hard to visualize sea ice, its textures, substance, and behaviour – except in relation to humans and their artefacts. Every museum dedicated to the Arctic is therefore a museum of representations of the Arctic, of human interactions with polar environments and histories of encounters with the Arctic. Polar museums share some of the often unstated, unacknowledged colonial histories of other natural history museums (Arends, 2020). They are, however, a particular subset of natural history museums, in that more than other natural history museums the geography, its inaccessibility, and the role of heroic adventure, knowledge, and the technologies for enabling it, are central to the stories of discovery being told.

Sea ice in the Museum of the Arctic is alluded to and come at from different angles and directions, but it is the one, central material object that cannot be included in the museum. In its place are representations and, most importantly, objects that have been in contact with it, which conjure the stories of Soviet exploits and events, such as Papanin's tent in which four Soviet men camped on Arctic sea ice and which was their floating research station. Some of these objects have value in themselves: as important inventions, patents, and technological discoveries. There are tools for measuring sea ice, for measuring and establishing the location of a ship in the ice, and tools for cutting through sea ice. They do not quite tell the story of humans' (especially Soviet heroes') engagement with sea ice, but they assume there is or was sea ice to be engaged with. Some of the objects in the museum are quasi-religious relics, especially the pieces of the *Yermak*. The grandfather of Soviet icebreakers, the *Yermak* was the first Arctic icebreaker, built for the Tsar by shipbuilders Armstrong Whitworth in 1889 on the River Tyne, in the northeast of England.

The 1930s is a decade of scientific discovery and technological advances in the Soviet Arctic. It is also a time when assemblages of technologies to work with sea ice become effective, for instance in the rescue of the *Chelyuskin* expedition in 1933, which used radio to guide aeroplanes landing on sea ice to evacuate the crew. As you enter the museum, you almost duck your head to avoid the aeroplane flying just above the entrance. The plane is coming to the rescue of the *Chelyuskin* crew stranded on Chukotka after the sinking of their ship in 1933, as it attempted to complete the journey along the Northern Sea Route without an icebreaker convoy. The plane creates an immediate sense of drama. You are going to be led through the adventures of Soviet heroes in this hostile environment. Analysing this object, we find evidence of what is known as 'the double address of spectacular aesthetics' (Ekström, 2012: 473) with a spectatorship oriented towards the disruptive experience of shocking and engaging audiences in the remediation through which disastrous events are disconnected from their origins.

The flimsy-looking canvas-winged aircraft establishes a human scale. It seems barely large enough to scoop up a couple of stranded polar explorers, but its skis and the fact that it is the original surviving aircraft signal its adaptation to the sea ice. The icy environment is assumed, not illustrated. We, the visitors, would be standing on it. If this were a film, the aeroplane flying over the entrance would be the opening shot, drawing the viewer into the themes of the film/museum: bravery, technological prowess, doughty Soviet ingenuity, and the conquest of the Arctic, and, more broadly, the taming of nature itself in the service of building communism. The opening shots of Alexander Dovzhenko's film *Aerograd* (1935), filmed from a small light aircraft flying low over northeastern Siberia, could have inspired this mise en scène at the entrance (Dovzhenko, 1935).

The *Chelyuskin* epic was an international media sensation. The journey of the *Chelyuskin* was intended to demonstrate the Soviet Union's pioneering spirit, bringing families to settle Wrangel Island, and the potential of the Northern Sea Route by sailing across it in one season in a non-ice-class vessel. Although the *Chelyuskin* sank, before completing the entire Northern Sea Route in one season, the failure was turned into a heroic success. The rescue of all but one of the crew (who died as the ship sank) encamped on sea ice represented the triumph of Soviet technology and heroism. Using radios to guide rescue aeroplanes, all the crew including women and children were brought home and met with great fanfare. The whole world watched. The survival and rescue enabled attention to be shifted, directed away from the sea ice as destructive force to sea ice as temporary home, and the opportunity to demonstrate to the world the technological and human prowess of the Soviet polar heroes.

Surprisingly, the current museum exhibition only briefly alludes to the *Chelyuskin* epic in a few images and objects. Disaster was gradually excluded, and today the museum is one of success stories. But this was not the case when it first opened: a large diorama, 'Schmidt's Camp', named after the leader of the expedition, Otto Schmidt, occupied a central space. A huge white ice floe split down the middle with a dark line of water in the rift was populated with small makeshift dwellings and little figures. The red flag flying on the tower built on the ice ledge in the left-hand corner was the only thing that stood out against the icy landscape. This diorama was lost during the war along with several others which had not been evacuated, and their fate during the Siege of Leningrad remains unknown. Although other dioramas were rebuilt after the war, such as the famous one of the bird colony at Novaya Zemlya by the well-known artist Mikhail Platunov, the Schmidt's Camp diorama was not.

In this representation of the *Chelyuskin* camp, the sea ice itself is the infrastructure of the Northern Sea Route. The sea ice becomes home to the crew, safe from the cruel, icy-cold ocean waters below, and, moreover, the ice is seen as an extension of Soviet territory into the ocean. The ability to build and maintain an airfield on the ice, as shown in the diorama, reinforces this feeling of the habitability of the ice and its prospective usefulness as infrastructure.

Aircraft enable a new perspective on the planet, especially the 'top' of the planet, the Arctic Circle and the North Pole (Bravo, 2019). By the 1930s aircraft had come to play a vital role in Arctic engagement and imagery. This aerial theme comes into focus in the central rotunda of the church: a large-scale relief model of the sliver of the globe above the Arctic Circle stands at waist height. It was designed by the artist V. V. Novodvorsky in 1936 based on the cartographical work of one of the most prominent geographers, Yulii Schokalsky, whose relief map of the Arctic was exhibited in Paris and London in 1935. The 3D relief shows areas of sea ice and political and physical geographical landmarks. The dome of the model stands under the dome of the neoclassical church, perhaps replacing a large

Figure 8.3 3402: Ruth Maclennan, Museum of the Arctic and Antarctica, St Petersburg.

font that might have stood there. Above the model North Pole fly Soviet-built aeroplanes, not angels. It recalls the final jubilant scene of the film *Aerograd* – when hundreds of pilots with parachutes tumble out of the sky over the site of the future city. The symbolism is powerful and multilayered and reinforces the overt narratives. Children and other visitors can stand at the edge of the Arctic Circle, and touch it. They can admire the vast Arctic wastes of the Soviet Union's landmass, coastline, and, by extension, sea area. The Soviet visitor can learn about the Arctic, and, by proximity and symbolically, feel a part of the Soviet heroic creation myths. The museum exhibitions enable visitors not only to see and know but also to be with the Soviet Arctic past (Ekström, 2019: 18). The visitor is included in the stories, and even invited to imagine themselves as joining in the future construction of the socialist Soviet Arctic – the endgame in the mastery of the Arctic.

The museum contains miniature icy spaces and models of Arctic environments, with which the visitor is invited to interact, to imagine the operations of Soviet scientists and explorers in the harsh environment. They are elaborate mechanical toys with moving parts which are no longer turned on for fear of them breaking. Each exquisite cabinet artfully imagines and reproduces a scene from the history of the Soviet Union's scientific and engineering conquest of the Arctic – a weather station, for instance, and an icebreaker with a layer of sea ice and the ocean below with swimming fish.

Friendly Arctic v. frozen Arctic

The main gallery contains several life-size dioramas of natural Arctic environments against the walls that were designed especially for the opening of the museum, and free-standing glass cabinets with Arctic wildlife. A stuffed polar bear represents the generic, symbolic polar fauna. The top Arctic predators are not displayed as a threat to humans. The bear is perched on a pale-blue cloth-covered plinth to represent an ice floe. It greets the visitors, and, recently, the old one that was petted so much its nose was partly rubbed away was replaced with a new one inside a glass case. The use of polar bears as an attraction in different kinds of exhibitions is traditional – for instance, a stuffed polar bear was put in the middle of the Russian section at the Great Fisheries Exhibition in London in 1883.

Further on in the main gallery a wolf at floor level stares through glass at latter-day Little Red Riding Hoods; birds and other wildlife are gathered picturesquely in dioramas and models. One raptor seems to have escaped and is seated high up on a classical pediment eyeing the scene below. What is the purpose of all these animals and dioramas if not to form the stage set in which the human drama of mastering the Arctic unfolds? There is a long history of showing wild nature by means of dead natural objects. And here this wilderness is shown as the starting point for transformation, as the natural world that will undergo a new environing. The objects of nature do not appear in conflict with humans, however. The animals taken out of their habitat look domesticated. The dioramas on the other hand seem designed to bring a real Arctic flavour into the museum, as a window into the wilderness. There is a diorama of a seashore colony of birds, a walrus colony, and scenes of the tundra in summer and winter with stuffed reindeer. This is an inhabited, friendly Arctic.

The friendly Arctic suggested by the animals displayed in the cabinets coexists with a conception of the Arctic environment as a frozen wilderness, empty of life. This is the place to find Soviet explorers, scientists, and engineers (all heroes); this is the backdrop for individual bravery and collective achievement. The real objects donated by explorers and scientists, or salvaged from rescue missions, have an aura about them that the stuffed polar bears and paintings lack, because they have played a part in a story. Polar bears are marked as ahistorical nature (albeit domesticated here), while the man-made objects occupy a place in history, as evidence of the successful deployment of Soviet science and engineering in the Arctic. Framed photographs, murals, the dioramas and paintings reinforce the stories, animate the objects (sometimes depicting them with the heroes who used them). Being in the presence of these relics the visitor can commune with and reflect on the lives of those who used them, who cooked tins of *tushenka* and boiled a kettle for tea here, who wrote their diaries by this lamp, and wrote down measurements of ice depth and temperatures in this logbook. This is the central Soviet human drama, which unfolds in and is entangled with the frozen elements of the Arctic, most notably on or against or through sea ice. This is the Arctic nature that is the backdrop, and the environment that is being transformed.

Friendly Arctic nature cedes its riches to valiant Soviet explorers, scientists, and engineers who brave the elements. The Arctic backdrop (with its duality) is also represented in murals and paintings. Many of the paintings were added later after World War II. Murals of Soviet Arctic history – in a roughly Socialist Realist mould – adorn the neoclassical arches of the church, along with framed oil paintings in a Romantic landscape tradition. Some of the framed paintings reiterate a vision of the Arctic as a sublime unpeopled landscape, familiar from nineteenth-century painting. But rather than inviting visitors to swoon in awe at the Arctic sublime, which is (conventionally) considered to be a catalyst for individual, spiritual transformation, these paintings seem to contribute to the domestication of Arctic nature. They are not like icons in a church so much as the backdrop to the range of human activities represented through objects, tools, and technologies. The activities – including the scientific study of sea ice, exploration, settlement, economic exploitation, and the military domination of the Arctic – are all aspects of the *osvoenie* or mastery of the north. Planning and strategy, and the role of the Party, are also represented with murals of gatherings of Soviet leaders and aviators, and framed historical maps are to be found in the display.

One of the murals filling an arch depicts the rescue of the stranded Papanin and his crew. It was painted in the 1970s by M. A. Kaneev. Four men in heroic upbeat poses stand on the ice holding a huge red flag, with an excited dog in the foreground. Two of the men face out towards us, as if there was an audience at the scene. The other two men are turned towards the rescue ship in the background off to the left where more giant red flags are flying. The ship's smoke wafts into the clouds suspended over our heroes. This mural in effect tells the story of the Papaninites as it is meant to be understood – in heroic terms – and reinforces the significance of the real Papanin tent which stands beneath it nearby.

The real-life tent used by the Soviet ice-drifting research station North Pole-1, under the leadership of Ivan Papanin, is the most important exhibit in this central room. For Papanin and his crew, and for their idea of drifting research stations, which follows Nansen's experience of drifting on the *Fram* and the unrealized dream of a drifting ice station, sea ice is a dynamic agent. It is a force to be reckoned with but also to be harnessed. In this view, humans land on the surface of sea ice, creating a temporary living space, an environment for conducting their scientific work. There is a sense of a light touch here: men 'alight on' the ice, like the bird and walrus colonies in the dioramas. The scientists observe the ice, measure it, move with the element rather than against it. Here the 'nature' of sea ice is to be implacable; winter ice is conceived of as a force of nature. Rather than battling with it, the scientists opportunistically hitch a ride on it, turning the power of sea ice to their scientific advantage.

The least convincing aspect of the Papanin display, and of all the life-size displays, is the ice itself: a rumpled sheet tinged with blue stands in for the ice floe that defined Papanin's mission and his and his colleagues' experience. But, like theatre, the audience is invited to suspend disbelief and imagine themselves holed up with Papanin in his tent. The movement and flow of sea ice is not depicted. Instead, the viewer is invited to engage not with the rhythms of Arctic seasons, of day and night, wind and weather, and the movement of sea ice, but with the story of Papanin as interpreted at the time, in newspapers, by Papanin himself in the (later, postwar) painting described above, and as mythologized and reinforced by the curators of the museum. The Papanin story and its representation is important for understanding the multiple roles and affordances of sea ice that become recognized and used in the Soviet Union of the 1930s. Sea ice is an agent – moving the ice-bound ship and the floating research station. It is both the environment and the object of scientific study, and with the pitching of a research station it can also be imagined as an extension of Soviet territory.

This view of the nature of Arctic sea ice that attaches to the Papanin expedition is counter to that heralded by the successful deployment of Arctic icebreakers (see below), which crash right through sea ice. Icebreakers 'defeat' sea ice, so it is no longer a foe, an obstacle, something to be taken into account – in fact, it can almost be ignored. This is in a sense the first 'loss' of sea ice. The second loss of sea ice comes now, with the decrease in sea-ice cover due to global warming. What can the 'loss' of sea ice in the 1930s say about today's loss? Mourning the loss of sea ice may take many forms. For the time being, in the current limbo of business as usual, the loss of sea ice, and mourning this loss, are disavowed.

Polar museums and their ice-bound heroes

Other polar museums of the same period present their own histories of the Arctic and human interactions with Arctic sea ice, and each represents a national and local narrative (Houltz, 2013). The Scott Polar Research Institute (SPRI) at Cambridge University was founded as a memorial to Robert Falcon Scott in 1920. The Polar Museum, which opened in its own building in 1934, is also based on epic exploration myths, in this case the failure – but brave, noble, heroic attempt – of Robert Falcon Scott's South Pole expedition, as well as Ernest Shackleton's earlier voyages. Scott's failure is attributed to many things – the wrong technologies, bad decisions, and bad luck primarily. Unlike the Chelyuskinite failure, modern technologies (radio, aeroplanes) and heroic pilots did not come to the rescue of Scott's party. However, the scientific discoveries and collections, the photographs, diaries, and other writings made during the expedition, are rescued, along with clothing and other relics, and these are on display in the museum. The UK is not an Arctic or even a near-Arctic state, so its relationship to the Arctic cannot be taken for granted and needs to be demonstrated and continually redefined in the light of changing priorities and perceptions of the UK's post-empire, post-Brexit place in the world. The Polar Museum emphasizes Britain's role in polar exploration, scientific research, and establishing international treaties. At the same time, the collections also include traded and purchased Indigenous historical and contemporary artefacts, which these days are the focus of renewed curatorial and historical research, and efforts at cultural decolonization. Geography, science and the study of natural environments, climate science, anthropology, history of science, and environmental humanities are all strong areas of research at the institute, which has a unique and important library and archive, and all these themes get featured in temporary exhibitions at the Polar Museum. But the historical polar expeditions of Scott and Shackleton are the founding stories, and collections, of the museum. Everything else follows in the wake of this strangely persistent national myth, with its colonial and class subtexts (Spufford, 1996).

The Stromness Museum in Orkney tells a quite different British polar story to the SPRI Polar Museum – about labour, seafaring, and the stories of Orkney islanders being hired by the Hudson's Bay Company to man whaling ships sailing to the Arctic. It is a history of islands whose economy is caught up in globalization, in colonial exploitation, but also the story of the labour of islanders and their contact with Indigenous people in the Hudson Bay area and Greenland. Many of the objects in the museum come from families on Orkney whose forebears worked at sea in whaling etc, and from Indigenous people who visited, or traded, or were the children of Orcadians. Like the Museum of the Arctic in St Petersburg, practices of engagement with the sea, and sea ice, are central to the stories on display. For instance, the museum tells the story of Dr John Rae from Orkney, who worked for the Hudson's Bay Company, and undertook several expeditions to find the Northwest Passage where he discovered the remains of Captain

John Franklin's failed expedition. He brought back relics from the Franklin expedition and traversed the final section of the so-called Northwest Passage (McGoogan, 2002). The museum has a life-sized model kayak with its local hero John Rae dressed in sealskins. The message is clear that, like Nansen and Amundsen, Rae learned to work the Arctic environment from Inuits – Indigenous, adapted technologies proving most successful for surviving and navigating the sea ice. Rae also had a second family with an Inuit woman. Orkney's local island narrative and its globetrotting, seafaring history is intertwined with Scandinavia, Greenland, and the Arctic. The museum presents a local version of a larger historical narrative that contradicts or at least complicates the grand myths about the English heroes of Arctic exploration.

The St Petersburg museum is very different from that as the voices of Indigenous people are missing. Their objects, their experiences of collectivization and residential/boarding schools (established in the 1920s), decidedly do not belong to the dominant narratives of the heroic Soviet conquest of the Arctic. A potential but unrealised way of experiencing their presence, however, is to see them as a powerful counter-narrative that unsettles the dominant colonial narratives of the Arctic of the 1930s. For example, the collection of drawings and graphic artworks produced by students of the School of the Peoples of the North, most of which are stored in the archives upstairs in the museum, other stories of the north, of Indigenous life-ways and relationships with Arctic environments, including sea ice, are powerfully articulated.

Arctic icebreakers

This brings us to the pivotal role played by Arctic icebreakers in the Soviet Union in the 1930s. We shall focus on these objects because of their critical role in transforming engagement with Arctic sea ice in the 1930s, and their continuing and evolving role in Russia's Arctic policy, and in the way the Arctic sea ice is imagined. Icebreakers transform Arctic sea ice both literally and figuratively. The icebreaker is a figure, a tool, a heroic object, a force of (human) nature, an architectural structure, an environment. Moreover, it could be considered as an example of the 'technological sublime', where technological achievements are lionised by people (Nye, 1994). Other nations have built iconic techno-figures, such as the glamourous but short-lived French-British supersonic aircraft Concorde, and the highly successful US Apollo space rockets, but in Russia, despite all the celebration of Soviet space exploration, the icebreakers were and are probably even more visible. Soviet and Russian icebreakers are world leaders that represent innovation, technology, and historic continuity with the heroic vehicles used to rescue Nobile and the Chelyuskinites.

Icebreakers are heroes themselves – they do not just carry heroes. They are personified, but luckily for them they do not suffer the fate of human heroes in the 1930s who could and did fall from favour. In *Podvig vo l'dakh*, by Nikolai Ivanov (1928), a book and then a film for young people, the icebreaker *Krasin* together with the pilot Chuknovsky come to the rescue of the survivors of the Nobile airship that crashed in the Arctic. The pilot is compared to Robinson Crusoe on ice, which signals of course the doughty survival credentials of the pilot, and the genre of the text. Like earlier ships, the *Krasin* is given a personality. In England and English, ships are gendered female, while in Russian they are male. The *Krasin* is not just a powerful creature pushing through the ice, oblivious to the elements, but 'feels' them. 'Enveloped in the breath of huge blocks of ice, the ship began to feel/sense the winter.'

Icebreakers, like aircraft, are important objects for engaging with and moving through Arctic sea ice. They are also powerful figures to be seen, and from which to view the Arctic sea ice through which they break a path. They are both technologically sophisticated, and objects of brute force. The preferred design is, still, a rounded bottom with a prow that slides up onto the sea ice, and then the weight of the ship breaks a passage through the ice. The Museum of the Arctic tells episodes from the history of their development from the first Arctic icebreaker, the *Yermak*, to the *Krasin* – which helped rescue Nobile and his team – to the *Lenin*, the first nuclear-powered icebreaker, to *Arktika* (1970s), and current icebreakers such as the huge nuclear icebreaker *Sibir*, currently being fitted out in the Baltic Shipyard. The icebreaker military-industrial-scientific complex is key to understanding relationships to Arctic sea ice in Soviet and post-Soviet Russia.

The relics from the *Yermak* are less evocative than the Papanin tent. The *Yermak* stayed in service until the 1970s when it was dispersed, and some pieces found their way into the museum, along with a model ship in a cabinet. The *Krasin* is moored across the Neva along the Lieutenant Schmidt Embankment next door to the Baltic Shipyard: a strong statement of continuity between 1930s Soviet Arctic shipping and the present. This a better place to understand the iconic status of icebreakers, and to learn about the role of icebreakers in Soviet and post-Soviet Arctic policy. In spring, icebreakers are brought to St Petersburg for a clean up and repairs. Across the river from the Baltic Shipyard are moored the diesel icebreakers of the civilian fleet used as supply ships for Arctic settlements. If you are lucky enough to visit St Petersburg in May, you might catch the Icebreaker Festival, and get to take a tour of one of these leviathans or watch from a bridge as tugboats 'waltz' in formation between the moored icebreakers.

Sea ice is a moving frontier. It was a worthy foe of Soviet explorers and engineers in the 1930s. Icebreakers enable sea ice to be imagined as a surmountable barrier, as a navigable system. Today, although Arctic sea ice itself is threatened, Arctic icebreakers still conjure this idea of the frontier, of strength and bravery, and the promise of vast riches to be won. Stock photographs of icebreakers cutting through sea ice are used on billboards to advertise financial services. The imagined Northern Sea Route has been as important since the 1930s as its actual use. Icebreakers and the sea ice on which they depend are key to conjuring this piece of infrastructure, the Northern Sea Route. Communities along the coast, however, are not connected by the Northern Sea Route – icebreakers avoid the coast. The successful deployment of ever more powerful nuclear and diesel icebreakers in the Arctic through Soviet coastal sea-ice ways is perhaps the most powerful role of sea ice that begins in the 1930s. The Northern Sea Route, like the Turkestan–Siberian railway (1930) or the Trans-Siberian railway, appears to connect the European Soviet republics with the Far East. The main economic purpose is to enable the extraction of oil, gas, coal, and other minerals that can be shipped along the route.

Without sea ice, there is no need for Soviet – and now Russian – icebreakers to lead ships along the route, and the top of the world becomes a free for all. It is important to note that the route itself exists only through the ice; if ice disappears there is no route, as you could drive ships everywhere. In fact, this is precisely what China and other 'near-Arctic' countries envisage as they shift their attention to the Transpolar Sea Route (TSR). 'The latest scientific models suggest that a route directly via the North Pole may open as early as the 2040s [...] As the TSR develops, two Arctic powers – Russia, a historic one, and China, an emerging one – will no doubt be countries to watch' (Bennett *et al.*, 2020).

Conclusion

Studies of the literature and films of the 1930s have shown how the colonization of the Soviet Arctic differed from that of other countries due to the 'grade of integration of the Arctic into the symbolical construction of national space', what is known as the normalization of the Arctic within the Soviet space (Frank, 2010: 22). In analysing the Museum of the Arctic's displays, we see this process in a longer temporal framework and with some important nuances. The exhibitions of the 1930s, developed in the immediate aftermath of the events of polar conquest, were focused on heroism, and sea ice played a central role. Ice was in turn an enemy to be destroyed, and a tamed space that could and should be inhabited and included in the Soviet realm. This, certainly, can be seen as part of the process of normalization. However, with constant gradual refurbishment and redesign of the museum spaces from the late 1940s to the early 1970s, encounters with sea ice became less prominent and visible. Icy landscapes were mostly moved higher up the walls, the 'haptic' visuality of dioramas often replaced by the plain visuality of paintings. The museum presented more and more historical exhibits rather than contemporary research and exploits. The struggle with sea ice too became a thing of the past: with the development of powerful icebreakers journeys in icy waters stopped being so risky; the work of drifting scientific research stations in a connected world after Sputnik became routine.

However, the legacy of the 1930s still lives on in the museum and in the overall Russian discourse about the Arctic. Certain important aspects of the history of the Soviet Arctic are still missing – the Gulag, and even the history of the purges of well-known polar explorers like Rudolf Samoilowitsch or the museum personnel. When it comes to sea ice, the museum is not open to new approaches to ice, towards recognizing sea ice loss as an environmental emergency. There is a paradoxical or at least contradictory position on climate change today. On the one hand, in the Soviet Union and especially under Stalin, there was a strong narrative of 'controlling nature' and taming the wildness (within humans as well as wild nature), and many claims to be succeeding at doing this. And on the other, probably as a counter approach, anthropogenic climate change denial still exists in Russia (Poberezhskaya, 2018).

Does this mean the Soviet experiment is recognized as having failed, and there is no control over nature – and therefore, climate change is just another natural cyclical phenomenon that humans cannot influence? Or can we have it both ways – control of nature but not of climate? What will the future of sea ice be? In reality, sea ice is disappearing. But bigger and bigger icebreakers keep on being built.

Current sea voyages in the Arctic are hard to narrate. They seem technologically insulated against the elements. The conditions for the types of stories of adventures with sea ice discussed above have disappeared: wealthy tourists can now charter an icebreaker to travel to the North Pole, as shown in a 'home video' on YouTube shot and narrated by a tourist on the Russian icebreaker *50 Years of Victory*. The predicted havoc of the effects of global warming is barely contemplated, although it is the reason to take an icebreaker to the North Pole before it is too late. The icebreaker is the future speeded up; as polar sea ice melts it keeps the Northern Sea Route open to enable fossil fuel extraction. The ship thus helps bring about its own obsolescence. For the icebreaker to keep going, as an icebreaker, it needs the sea ice to be the infrastructure it is imagined to be. The Arctic waters must keep freezing; carbon must stay in the ground.

References

Andreev, A. O., V. I. Boiarskii, and M. V. Dukal'skaia (2013) *Rossiskii gosudarstvennyi muzei Arktiki i Antarktiki: Stranitsy istorii'*. St Petersburg: Aurora.

Arends, B. (2020) 'Decolonising Natural History Museums Through Contemporary Art', in C. Rossi-Linnemann and G. de Martini, eds, *Art in Science Museums: Towards a Post-Disciplinary Approach*. London: Routledge, 213–223.

Bennett, M., S. Stephenson, K. Yang, M. Bravo, and B. De Jonghe (2020) 'The Opening of the Transpolar Sea Route: Logistical, Geopolitical, Environmental, and Socioeconomic Impacts', *Marine Policy* 121. https://doi.org/10.1016/j.marpol.2020.104178.

Bravo, M. (2019) *North Pole: Nature and Culture*. London: Reaktion Books.

Bugaeva, L. (2018) 'Arkticheskii mif v sovetskoi kul'ture 1930- kh godov i ego vozrozhdenie', *Zvezda* 8. Available at: https://magazines.gorky.media/zvezda/2018/8/arkticheskij-mif-v-sovetskoj-kulture-1930-h-godov-i-ego-vozrozhdenie.html (accessed 25 March 2021).

Chlenova, M. (2017) 'Soviet Museology During the Cultural Revolution: An Educational Turn, 1928–1933', *Histoire@Politique*, 33(September–December). Available at: www.histoire-politique.fr (accessed 21 March 2022).

Dovzhenko, A. (1935) Аэроград (*Aerograd*) [film]. USSR: Mosfilm and VUFKU.

Eklund, N., J. Lajus, V. Borovoy *et al.* (2019) 'Imageries and Historical Change in the European Russian Arctic', in C. E. Keskitalo, ed., *The Politics of Arctic Resources: Change and Continuity in the 'Old North' of Northern Europe*. New York: Routledge, 200–220.

Ekström, A. (2012) 'Exhibiting Disasters: Mediation, Historicity and Spectatorship', *Media, Culture & Society* 34(4): 472–487.

Ekström, A. (2019) 'Walk-in-media: International Exhibitions as Media Space', in K. Drotner, V. Dziekan, R. Parry, and K. C. Schrøder, eds, *The Routledge Handbook of Museums, Media and Communication*. London, New York: Routledge, 17–30.

Frank, S. (2010) 'City of the Sun on Ice: The Soviet Counter-Discourse of the Arctic and its Western Equivalents in the 1930s', in J. Schimanski, A. Ryall, and H. Waerp, eds, *Arctic Discourses*. Cambridge: Cambridge University Press, 106–132.

Gavrilova, S. (2019) 'Nasledoe sovetskikh teoreticheskikh i ekspozichionnykh metodov v sovremennykh kraevedcheskikh museiiakh', in A. Zavadsky, V. Sklez, and K. Suverina, eds, *Politika affekta: Muzei kak prostranstvo publichnoi istorii*. Moscow: Novoe literaturnoe obozreniie, 174–194.

Houltz, A. (2013) 'Displaying the Polar Nation: Nordic Museum Exhibits and Polar Ambitions', in S. Sörlin, ed., *Science, Geopolitics and Culture in the Polar Region: Norden Beyond Borders*. Farnham: Ashgate, 293–327.

Ianvarsky, A. (1936) 'Musei Arktiki', *Poliarnaia Pravda* 98(2863), 28 April.

Ivanov, N. (1928) *Podvig vo l'dakh* [film]. Sovkino, directed by Georgii and Sergei Vasiliev.

Kaganovsky, L., S. MacKenzie, and A. W. Stenport, eds (2019) *Arctic Cinemas and the Documentary Ethos*. Bloomington, IN: Indiana University Press.

Lajus, J. (2013) 'In Search for Instructive Models: The Russian State at a Crossroads to Conquering the North', in D. Jorgensen and S. Sörlin, eds, *Northscapes: History, Technology, and the Making of Northern Environments*. Vancouver: University of British Columbia Press, 110–136.

McCannon, J. (1998) *Red Arctic: Polar Exploration and the Myth of the North in the Soviet Union, 1932–1939*. Oxford, New York: Oxford University Press.

McCannon, J. (2007) 'The Commissar of the Ice: The Main Administration of the Northern Sea Route (GUSMP) and Stalinist Exploitation of the Arctic, 1932–1939', *Journal of Slavic Military Studies* 20(3): 393–419.

McGoogan, K. (2002) *Fatal Passage: The True Story of John Rae, the Arctic Hero Time Forgot*. Toronto: Harper Collins; USA: Carroll and Graf.

Muzei Arktiki (1931) *Biulleten' Arkticheskogo instituta SSSR*, 1–2. Leningrad, 14.

Nye, D. E. (1994) *American Technological Sublime*. Cambridge, MA: MIT Press.

Poberezhskaya, M. (2018) 'Blogging about Climate Change in Russia: Activism, Scepticism and Conspiracies', *Environmental Communication* 12(7): 942–955.

Shpanov, Nikolai N. (1930) *Podvig vo l'dakh: dlia detei starshego vozrasta*. Moscow: Molodaia Gvardiia.

Sörlin, S. (2019) 'Cryo-history. Ice, Snow and the Great Acceleration', in J. Herzberg, C. Kehrt, and F. Torma, eds, *Ice and Snow in the Cold War: Histories of Extreme Climatic Environments*. New York, Oxford: Berghahn Books, 20–45.

Sörlin, S. and J. Lajus (2013) 'An Ice Free Arctic Sea? The Science of Sea Ice and Its Interests', in M. Christensen, A. E. Nilsson, and N. Wormbs, eds, *Media and the Politics of Arctic Climate Change: When the Ice Breaks*. New York: Palgrave Macmillan, 70–92.

Sörlin, S. and N. Wormbs (2018) 'Environing Technologies: A Theory of Making Environment', *History and Technology* 34(2): 101–125.

Spufford, F. (1996) *I May be Some Time: Ice and the English Imagination*. London: Faber.

Steinberg, P. E., J. Tasch, and H. Gerhardt, with A. Keul and E. Nyman (2015) *Contesting the Arctic: Politics and Imaginaries in the Circumpolar North*. London: IB Tauris.

Strukov, V. (2021) 'The Arctic on Display: Museums, Art and Haptic Visuality of the North', in M. Lehtimäki, A. Rosenholm, and V. Strukov, eds, *Visual Representations of the Arctic: Imagining Shimmering Worlds in Culture, Literature and Politics*. New York, London: Routledge, 257–276.

Youngs, T. (2010) 'The Conquest of the Arctic: The 1937 Soviet Expedition', in J. Schimanski, A. Ryall, and H. Waerp, eds, *Arctic Discourses*. Cambridge: Cambridge University Press, 132–150.

9

Waiting and witnessing at Larsen C Ice Shelf, Antarctica

Jessica O'Reilly[1]

> I have never in my life seen so much never-ending attention paid to a single iceberg floating around in the ocean.
>
> Mike MacFerrin (@IceSheetMike), Twitter, 30 January 2021

During the Northern Hemisphere winter of 2020–21, in which much of humanity was hiding from the coronavirus pandemic, a huge iceberg travelled up the Southern Ocean's 'iceberg alley' and came perilously close to the coast of South Georgia Island. The huge block of ice could impede penguins' and seals' travel routes from their land nesting and calving grounds to their marine feeding places, scour the diverse seabed floor, and cause freshening in a marine protected area (Gibbens, 2020). The threat gave us something to watch aside from the spiking virus caseload, but eventually diminished as the berg cleaved and shattered in late January or early February 2021 (ESA, 2021). Mike MacFerrin, a glaciologist who regularly posted updates and insights into the iceberg's movements, noted the extreme intensity of attention paid to this chunk of ice in the tweet above. The iceberg, named A-68, came to its end after a closely observed, years long journey that began with a rift on the Larsen C Ice Shelf.

A series of ice shelves wedge up against the land on the east coast of the Antarctic Peninsula, the strip of the continent that points towards the bottom of South America. Named after Norwegian whaling captain Carl Anton Larsen, the four Larsen ice shelves, practically named Larsen A, B, C, and D, sit alphabetically from north to south. As of early 2021, only Larsen D remains intact, as the more northerly ice sheets have disintegrated. Larsen A broke up in 1995. Larsen B collapsed in 2002, in a spectacularly rapid event over the course of six weeks. The glacier behind Larsen B accelerated into the Weddell Sea. This moment became significant among scientists and in popular culture. Larsen B is featured in the opening scene of the 2004 climate disaster movie *The Day After Tomorrow*, two songs, and numerous climate science PowerPoint presentations (O'Reilly, 2017). The expert and popular mediations of the Larsen B collapse set the stage for increased anticipatory scrutiny on the remainder of the ice shelves.

Larsen C's break-up event, which saw about 10 per cent of the ice shelf calve, took longer than its northern counterpart. A stark, 110-kilometre-long crack emerged in 2016, detected by satellites and later by airplane as scientists flew closer to the rift. The precipice of the break-up of Larsen C lasted seven months, as the crack edged towards open water and formed ancillary breaks.

Ice experts tracked the satellite readings closely, looking for signs of the looming break-up. Some did so in public, in real time. People invested in the event were on tenterhooks, especially as Larsen C hung on to the continent by less than three miles of ice for almost a week, before finally calving. The slowness of Larsen C's break-up in contrast to that of Larsen B was an epistemic surprise; since Larsen B had disintegrated so rapidly, the tentative, meek behaviour of its southern counterpart seemed to challenge the idea of accelerating ice loss in the region. Larsen C's hesitation also was one that allowed scientists to ponder extensively over the minute changes visible in the ice and to create a narrative catalogue of speculative possibilities for the imminent break-up of the ice sheet. The spectacle of the iceberg calving, and the breathless reporting on each change ahead of the event, demonstrates how planetary visualizations relate to concerns over environmental change. As ice experts – predominantly based in the United Kingdom, Europe, and the United States – observed and reported on Larsen C's behaviour, they sought to – or fulfilled expectations to – enfold their expert interpretations in broader stories about Antarctic climate change.

How does the translation of expert knowledge fare in near real-time environmental events? In this case, the ice spoke, but slowly, relative to Larsen B's timeline – a meandering, obfuscated sign of environmental change. Nonetheless, the physical, material event of ice calving off of Antarctica has a resonant power unmatched by climate projections more removed from the tangible, earthly register. The eventization of ice happenings is an emerging narrative in cryo-history and climate change generally, including the annual announcement of the Arctic's sea ice minimum (Sörlin 2018; Vardy 2020) To capture this power, experts explained the images and indicators of Larsen C's break-up, producing knowledge that can be repackaged for broader public consumption. In particular, scientists used the Larsen C event as a launching point to highlight examples of Antarctic climate changes.

Puzzling out the spectacle at the icy edge

Climate change, as a major social, economic, political, and environmental problem challenges human notions of scale and time. For example, political time frames are not commensurate with geological time by any stretch and imagining the global scale of atmospheric carbon dioxide levels is so challenging that only powerful models can approximate the complexity and potential futures of this planetary crisis. Scientific notions of nature, too, reify Western concepts of space. For example, the 'ice edge' encountered below promotes the idea that there is a stable interface between solid and liquid water, a fixed boundary between states (Steinberg and Kristoffersen, 2017). The facile use of 'ice edge' hides the transition zones as well as the give and take between frozen and liquid water over various time frames. The three dimensions of the ice shelf are also hidden, glossed over with two-dimensional maps and images, concealing the ice shelf's depth, volume, and architecture, as well as its interaction with the liquid sea under it and at its face.

Connecting perspectives on new materialisms (Coole and Frost, 2010) in feminist and science studies with climate change is what I call 'the substance of climate' (O'Reilly, 2018). Works like Jane Bennett's *Vibrant Matter* repositions material into active roles, engaging with people instead of simply enacted upon by them (Bennett, 2009). New materialisms can inspire how scientists communicate their findings as well, thinking about resonances, mediation, and enactment between facts, experts, and audiences (Coopmans *et al.*, 2014; Messeri, 2017). Ice as a material of climate change offers considerations of its liminality, its role as a repository of history (particularly the history of Earth's atmosphere), and the way it mediates the frontier of anthropogenic climate change (Sörlin, 2015; Antonello and Carey, 2017; Salazar 2018). The shifting ice edge is the front that delimits a powerful imaginary of climate change, that of virtuous nature in contrast to maladaptive human practices. Ice also engages, too, in predictable and surprising behaviours that require expert understanding to explain the signals the ice sends. It holds on longer than expected or breaks off more rapidly than previously observed; it gets stuck and unmoored from pinning points and other topographical features.

Scientific observations of natural phenomena require particular relationships with the environment. These relationships require sensory engagements with nature, either directly through field observation or mediated through technological tools, like satellites or radar. Researchers consider their ability to distil facts from nature as something between skill and gift, enabling scientists to make discoveries (Keller, 1983; Helmreich, 2009; Hustak and Myers, 2012). Knowing how to interpret data – from model outputs to satellite images – is a core component of scientific expertise that sets scientists apart from laypeople, who do not have the necessary knowledge to analyse the raw data at hand. In addition to being an embodied practice where people learn to observe, expertise is also a performance. Expertise enables particular forms of judgement (Lamont, 2009; Oppenheimer *et al.*, 2019), with contestation and scepticism being hallmarks of expert judgement.

Scientific practice is generally a set of skilled work that helps experts reason out a puzzle. Expert judgement under uncertainty – which we can also call expert intuition – is a mix of skill, practice, memory, and learning from previous judgements, if it exists at all. (Kahneman, 2011). This disciplined intuition is honed not just in everyday scientific practice but in moments of epistemic surprise, particularly among 'radical skeptic' scientists (Guyer, 2013: 286). While there have been plenty of forays into understanding how scientists grapple with uncertainties, including the deep uncertainty of things unknown or unknowable, we have paid less attention to what happens when things scientists thought they knew were instead surprising. As Guyer claims, the surprise eventually becomes known through the processes of reason (2013). For the purposes of this chapter, however, we will spend time on the phenomenon of framing science as suspense and surprise and less with the intervening time in which scientists make events reasonable in hindsight. Such framing takes place through the expert interpretation of data: in this instance, the interpretations of airborne and satellite images.

Satellite images and airborne photography imbue the spectacle with a particular aesthetic sense; the images let our eyes glide over the scene as a whole, removed from human scale situatedness at the site. They require expert knowledge to interpret; the image is loaded with codes, colours, and nuance that a casual observer cannot make sense of (Wormbs, 2013). Writing about portrayals of disaster in science fiction film, Sontag describes this as 'the aesthetics of destruction, with the peculiar beauties to be found in wreaking havoc, making a mess' (1965: 44). She continues, explaining that film allows a certain view of disaster, 'a dispassionate, aesthetic view of destruction and violence – a *technological* view' (45). Satellite visualizations, rapidly and partially interpreted in real time on social media, turn scientific observation into suspenseful, filmic events, capturing the sublime beauty of wreckage. The ice itself – moving glacially – does not immediately lend itself to ecological mediation, but instead offers cues for disaster narratives in a broader sense.

Cinematic visual cues are part of a familiar script, evidence of a precipice followed by a powerful disaster. Climate scripts are well-worn; those of us thinking about climate regularly consume violent imagery of floods, drought, death, and societal collapse. News outlets often follow typological and chronological patterns, as Button (2010) has pointed out: there are 'miracle' reports after days of terrifying reporting, interviews with religious leaders on the Sunday following a calamity, and so on. Reporters and experts present news in the form of 'moments' that serve as touchstones in the broader media event of climate change (Christensen *et al.*, 2013). Indigenous resilience to climate disaster is a well-known script, in which consumers are expected to celebrate Indigenous knowledge and innovation while ignoring how colonial power makes 'resilience' the only option as traditional livelihoods are increasingly diminished in rapidly degrading environments (Reid, 2019). Climate scripts also usually have a trodden end point, usually some vague 'hope' or call to action that is increasingly ridiculous in light of the deep problem we face.

Working with and against these scripts, how do experts situate events into the universe of what is known? In reading the signals of an active planet, particularly ice as ecological and political media, the scientists below perform acts of sensing, witnessing, and interpreting environmental change. These expert acts turn spectacles into indicators and surprise into sense.

Reading the precipitous image

An ice rift had been detected on Larsen C since the earliest satellite images of the ice shelf, in 1963 (Glasser *et al.*, 2009). The rift lay fairly static until 2016. Between March and August of that year, observers had not been able to get a look at the rift. While satellite imagery can look through the dark, the cloudiness that goes along with the Antarctic winter, and the Antarctic Peninsula generally, meant that images of the ice shelf were not updated for five months. When the new images came through, the scientists were excited to see that the rift length had expanded 22 kilometres. It looked like the next Larsen disintegration was on its way.

Project MIDAS (Managing Impacts of Deep-Sea Resource), a research team based in the United Kingdom, had been studying Larsen C ahead of the break-up, noting the rapid expansion of the rift in 2014 (Jansen *et al.*, 2015). MIDAS researchers were among the most vocal experts on social media reporting on the calving event. MIDAS drove much of the social media narrative. Adrian Luckman, the Principal Investigator for Project MIDAS, was particularly active. Additionally, Stef Lhermitte, a professor at the Delft University of Technology in the Netherlands (and not affiliated with MIDAS) regularly updated his Twitter page with recent satellite imagery, interpreting changes in the ice shelf and what various rifts indicated for its future.

During the subsequent Antarctic summer season, NASA's Project Icebridge and the British Antarctic Survey flew aircraft over the rift, photographing the jagged line extending out into the white expanse (Figure 9.1). A close-up photograph taken by John Sonntag from Project Icebridge captured the shadows of the rift:

I asked Professor Adrian Luckman to explain what the photograph depicted, as one side looked like a sheer edge and the other slumping and disorderly.

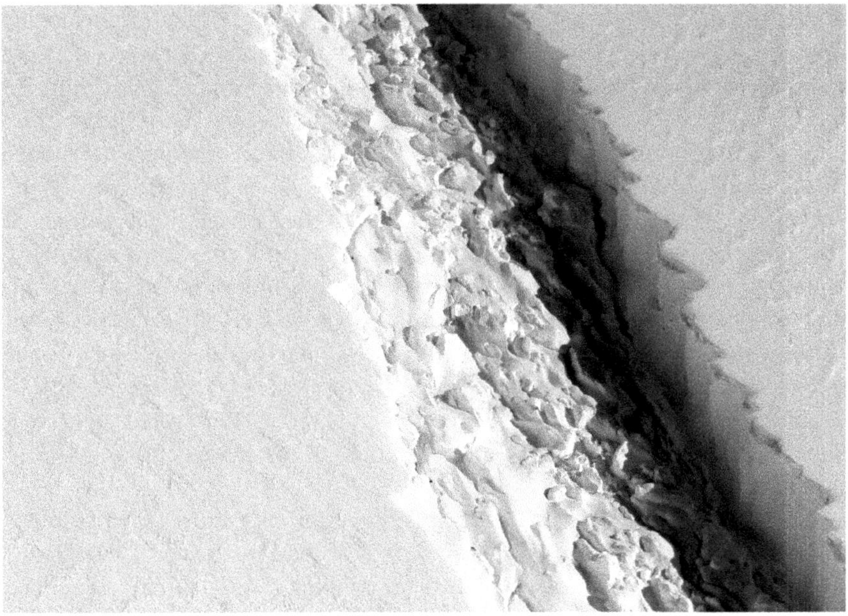

Figure 9.1 Project Icebridge overflight of the Larsen C rift. Grayscale added by author.

My inexpert interpretation of the image was wrong; I did not know how to interpret the photographed image with my minimal Antarctic icefield experience. Technology renders new patterns of visuality that can eclipse *in situ* expertise. Aporta *et al.* (2005) noticed this as they wrote about Inuit wayfinding changes with the introduction of GPS devices – while the technology could draw a straight line to a spot, it couldn't read the vagaries and risks of travel over the ice. Luckman explained to me:

> This picture presents a bit of an illusion. What you are seeing is the ice shelf, which is quite flat and featureless, cut through by the rift, which is full of sea water on which is floating a fairly flat mess of ice and snow. There is no slump, although the lighting makes it appear so. You can see one vertical edge of the rift on the right which is casting a shadow onto the floating ice. The other vertical edge is not visible because of the angle of the shot but starts where the featureless ice shelf ends on the left. The vertical 'freeboard' from the shelf top to the floating ice melange, is probably 30 or 40 meters, showing the full ice thickness is 300–400 m, so most of the ice shelf thickness is hidden (personal communication, 28 February 2020).

Luckman's countless hours poring over images of the rift from aircraft and satellite allowed him to explain how the angle of the photograph revealed texture on one side of the crack while the shadow covers the further details on the right. Importantly, explaining the scale of the photograph – that the dark, icy water visible in the rift is about 90–120 feet below the surface – helps make spatial sense of the white, often featureless Antarctic expanse.

The rift continued to lengthen during the summer season. In January 2017, Project MIDAS observers updated their map twice to depict the lengthening rift as it approached the far side of the ice shelf. Luckman noted that Larsen C was now 'poised to calve'. My Antarctic environmental NGO friends began preparing press briefings for the imminent event, sending emails with drafts to edit as we waited for the calving to occur. MIDAS, the European Space Agency (ESA), the National Aeronautics and Space Administration (NASA), and the National Snow and Ice Data Center (NSIDC) all had press releases ready to go immediately after the event, suggesting similar activity elsewhere. They prepared their public statements and waited for the ice to move.

In mid-February, Larsen C observers noted another short lengthening of the rift, but then, as the Antarctic summer came to a close, extensions ceased. The length held steady. However, while the ice shelf still remained connected to the continent, the rift itself began to widen, upwards of a metre a day. In May, observers noticed that while the rift had not lengthened, a new branch had formed, and after the branching the widening accelerated. The shelf was floating out to sea even as it remained attached at the north. Synthetic Aperture Radar (SAR) interferometry from the European Space Agency's Sentinel-1 satellites clearly depicted the acceleration of ice surface speed as the massive future iceberg moved (Figure 9.2). The rainbow shading of the satellite data emphasized the speeding up near the rift as the rest of the ice shelf flowed more slowly behind it.

In response to this accelerating movement, the ice near the rifting edge began to slump, creating features so large they were visible on the satellite images. Both the iceberg side and the ice shelf side slumped as the edges lost contact with each other. As the MIDAS scientists explained, 'Larsen C begins to react to the rift', as the ice responded to the changing landscape by filling in newly empty space.

The update on 31 May was a nail biter. The rift had lengthened again, without detaching. The rift was now 13 km (about 8 miles) from the edge. Luckman wrote, 'the rift has now fully breached the zone of soft "suture" ice originating at the Cole Peninsula and there appears to be very little to prevent the iceberg from breaking away completely' (Luckman *et al.*, 31 May 2017).

In June, the berg was clinging onto the peninsula at the northern tip as the ice surface speed of the detaching ice continued to accelerate, tripling in speed as it moved away from the peninsula. Towards the end of June

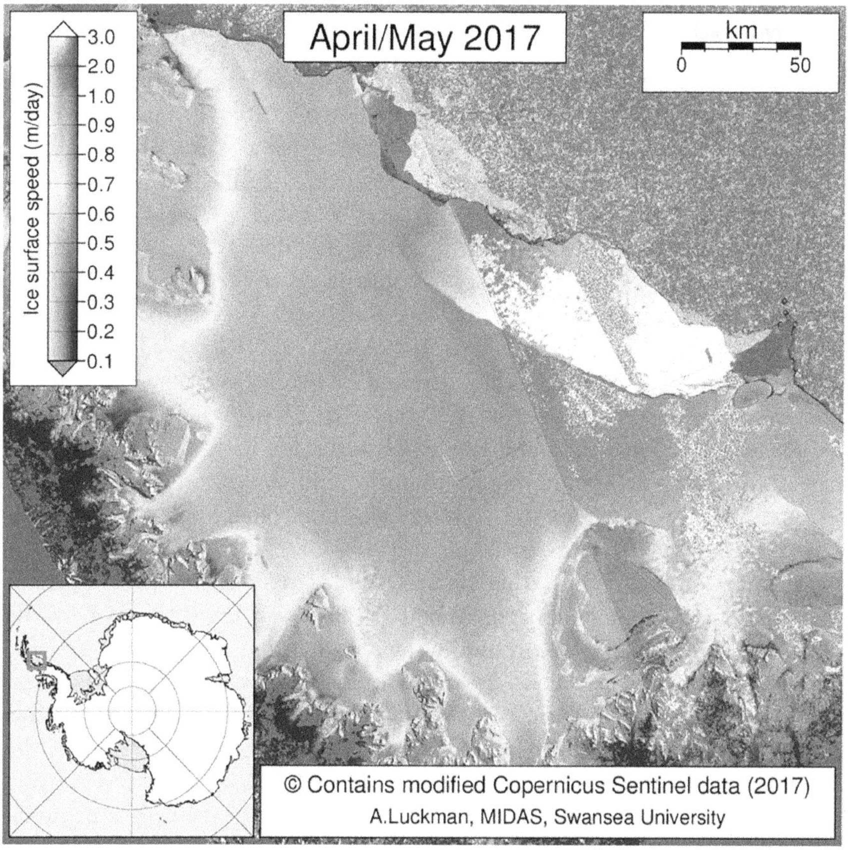

Figure 9.2 Synthetic aperture radar interferometry from the European Space Agency's Sentinel-1 satellites. The tones near the centre top show where the future iceberg's speed is accelerating away from the Antarctic Peninsula.

and into the beginning of July, the rift tip began to branch, fracturing in multiple directions as the attachment weakened. By 6 July, the ice was holding on with a connection of just 4.5 km, or 2.1 miles.

Sometimes the ice was presented as a character, personified and projecting the human experiences of observers. 'It's stressful being an iceberg hanging on by a thread', wrote Brian Kahn for the LiveScience blog (8 July 2017), capturing the anticipation of the ice shelf watchers along with the stress-related rifts observed in the days ahead of calving.

On 12 July, Landsat – Land Remote-Sensing Satellite (System) – data, compiled into images by NASA Earth Observatory's chief visualizer, geographer Joshua Stevens, revealed that the Larsen C iceberg had finally detached.

The images, taken over the dark, midwinter Antarctic Peninsula, illuminate the warmer liquid water in between the colder ice in iridescent rainbow shades. The thermal signature of the water is finally clearly visible through the entire rift. The visibility of water in between the rifting sides of ice allowed scientists to more precisely measure the progression of rifting rate.

The iceberg, named A-68 by the United States NSIDC, was massive: 1 trillion tons of ice. On social media and in press releases, this almost unimaginable size was translated into more familiar spaces, often related to the nationality of the reporting. The US-based environmental NGO the National Research Defense Council compared it to the state of Delaware. The European Space Agency noted that A-68 was 'twice the size of Luxembourg'. CGTN (China Global Television Network), the state-run Chinese television network compared it to the size of Shanghai. Project MODIS (Moderate Resolution Imaging Spectroradiometer), with Welsh leadership, had anchored their update images to scale with Wales, and described the volume of the iceberg as twice that of Lake Erie, one of the smaller Great Lakes in the United States.

A major finding of the earlier Larsen B break-up was the speeding up of the ice on land that had been pinned behind, and then released by the collapse of that ice shelf. And unlike the ice shelf that already was floating over water, ice from the land contributes to increased sea levels. Researchers continue to study the response of this ice after the calving event, and at this point, the ice behind A-68 has accelerated, but not at the rate from the Larsen B event (Hogg *et al.*, 2017). Instead, scientists have observed a rate of about 0.1m/ day faster movement, an increase of 3–5 per cent (Luckman, personal communication, 28 February 2020). This ice moving is still remnants of the ice shelf, so the response of land-based ice has not yet been detected.

Throughout the calving event, experts pointed out and read indicators of ice shelf behaviour, communicating suspense. Expert discernment, along with occasionally transforming the ice into a personified character, amplified the experience of ice-watching as something precipitous.

The spectacle of ice, the climate ask

The potential dramatic spectacle of ice shelf collapse is an attractive script for depicting environmental change. While the calving event of Larsen C ended up being agonizingly slow compared to the rapid break-up of Larsen B the decade before, it nonetheless contained the idea of the ground disintegrating beneath our feet, an apocalyptic consequence of our behaviour. Reporters covering the 'Antarctic beat' – science and environment writers – spent several months of 2017 with laser-like focus on Larsen C. However, other parts of the Antarctic continent were sending out their own, more urgent, and clear climate signals. Thwaites Glacier – now called the Antarctic 'canary in a coal mine' repeatedly enough to be considered a conventional trope – has been rapidly thinning and accelerating towards the sea. And since Thwaites is on land, this behaviour does ultimately raise global sea levels. Attention stayed over at Larsen C anyway, even though some of the scientists interviewed encouraged their readers and listeners to places of more significant Antarctic climate activity.

Taken on its own, the breaking off of an ice shelf is not an indicator of climate change: it is typical ice behaviour. Ice flows slowly, it is viscous, and it interacts with the topography and water in which it is situated. An ice shelf sits over water, bound to a coast. It does not contribute a significant amount to global sea levels, as the ice is already part of the sea: a change in its stage from solid to liquid does not add or subtract water from the seas. However, ice shelves often act as plugs, holding back continental ice – in smaller, flowing terms a glacier, and as a system of glaciers, an ice sheet. When the ice shelf disintegrates, the land ice behind it may (but not always) flow more rapidly into the ocean. When ice on land flows into the ocean, this does eventually raise global sea levels – and when snow accumulates on top of ice sheets, sea level decreases.

While the dynamics of ice shelves breaking is part of typical ice sheet behaviour, it is worth noting that each of the Larsen shelves have been sequentially more southern. Attached to the eastern side of the Antarctic Peninsula, the region is the most rapidly warming part of the Antarctic. Marambio base, an Argentinean research station, reached a record high Antarctic temperature of 20.75°C on 9 February 2020, flowering plants have increased their ranges, and penguin species, who inhabit specific temperature bands, have shifted their colonies south to resituate themselves in the microclimate they find comfortable (Turner *et al.*, 2009). People living and working in Antarctica are increasingly concerned about – and finding evidence for – environmental tipping points in Antarctic natural systems. However, this concern stands in contrast with popular portrayals of climate catastrophes, which most scientists carefully distance themselves from, to avoid their portrayal as biased or 'alarmist' (Brysse *et al.*, 2013).

In media accounts, talking about climate in relation to Larsen C seemed to be irresistible even though the causality does not exist or is at least quite weak. While the ice movements at Larsen C were less significant scientifically and in terms of environmental impact, the attention given to that moment could be parlayed into broader conversations about the Antarctic ice sheets and climate change. The scripts and surprises that Larsen C afforded demonstrate an impulse to think about climate change as a precipice, a tipping point, a waking giant, or an alarm signal. It is captivating and terrible, so much livelier than model projections or cautious scientific talk about carbon budgets and climate scenarios. Larsen C displayed the aesthetics of disaster in its slow rifting. That the narrative inertia of Antarctica has shifted to climate, too, means that scientists expect a 'climate ask' from the public. Experts here tried to explain the Larsen C phenomenon carefully, delineating between natural and anthropogenic change and redirecting towards other Antarctic sites more responsive to climate change.

Conclusion

Everything we know about the calving of A-68 from Larsen C had to be mediated and interpreted from captured images of the ice by experts. These experts in turn had to translate their knowledge about the textures and shadows on the images alongside their experiences on the ice itself. Looking over the Antarctic horizon, the landscape stretches out blank, white, and flat. Travelling over that expanse, however, is perilous, broken up by crevasses hidden by ephemeral bridges of snow.

The transience of ice helps people faraway see the Antarctic as material, tangible, and present. Events like the calving of Larsen C and its neighbours to the north animate notions of a dramatically changed planet in response to human activities, even though ice shelf collapse is not correlated to anthropogenic warming. The spectacle drives a script of a planet responding to our carbon combustion that is more powerful than careful and accurate climate models.

Antarctic scientists work with the substance of climate change, living it in, turning it into data, and situating their research into the bigger stories of climate change. Larsen C, disappearing penguin colonies, and charismatic discoveries, achievements, and tragedies manage to cast some public attention towards the bottom of the world. Scientists (and spokespeople for national Antarctic programmes) are increasingly savvy at leveraging Antarctic events into opportunities to communicate about the continent.

The collapse of the Larsen C Ice Shelf made apparent the leaps in satellite visualization, even in the intervening decade since the collapse of Larsen B, as scientists, reading satellite images, were able to show us how the event was unfolding in near real time, in an increasingly dark and cloudy landscape. Improved, publicly available visualizations let us be spectators from our phones and computers, witnesses to a major icescape change. Expert narration through social media and journalism helped translate the difficult to scale and challenging to understand images available.

There is a tangible and tactile immediacy to massive chunks of ice breaking off the continent; these moments provide an opportunity to fuse

Hollywood-style disaster scripts with scientific expertise. However, these dramatic moments are not really what climate change looks like or feels like: it is a slow onset disaster of already marginal farms no longer arable, of conflict and migration over increasingly scarce water, of disease vectors moving readily into places that used to be too cold. And the spectacle of Larsen C is not easily correlated to anthropogenic climate change. However, this icy moment did become a powerful vantage point from which to recognise that ice will continue to be responsive to ongoing and intensifying environmental change.

Note

1 The author thanks Klaus Dodds, Steff Kane, and Ilana Gershon for comments on chapter drafts, and Adrian Luckman for his ice expertise.

References

Antonello, A. and M. Carey (2017) 'Ice Cores and the Temporalities of the Global Environment', *Environmental Humanities* 9(2): 181–203.

Aporta, C., E. Higgs, D. Hakken *et al.* (2005) 'Satellite Culture: Global Positioning Systems, Inuit Wayfinding, and the Need for a New Account of Technology', *Current anthropology* 46(5): 729–753.

Bennett, J. (2009) *Vibrant Matter: A Political Ecology of Things*. Durham, NC: Duke University Press.

Brysse, K., N. Oreskes, J. O'Reilly, and M. Oppenheimer (2013) 'Climate Change Prediction: Erring on the Side of Least Drama?', *Global Environmental Change* 23(1): 327–337.

Button, G. (2010) *Disaster Culture: Knowledge and Uncertainty in the Wake of Human and Environmental Catastrophe*. Walnut Creek, CA: Left Coast Press.

Christensen, M., A. E. Nilsson, and N. Wormbs, eds (2013) *Media and the Politics of Arctic Climate Change: When the Ice Breaks*. New York: Springer.

Coole, D. and S. Frost (2010) *New Materialisms: Ontology, Agency, and Politics*. Durham, NC: Duke University Press.

Coopmans, C., J. Vertesi, M. E. Lynch, and S. Woolgar, eds (2014) *Representation in Scientific Practice Revisited*. Cambridge, MA: MIT Press.

European Space Agency (ESA) (2021) *Is This the End of the A-68A Iceberg?* European Space Agency, 3 February. Available at: www.esa.int/Applications/Observing_the_Earth/Is_this_the_end_of_the_A-68A_iceberg (accessed 21 March 2022).

Gibbens, S. (2020) 'Huge iceberg breaking up off South Georgia Island is still a threat', *National Geographic*, 28 December. Available at: www.nationalgeographic.com/environment/2020/12/world-largest-iceberg-a68-heading-for-south-georgia-wildlife/#close (accessed 21 March 2022).

Glasser, N. F., B. Kulessa, A. Luckman *et al.* (2009) 'Surface Structure and Stability of the Larsen C Ice Shelf, Antarctic Peninsula', *Journal of Glaciology* 55(191): 400–410.

Guyer, J. I. (2013) '"The Quickening of the Unknown" Epistemologies of Surprise in Anthropology: The Munro Lecture, 2013', *HAU: Journal of Ethnographic Theory* 3(3): 283–307.

Helmreich, S. (2009) *Alien Ocean: Anthropological Voyages in Microbial Seas*. Berkeley, CA: University of California Press.

Hogg, A. E. and G. H. Gudmundsson (2017) 'Impacts of the Larsen-C Ice Shelf Calving Event', *Nature Climate Change* 7(8): 540–542.

Hustak, C. and N. Myers (2012) 'Involutionary Momentum: Affective Ecologies and the Sciences of Plant/Insect Encounters', *Differences* 23(3): 74–118.

Jansen, D., A. J. Luckman, A. Cook *et al.* (2015) 'Brief Communication: Newly Developing Rift in Larsen C Ice Shelf Presents Significant Risk to Stability', *Cryosphere* 9(3): 1223–1227.

Kahn, B. (2017) *The Larsen C Iceberg is Expected to Have Company*. LiveScience, 8 July. Available at: www.livescience.com/59732-larsen-c-iceberg-swarm-expected.html (accessed 21 March 2022).

Kahneman, D. (2011) *Thinking, Fast and Slow*. New York: Macmillan.

Keller, E. F. (1983) *A Feeling for the Organism*. San Francisco: Freeman.

Lamont, M. (2009) *How Professors Think*. Cambridge, MA: Harvard University Press.

Luckman, A. (2020) Personal communication with author, 28 February.

Luckman, A., M. O'Leary, and Project MIDAS (2017) 'Larsen C Takes Another Step Towards Calving', Project MIDAS blog (now offline), 31 May.

Messeri, L. (2017) 'Resonant Worlds: Cultivating Proximal Encounters in Planetary Science', *American Ethnologist* 44(1): 131–142.

Oppenheimer, M., N. Oreskes, D. Jamieson *et al.* (2019) *Discerning Experts: Understanding Scientific Assessments for Public Policy*. Chicago, IL: Chicago University Press.

O'Reilly, J. (2017) *The Technocratic Antarctic: An Ethnography of Scientific Expertise and Environmental Governance*. Ithaca, NY: Cornell University Press.

O'Reilly, J. (2018) 'The Substance of Climate: Material Approaches to Nature under Environmental Change', *WIREs Climate Change* 9(6): e550.

Reid, J. (2019) 'Narrating Indigeneity in the Arctic: Scripts of Disaster Resilience Versus the Poetics of Autonomy', in N. Sellheim, Y. V. Zaika, and I. Kelman, eds, *Arctic Triumph*. Cham, Switzerland: Springer, 9–21.

Salazar, J. F. (2018) 'Ice Cores as Temporal Probes', *Journal of Contemporary Archaeology* 5(1): 32–43.

Sontag, S. (1965) 'The Imagination of Disaster', *Commentary* 40(4): 42–48.

Sörlin, S. (2015) 'Cryo-history: Narratives of Ice and the Emerging Arctic Humanities', in B. Evengård, J. N. Larsen, and Ø. Paasche, eds, *The New Arctic*. Cham, Switzerland: Springer, 327–339.

Sörlin, S. (2018) 'Cryo-history: Ice, Snow and the Great Acceleration', in J. Herzberg, C. Kehrt, and F. Torma, eds, *Ice and Snow in the Cold War: Histories of Extreme Climatic Environments*. New York, Oxford: Berghahn Books, 20–48.

Steinberg, P. and B. Kristoffersen (2017) '"The Ice Edge is Lost … Nature Moved It": Mapping Ice as State Practice in the Canadian and Norwegian North', *Transactions of the Institute of British Geographers* 42(4): 625–641.
Turner, J., R. Bindschadler, P. Convey *et al.* (2009) *Antarctic Climate Change and the Environment*. Cambridge: Scientific Committee on Antarctic Research.
Vardy, M. (2020) 'Relational Agility: Visualizing Near-real-time Arctic Sea Ice Data as a Proxy for Climate Change', *Social Studies of Science* 50(5): 802–820.
Wormbs, N. (2013) 'Eyes on the Ice: Satellite Remote Sensing and the Narratives of Visualized Data', in M. Christensen, A. E. Nilsson, and N. Wormbs, eds, *Media and the Politics of Arctic Climate Change*. London: Palgrave Macmillan, 52–69.

Part III

Thinking with ice

10

Imperial slippages: encountering and knowing ice in and beyond colonial India

Thomas Simpson

Most Europeans didn't anticipate encountering ice during their time in India. If they did come into contact with it, they struggled to get any purchase on this slippery stuff epistemologically as well as physically. This chapter makes forays onto three distinctive types of ice in order to understand how, why, and to whom frozen matter mattered in the nineteenth-century subcontinent and its mountain fringes. The first form was the least durable: the consumable ice of lowland northern India. This substance congealed a diverse mixture of globe-spanning trade connections, everyday practices of racialized sociability, and innovative enquiries into atmospheric physics. Europeans encountered the second variant considered in this chapter during the first half of the nineteenth century as they began to penetrate portions of the high Himalaya bounding Britain's Indian Empire to the north. I focus on conjoined processes of experiencing and understanding ice, both of which were shaped by Asian actors and knowledge even as they related to distinctively Euro-Western dynamics. The third kind of ice was the expansive and deep, moraine-laden and crevassed masses in the still-higher Karakoram and borderlands of Sikkim and Tibet. On the world's largest extra-polar glaciers during the late nineteenth and early twentieth centuries, timescales and processes of high imperialist geopolitics collided with those of geophysics as the global 'Little Ice Age' drew to its regionally uneven close (Hewitt, 2014: 300).[1] This ice became 'vital matter' for performing race and masculinity, extending the reach of imperial influence, and developing conjectures of climate change (Bennett, 2010).

A number of core themes span these distinct substances, locales, and junctures. First, Europeans configured consuming, climbing, and categorizing ice as markers of racial and gendered distinction. However, these activities relied upon Asian people and infrastructures – and entailed physical and reputational risks – in ways that could make a mockery of elitist self-assertions. Second, multiple temporalities and types of motion coalesced and clashed on the ice, ranging from humans moving in different spaces and different rhythms, to the assumed actions of powerful and mysterious glaciers and gods, to instruments and texts precariously generating and bearing information. In this flux, knowledge and practices of power might congeal for a time; but only on rare occasions, and with difficulty, could they be made to stick durably.[2] Finally, all forms of Asian ice appeared sufficiently distinctive to many Europeans to prompt or substantiate a vast range of scientific investigations and theories across natural and physical sciences. Difficulties in getting ice – both the actual substance and representations – to travel well enough to substantiate knowledge claims provided opportunities as well as problems to imperial agents and their interlocutors.

A fluid commodity

Like the majority of his European contemporaries who spent time in a subcontinent increasingly dominated by the British East India Company, the Commander-in-Chief of the Company's army in the early 1770s, Robert Barker, 'never saw any natural ice … in that quarter of the globe'. Yet, he reported, 'it was usual to collect a quantity [of ice] every morning before Sun-rise' at sites across northern India during the cold season. Taken 'to the grand receptacle or place of preservation … it is beat down with rammers, till at length its own accumulated cold again freezes and forms one solid mass' (Barker, 1775). Consumable ice had arrived in this region with the Mughal Empire 250 years earlier. Accustomed in Kabul to a ready supply of chilled beverages, the dynasty's founder Babur complained that there was 'no ice or cold water' in his newly conquered territories (Babur, 2006: 131 and 275). Among the Mughals' imports were the technologies that Barker saw – earthenware pans in which ice formed and 'houses' where it was sequestered. By the turn of the seventeenth century, the imperial court also enjoyed ice brought down from the Himalaya year-round (Eraly, 2007: 66).

Barker's account in *Philosophical Transactions*, the prestigious journal of London's Royal Society, marked the moment that an Indian luxury started to become a significant component in European theories of atmospheric physics. The methods and materials of ice manufacturing when the ambient temperature remained above freezing were, Barker stated using a keyword of European natural philosophy of the eighteenth century, a 'discovery' – but one distinctive to 'the Asiatic (whose principal study is the luxuries of life)' (Barker, 1775: 256). Barker may have qualified his insistence on the scientific significance of Indian ice making with a pejorative slur, but over the following century the British came to consider this frozen matter good for drinking and for thinking.[3] As an antidote to atmospheric heat and fever, it was perceived crucial to the health of white bodies (Harrison, 1999). In her widely read memoirs of colonial life between the 1820s and 1840s, Fanny Parkes claimed possessively that 'ice is our greatest luxury' (Parkes, 1850: I: 83–84). Corporeal anxieties underlay increasingly urgent British efforts to control the production and distribution of Indian ice. Existing manufactories came under European management, and a slew of new ones opened. One such occurrence, the entry in the early 1830s of the British doctor Thomas Wise into ice making at Hugli in Bengal, prompted a burst of intensive discussion about ice among the European community of Calcutta. The short-lived popular scientific periodical *Gleanings in Science* and more highbrow *Journal of the Asiatic Society of Bengal* published articles combining primers on Euro-Western ideas of ice formation in the wake of Barker's article alongside practical pointers for readers to instruct their *aubdar*s (beverage-cooling servants) in the art of chilling wine (G., 1831; Anon, 1832a; Anon, 1832b; Dickason, 1991: 67). The process of ice formation and the pleasures of its consumption were equal sources of fascination.

While drawing in their explanations of ice in lowland India on prestigious works such as the American William Wells's 1814 *An Essay on Dew* (1866), scholars in the colony tended to claim that the frozen matter they had seen, touched, and ingested undermined universalizing theories. We might say that knowledge derived from the West's ice was 'provincialized' in India (Chakrabarty, 2008), as ostensibly global explanations faltered in the face of regionally distinctive icy materials and practices. Uncertainty and amazement regarding congelation in a 'tropical' region worked in tandem with British eyewitnesses' thick descriptions of the vessels, people, and environs of ice making, which emphasized minute variations in circumstances between sites (Williams, 1793; G., 1831; Parkes, 1850: I: 78–82). Thomas Wise invoked three knowledge traditions in producing and explaining his 'Hugli ice'. He drew on North Indian technology and terminology, including the conceptual distinction between solid ice (*pakka baraf*), thinner ice that began to congeal later into the night (*paperí*), and the thinnest ice that congealed only shortly before dawn (*phúl baraf*). He also recounted undertaking 'experiments' that affirmed Wells's key claim (1866: 108–13), in opposition to Robert Barker's argument (which had been taken up by metropolitan luminaries including Humphrey Davy in the meantime), that ice formed by radiation rather than evaporation. Finally, he found a place for his situated knowledge of Hugli's climate and environment, placing great significance on the direction of winds and 'the influence of soil and the elevation of the dry ground, its inland situation and free exposure to the sky' (Wise, 1833). In Wise's icy theories and practices, Indian artisanal practice, Euro-Western natural philosophy, and European residents' experience of the colonial subcontinent mutually informed and transformed each other. Here was one instance among many in which Indian and European ice did not necessarily seem to be the same stuff.

Within months of the publication of Wise's account, an interloper threatened his product. In September 1833, Calcutta's Diamond Harbour received 'clearest crystal ice' that had departed Boston four months earlier. The Asiatic Society gleefully reported that 'about one hundred tons' – a similar quantity to Wise's annual production in a good year at Hugli (Anon., 1832a) – had made it 'from one side of the globe to the other, crossing the [equator] line twice'. Backed by the capital of Boston businessman Frederic Tudor, who had already established a profitable ice trade to the southern United States and Caribbean, the American upstart seemed to melt into air both 'Hugli ice' and a venture in Calcutta to freeze water using steam pumps (Anon. 1833, 491). Even Fanny Parkes, whose husband had managed an ice-manufactory at Allahabad, judged the 'fine clear ice from America, sent in enormous blocks' to be 'greatly superior' to the 'beaten up' stuff made in the subcontinent (Parkes, 1850: I: 82). The combination of 'an insulated house' on his ships and an 'ice-house' for its 'reception and preservation' in Calcutta enabled Tudor's product to avoid the fate of an earlier shipment of American ice arranged by residents of Calcutta (Anon., 1833: 492; Anon., 1835). In the words of one periodical, on this previous occasion 'the commodity became fluid' (Anon., 1835); by contrast, Tudor's commodity seemed to *flow* primarily in the sense of '"pure" movement' that anthropologist Stuart Rockefeller has recently shown to be vital to the self-perception of global capitalism (Rockefeller, 2011: 558).

Despite swiftly becoming established as 'a regular trade' (Anon., 1835), Tudor's ice was not entirely frictionless.[4] In transit it 'melted and subsided bodily' between each individual block within the stack, rather than only from its exterior surfaces as had been anticipated, losing around a third of its original mass (Anon., 1833, 492). Once in India, American ice initially shocked many of those who experienced it. Much to the surprise of Westerners engaged in the trade, some of these sticking points involved ice and its associated paraphernalia being hot. Some Hindus avoided ice according to the Ayurvedic understanding of it as a substance that heated the body while children reportedly experienced a burning sensation when touching it with 'their nearly paralyzed little fingers' (Grant, 1862: 36–37; Dickason, 1991: 73). Moreover, in 1836 the dried bark from tanners' pits used to insulate the Calcutta ice-house fermented when it became wet, almost catching alight and releasing gases that killed a man who ventured into the chamber (Anon., 1836).

Notwithstanding these conceptual and corporeal difficulties, by the 1860s 12,000 tons of ice annually made it intact from New England to entrepots across the subcontinent (Dickason, 1991: 73–75). 'Like most other conveniences which habit renders familiar', as one British resident of Calcutta put it, ice became an expected rather than exceptional element of affluent life across communal lines in the major cities of colonial India (Grant, 1862: 36–37). For four decades, Boston ice enhanced racialized and class-bound cultures of sociability that had already developed around locally produced ice, especially British attempts to keep the best quality product for themselves and away from supposedly perfidious servants' attempts to sell it on to Indian consumers (Parkes, 1850: I: 81). Then, from the 1870s, it rapidly receded owing to a combination of the advent of artificial refrigeration techniques in South Asia and milder winters accompanying the end of the Little Ice Age in New England. Ice's decline in the northeastern United States resulted from a combination of natural cycles and coal-fired heavy industry – much like the carbon-intensive ice-manufactories that subsequently spread through the colonial and postcolonial subcontinent (Dickason 1991: 79–83).

Making (it) up glaciers

British encounters with Himalayan ice resulted from aggressive imperialism, but at the same time were thoroughly reliant on Asian personnel, infrastructure, and knowledge.[5] East India Company soldier-surveyors ventured into Kumaon and Garhwal (present-day Uttarakhand) from 1807 onwards to locate the sources of the River Ganges while also countering Nepalese expansion and, subsequently, establishing colonial authority. Their first encounter with ice came as they travelled upstream along the banks of the two main branches of the Ganges in the area, the Bhagirathi and the Alaknanda. 'The bleak aspect of the country', the lead officer reported, 'gave the appearance and sensation of the depth of winter, in a much more northern latitude' (Raper, 1810: 523). While inimical to the widespread European notion that latitude and climate were analogous, the region's ice had been incorporated into Hindu writings and rituals (Colebrooke, 1810; Raper, 1810: 483–506). Although the British surveyors relegated geographical information contained in the shastras and puranas (ancient Hindu texts) to 'fable' and the 'entirely fabulous' (Raper, 1810: 484 and 524), sacred toponyms and pilgrims' paths to monasteries and spiritually significant sites structured their exploration.

In 1817, a British surveyor, John Hodgson, managed to get beyond the Gangotri monastery and onto what he termed the 'grand snow bed' from which the Bhagirathi issued. The vertical variability of this ice presented a particular challenge.[6] Rather than a solid mass, Hodgson found it a 'hollow and treacherous compound of snow and rubbish' into which he and members of his party 'sunk up to [their] neck' (Hodgson, 1822:

117–120). Comprehending the Bhagirathi's frozen source proved as arduous as traversing it. He speculated that 'deep rents' in the formation might result from earthquakes or hot springs, but also turned to Hindu nomenclature to try to get a conceptual grip on the ice (Hodgson, 1822: 127–128). Like the Western explorer-scientists in another Asian borderland some decades later analysed by anthropologist Erik Mueggler, imperial surveyors in the Himalaya found that 'the visions of others inflected their experience' (2011: 18). At the suggestion of a high-caste monk who accompanied him from the monastery to the river's source, Hodgson thought it 'a good idea' to label the 'icicles' descending over the arch from which the river emerged 'Mahadeva's [i.e., the god Shiva] hair, from whence as [the Brahmin] understood, it is written in the Shastra, the Ganges flows' (Hodgson, 1822: 117–118). Even though he journeyed further up the Bhagirathi than other British travellers had managed over the preceding decade, he followed in their footsteps in paying heed to Shiva's presence in the icescape. The surveyor William Webb had previously labelled the mountain above Gangotri 'Mahádéva-ca-linga' on the advice of his Indian assistant. And the Scottish traveller and artist James Baillie Fraser credited obtaining a clear view of the same peak to bathing in the river by the monastery in the fashion of a Hindu pilgrim, after which 'the day cleared up' – an event 'pronounced to be a very special sign of favour from the mighty deity there enthroned' (Fraser, 1820: 479). Even as these visual representations forced Himalayan ice into the ambit of imperial knowledge, other substantive and imaginative qualities of frozen water remained influential.

Three decades after Hodgson's '2½ hours hard labour of wading, and floundering' on and in the Gangotri ice (Hodgson, 1822: 121), it was discovered that he had been the first European to discover a Himalayan glacier.[7] Hodgson had neither made it up the glacier to its head, nor, with his lexicon of snow and shastras, had he quite made up a glacier. But, in 1847, the soldier-explorer Richard Strachey reproduced and annotated extensive excerpts of Hodgson's published description, which, Strachey asserted, 'clearly prove[d]' that the Bhagirathi's source was a glacier (Strachey, 1847: 807). Reinterpreting sloppy 'snow-beds' as robust bodies of ice was entangled with intensified efforts during the late 1840s to project imperial power and knowledge into what administrator George Traill had described only fifteen years earlier as the 'inaccessible … interior of the Himalaya' (1832: 3). The concept of durable glaciers emerged alongside British attempts to create durable boundaries in the western Himalaya, with the same personnel and survey instruments – theodolites and stakes planted in rock and ice – responsible for both (Fleetwood, 2019: 131–133; Gardner, 2021: 71–75).

The cryospheric conjunction of science, sovereignty, and the sacred that emerged in the mid-nineteenth century has enduring significance in the postcolonial period: for instance, the first Indian expedition to the Antarctic in 1981–82 and the research station it established were both codenamed 'Gangotri' (Qasim, 1983). In the mid-nineteenth century Himalaya, however, making the term 'glacier' stick to icy agglomerations was as laborious as ascending from snout to névé. Glaciers of the European Alps were well known and widely discussed in scientific and artistic circles by the early nineteenth century, but it was only from the mid-1840s that Strachey and others transplanted the concept to High Asia. As historians Lachlan Fleetwood and David Arnold show, this belatedness was due in part to the conceptual difficulty of squaring permanent masses of ice with India's supposedly 'tropical' climate, thought inimical to substantial and permanent masses of frozen water (Arnold, 2005a; Arnold, 2005b: 99–104; Fleetwood, 2019: 129–134). Also important was the difficulty of getting ideas of ice to move, owing to the substance's material qualities as well as the dynamics of scientific institutions and publications in colonial India. Unlike the fossils they prised from Himalayan rocks, the flowers and plant cuttings they pressed between paper, and animal skeletons and skins they preserved and packaged, sending samples of ice over long distances was not viable for mid-nineteenth-century scientific explorers. The tools and techniques of natural historical collection were nonetheless key elements in affirming the existence of glaciers in the Himalaya.[8] Edward Madden, for example, carried 'a hatchet' to the Pindari Glacier in Kumaon in 1846 with which he 'frequently broke off fragments' of the ice. These shards were put in motion not to Calcutta or London, but only as far as Madden's eyeline. Thus separated from the vast congelation beneath, he compared them with his recollections of ice in the European Alps: 'the only difference perceptible, or that I can remember, … being a coarser granular texture here' (Madden, 1847: 257–258).

Madden's account of his engagement with Himalayan ice featured two aspects that were crucial to conceptualizing glaciers in the 1840s. First, like knowledge of ice in alpine Europe (Hevly, 1996), explorer-scientists tended to represent discovering glaciers in High Asia as a heroic venture entailing mastery over various elements: one's own body and mind, an instrumental repertoire, the ice itself, and the other men in the exploration party. But ideologies of race set apart knowledge production in the imperial borderlands of the Himalaya. Some Europeans acknowledged the critical roles of Asians as informants, guides, and porters, but others occluded them and resorted to pejorative stereotyping (Madden, 1847: 249; Fleetwood, 2019: 133–134). Violence and reliance often intermingled conspicuously in harsh icy environs, as when the soldier-explorer Alexander Gerard 'pelted' a crucial member of his retinue 'with a few snow-balls' after the latter threatened to abandon the party unless it turned back from an arduous glacial pass (Gerard, 1841: 242). Second, Madden's emphasis on his recent experience of 'the Mer de

Glace and several glaciers of Switzerland' (1847: 252) was one of many instances in which European ice constituted the elemental and epistemological standards by which to assess the Himalayan variety. Madden further claimed that the 'beautiful curves' of the Pindari Glacier provided 'a striking illustration of Professor Forbes's *Viscous Theory*'. The Scottish physicist James David Forbes, who conjectured that glaciers moved as viscous liquids, profoundly influenced British men's experiences of ice in the Himalaya from the mid-1840s.[9] Forbes's 1843 book *Travels through the Alps* informed Himalayan fieldwork, sometimes via retrospective readings that compelled scientific explorers to revisit earlier observations and conclusions (Strachey, 1847: 795; Thomson, 1852: 476; Hooker, 1854: II: 133–134).

For men of this era, calling Himalayan ice formations 'glaciers' necessitated establishing material similitude with ice of the European Alps by invoking either first-hand experiences or Forbes's work. It also required critiquing their predecessors' narratives of encountering Himalayan ice and sometimes, as already discussed in Richard Strachey's case, revisiting these accounts in order to find glaciers nestled in books and journals as well as in high mountain valleys. Discovery was often a matter of textual hermeneutics as much as endeavours in the field. Madden combined complaint and indebtedness in recasting the engineer J. A. Weller's (1843) descriptions of 'fantastic castles, walls, &c.' of ice to 'denote the junction of a lateral glacier'. Weller, Madden pointedly noted, seemed unaware 'there was such a thing as a glacier; at least he never employs *the word*' (1847: 257–258). New words were a key part of making up glaciers; so too were new images. Accounts of the Himalaya in Calcutta journals and London-published books of the later 1840s contained notably more illustrations than their predecessors, and these were vital to making Asian glaciers gain traction among audiences in colonial and metropolitan hubs far removed from the icy heights. The visual repertoire for Himalayan ice that developed in conjunction with the verbal one equally bore marks of Forbes's influence. Renowned botanist Joseph Hooker's depiction in his 1854 *Himalayan Journals* of a block of gneiss perched on the ice (Figure 10.1) clearly resembled the rendering of a 'glacier table' on the Mer de Glace that formed the frontispiece to Forbes's *Travels* (Figure 10.2). Hooker's image did analytical work, relating glaciers in Sikkim to the archetype in Savoy. Forbes approved, commending Hooker's 'most complete illustrations' (1856: 66). Similarly, although the Asiatic Society of Bengal lacked access to the expensive colouring technologies of Forbes's Edinburgh publishing house, Richard Strachey's bird's eye and cross-sectional diagrams also appear to have taken cues from Forbes in combining multiple perspectives and using contour lines to denote topology and motion.

However, just as imperial power in the high mountains departed from models of exclusive sovereignty derived from the system of European states

Figure 10.1 Joseph Hooker (1854).

Figure 10.2 James Forbes (1843).

(Bergmann, 2016), images and descriptions of Himalayan glaciers in the late 1840s and early 1850s did not simply follow metropolitan conventions. As is also evident in the cases discussed in the next section, the Alps were construed as a point of difference as much as one of reference. Soldier-explorer Henry Strachey, for instance, believed that certain Tibetan categories for different icy formations did not translate into European concepts. He noted that '*kangri*' covered both glaciers and mountains covered in perpetual snow, and confessed that Tibetan understandings of '*dar*', or apparently permanent snowfields, surpassed his own (Strachey, 1853: 52–54). Having with conspicuous effort made up Himalayan glaciers partly through textual and visual analogies to European archetypes, explorer-scientists of the later nineteenth century increasingly insisted on the peculiarity of High Asia's ice.

Climbs and climes

Western Tibet's 'great variety of temperature-climates' eluded Henry Strachey's understanding every bit as much as its ice (Strachey, 1853: 58). The frozen regions of High Asia had long disrupted climatic expectations of Asians and Europeans alike. Crossing the Pir Panjal range into the Kashmir Valley in 1665, the court of Mughal Emperor Aurangzeb was caught in a snowstorm. Among the party was French physician François Bernier, whose disorientation at being suddenly in 'the midst of frozen snow' prompted him to undertake 'philosophical speculation'. Along with investigating the atmospheric agency of 'an aged hermit' credited locally with 'raising storms of wind, hail, snow, and rain', Bernier formulated a theory of circulating air around the mountain (Bernier, 1891: 408–409). During his subsequent stay in Kashmir he gleaned information on the surrounding uplands from itinerant wool merchants, taking particular interest in a pass across the Karakoram 'where, in every season, you must go a quarter of a league over ice' (Bernier, 1891: 427). Bernier's account is an early instance of encounters with Asian waters and winds prompting Europeans to advance climatic conjectures, akin to what environmental historian Sam White identifies as the 'new science' that emerged in early modern European colonists' attempts to 'unpuzzle' the climate of North America (White, 2015).

Bernier's ice long influenced Euro-Western knowledge of upland Asia's climate. The British translator and publisher of the 1891 English edition of Bernier's *Travels* compared its information on the Karakoram with recent European scientific explorers' accounts. 'Owing to changes in the ice' over the intervening two centuries, he noted, the route had become 'no longer passable' (Bernier, 1891: 427 footnote). Knowledge gleaned from Kashmiri traders in the seventeenth century was thereby harnessed to growing awareness among Europeans in the late nineteenth century that the ice of High Asia was liable to change rapidly. Recent historical climatology has shown that the global Little Ice Age continued unusually late in the Karakoram. Enhanced British efforts from the 1860s to gain footholds in the political and trading worlds of the contiguous locales of Ladakh and Turkestan (Gardner, 2021: 92–132) coincided with the growth of many of the largest extra-polar glaciers in the world (along with surprising and much-discussed events such as rapid surges and glacial lake outburst floods). These processes of geopolitical and geophysical expansion converged to render the Karakoram a key focal point for European speculation and adventure on ice.[10]

As early as the 1830s and 1840s, British administrator-scholars obtained information from upland communities and travellers suggesting that Himalayan ice could move appreciably within living memory (Traill, 1832: 3; Strachey, 1847: 802–804). However, they continued to assume Himalayan and Tibetan climate was stable over time if complex and variable across space, extrapolating present-day conditions to make claims about the distant past. James Gerard suggested that the Tibetan plateau's aridity rendered it the most likely location for intact 'antediluvian relics of the human skeleton' (Gerard, 1833: 256–257). Edward Madden, meanwhile, opined that first-hand experience of Himalayan weather enabled sympathetic understanding of the shastras' rendering of the region as the abode of gods (Madden, 1847: 239–240). Assumptions of climatic consistency melted away over subsequent decades as scientist-explorers encountered glaciers in Sikkim and the Karakoram. Not only larger than Himalayan glaciers elsewhere, these also left conspicuous signs of growth and recession. When Joseph Hooker witnessed 'a giant wall of rocks, piled – as if by the Titans' 10,000 feet up in Sikkim, his prior experience in Antarctica initially inclined him to trace their presence to 'a period when a glacial ocean stood high on the Himalaya'. After closely reading Forbes, however, he instead conceived of this and other 'vast old moraines' of the Kanchenjunga region as evidence that glaciers had descended much lower 'at some antecedent period': the ice, rather than topography, had changed (Hooker, 1854: I: 231–232). Images were every bit as vital to Hooker's point as romanticist-scientific prose: his *Himalayan Journals* depicted 'ancient moraine' in landscape and map forms.

Shortly after the publication of *Himalayan Journals* in 1854, the notion that glaciers had undergone substantial change in recent times also appeared further west along the mountain chain. Two expeditions were crucial: the German Schlagintweit brothers' travels in and beyond the northwestern Himalaya between 1854 and 1857; and the colonial-state-sponsored trigonometrical survey led by Henry Godwin-Austen in the Karakoram in 1861. As Moritz von Brescius's recent history shows, alongside their primary task of surveying magnetism the Schlagintweits executed measurements, maps, and sketches that 'contributed enduringly to the field of Himalayan glaciology' (2018, 139). Importantly, they identified 'most evident proofs' that nearly all of the region's glaciers had shrunk. The brothers went further than Hooker in suggesting that 'a diminution in the extent of *some* of these glaciers is still going on at present'. They also proclaimed the distinctiveness of these processes in Asia from the Ice Age theory then hotly debated in Europe and North America, on the basis that Himalayan 'ancient moraines ... are in situations very different' from the 'erratic blocks' that constituted the primary evidence in Europe (Schlagintweit and Schlagintweit, 1856: 123–125). As in the case of the lowland consumable variety, here again the subcontinent's ice appeared to be a very different substance to that in the West. The ongoing controversy around primary mechanisms of glacial changes in Europe coupled with the supposedly distinct situation in High Asia left the Schlagintweits uncertain about whether 'changes of climate' or 'general subsidence or elevation' drove glacial shrinkage. This marked the start of Himalayan evidence being invoked on both sides of debates over whether Ice Ages initially identified in the North Atlantic region were globally extensive (Campbell and Medlicott, 1877). The mystery of shifting Himalayan ice also contributed to the Schlagintweits' insistence that studies of weather and climate had to be systematized in the colonial subcontinent – a key influence on the foundation of the Indian Meteorological Department between 1867 and 1875 (von Brescius, 2018: 304–310).

In contrast to the Schlagintweits' apprehensions of dwindling ice, Godwin-Austen experienced growing glaciers in the Karakoram and was convinced that these indicated significant recent climate change. His account foregrounded the physical and geopolitical challenges involved in getting a grip on the vast icescapes of the Karakoram. The glaciers fascinated and frustrated him, presenting a kaleidoscopic array of colours along with extraordinary features and noises that he rendered in both anthropomorphized prose and watercolour sketches. Reliable and adaptable go-betweens with prior experience of travelling throughout the region were vital.[11] Like the Schlagintweits, to secure the required assistance Godwin-Austen relied on upland communities only loosely controlled by Kashmir, itself a tributary state of British India. The son of a potentate in Shigar, a state immediately southwest of the Karakoram that paid tribute to Kashmir, named Muhammad played numerous essential roles in Godwin-Austen's survey. The colonial state's inveiglement into the layered and overlapping forms of sovereignty in the mountains relied heavily upon the presence and actions of individual agents of empire (Bergmann, 2016; Gardner, 2021). Karakoram glaciers were important but unstable arenas for performances of British power. Far from infrastructures of colonial control, Godwin-Austen acknowledged that 'some man of influence accompanying the party' – like Muhammad – was essential to enforce discipline among inadequately equipped and overburdened porters, particularly during arduous stretches of walking and camping on ice (Godwin-Austen, 1864: 41). Muhammad played a further critical part in Godwin-Austen's work by informing him of 'the continual change going on in these regions', sharing the observation that ground free of ice during a previous visit to the area six years earlier had been covered by advancing glaciers and moraine. The British surveyor also placed great stock in traders and farmers' reports that glacial growth had closed off major passes and reduced two annual crops to one. He argued that these accounts complemented physical evidence such as 'overthrown trees and bushes' carried along at the snouts of glaciers, indicating 'considerable amounts of change of temperature within what we may call our own times' (Godwin-Austen, 1864: 51).

Whereas making up glaciers in the 1840s subjected ice to the instruments and close-focus visual techniques of natural history and surveying, making glaciers indices of climatic change in the 1860s involved zooming out spatially and temporally. As Anya Zilberstein has commented in the context of late eighteenth-century America, it was far harder to 'isolate, objectify, and circulate meteorological or atmospheric phenomena, let alone climatic patterns as a whole', than specimens of minerals, plants, and creatures (Zilberstein, 2016: 6). Along with material markers of change and oral traditions from both mobile and sedentary Himalayan communities, writings and images of earlier European explorers were crucial to rendering glaciers diachronic objects. The slew of climatic theories that emerged from surveys of Karakoram glaciers during the decades around the turn of the twentieth century and again in the 1930s established Godwin-Austen's work as the standard baseline from which to assess later movements (Conway, 1892: 753; Auden, 1935: 401). In the twenty-first century, his maps, texts, and watercolour landscapes (Figure 10.3) – which one glaciologist recently praised as 'detailed enough to compare with later photographs' (Hewitt, 2014: 305) – continue to be rendered as resources for analysing the 'Karakoram anomaly' of relatively stable glaciers at a time of rapid ablation elsewhere across the globe.

European fascination with the Karakoram from the late nineteenth century derived both from the monumental extent and unexpected behaviour of Asian ice. In Godwin-Austen's wake, visitors were attuned to evidence of rapidly shifting glaciers and saw this not just as a prompt to climatic theorizing, but also as lively matter for imperial adventuring. As well as seeking to quantify glacial shifts, alpinists and explorers in the later decades of the nineteenth century evoked ice in motion in impressionistic terms that gave prominence to their own physical labour. For instance, administrator-explorer

Figure 10.3 Henry Godwin-Austen, 'The End of the Punmah Glacier, Baltistan' (1861).

Robert Shaw's popular 1871 account of his journeys to project British influence into Turkestan, a region subject to Russian and Chinese expansion, depicted his party in the Karakoram avoiding a lake burst precipitated by sudden glacial movement. As with many other contributors to 'the Great Game' myth, Shaw imbricated geopolitical and geophysical dangers.

As well as being a foil to late Victorian heroism, Karakoram glaciers crystallized racial identities of high empire. What soldier, explorer, and mystic Francis Younghusband in his account of traversing the Karakoram in 1887 repeatedly termed 'pure white' ice was stuff on which the supposed distinction of having white skin could be enacted (1896: 188–213). Meanwhile, mountaineer Martin Conway's perception that the ice of the Biafo Glacier 'is as clear as that which freezes on the surface of a still pool of water in an English park' was one of many instances in which unsullied ice in Asia was imaginatively transplanted to Europe, even as explorers emphasized that the vast scale of Karakoram glaciers set them above their Alpine counterparts (Conway, 1894: 393–394). Exploration narratives routinely couched appreciating and mastering 'pure' ice as a marker of European distinction. Conway, for instance, claimed that when traversing

Figure 10.4 Robert Shaw, 'Escape from Inundation Caused by the Melting of a Glacier' (1871).

glaciers 'the foolish coolies greatly prefer stones to ice', whereas he and his European companions saw it as 'a natural highway' (Conway, 1893: 290; Conway, 1894: 537). This comment, made during an era when British power in the mountains depended on controlling actual roads (Gardner, 2021: 92–132), indicates how the British had become much more adept during the decades since John Hodgson's slips and falls at turning Himalayan ice to their advantage.

In 1910, at a time when he and other Westerners were scouring snows and sands at the heart of Asia for evidence of secular environmental shifts (Forêt, 2018), Swedish explorer Sven Hedin pronounced that 'the periodic oscillation of glaciers is ... very important on account of the light it throws on periodical climatic changes' (1910: 191). By the turn of the twentieth century, the nexus of climbs and climes in colonial India's icy borderlands was established not only in the Karakoram but also the eastern Himalaya. Increasingly intensive British interference in Sikkim precipitated a new phase of encountering and measuring glaciers. British mountaineer-geographer Douglas Freshfield emphasized the 'peculiarities' of Sikkim's ice, including 'yellow pools' of meltwater and a lack of crevasses on glaciers' surfaces. He claimed to comprehend these features with the feet of a climber – they 'delayed, but seldom stopped' movement across the glacier – and the 'trained eye' of a scientist, surmising that the ice 'has been in recent geological times a lot lower than it is now' (Freshfield, 1902: 460–461).

Like Conway in the Karakoram, Freshfield criticized colonial cartographic representations of glaciers as inadequate for the work of both mountaineers and physical geographers (Conway, 1894: x–xi; Freshfield, 1902: 460–461). The inclusion of renowned Italian photographers Vittorio and Erminio Sella in his party was one part of the attempt to rectify this practical and aesthetic shortcoming. Vittorio Sella's images of the glaciers around Kanchenjunga and, as part of an Italian climbing party seven years later, the Karakoram (Figure 10.5) were widely heralded as revolutionary in aesthetic terms and as a reliable index from which to assess subsequent glacial change in Asia and beyond (de Filippi, 1911: 26–28). Like Godwin-Austen's sketches, they continue to be used by present-day climatologists (Hewitt, 2014: 303). But even as these new visions inflected climatic understandings – and made new demands on Asian porters required to lug Sella's heavy photographic glass plates (de Filippi, 1911: 24) – older ways of positioning the Greater Himalaya's glaciers as objects in time persisted. Following the paper to the Royal Geographical Society at which Sella's Karakoram photographs met with great acclaim, British-Hungarian archaeologist Aurel Stein, a key figure in debates over environmental change in Central Asia's deserts, referred back to the account of trade caravans crossing the Karakoram that Kashmiri

Figure 10.5 Vittorio Sella (1909) 'The Snout of the Biafo Glacier' (1911).

merchants gave François Bernier. This, Stein argued, constituted evidence that some of the region's glaciers had grown significantly since the seventeenth century (de Filippi, 1911: 29). Diverse data from recent and distant climbs into the Karakoram and the Sikkim Himalaya became crucial components of high imperial – and still influential – theories of icy climes.

Conclusion: vital matter

As slush in drinks, Shiva's hair initiating a Subcontinent-spanning river, and rock-laden giants surging and retreating, ice in colonial India was vital in two senses. It was important stuff, crystallizing racialized and gendered identities while recasting extant Euro-Western understandings of atmospheric processes and climate. And it was a lively substance, in motion on various temporal and spatial scales from the race-against-entropy of New England's solidified and commodified ponds to the rapid surges and bafflingly consistent mass of Karakoram glaciers. We should read the diversity and intensity with which imperial agents described and depicted these frozen forms as an indication that ice appeared 'vital' or 'vibrant' in Jane Bennett's terms: 'as quasi agents or forces with trajectories, propensities, or tendencies of their own' (Bennett, 2010, viii). Ice penetrated the bodies and lodged in the minds of Europeans in South Asia, at times soothing and revivifying, at others baffling and wounding. Just as they did things to and with ice, ice also did significant things to them.

Looking through an icy lens helps us see how imperial sovereignty, sciences, and selfhood were far from stably congealed entities by the era of high empire in the late nineteenth century. A great deal of work on colonial India over the past four decades has emphasized the discursive and material forms that supposedly rooted in place rulers and ruled (Simpson, 2021: 5–11). Deploying a lexicon of icy rather than earthy metaphors enables an alternative and more insightful understanding of how structures of power, information, and identity operated in nineteenth-century India. Instead of processes of steady accretion leading to solid foundations, we should think of colonial power and knowledge as the always-provisional results of myriad slippages between varied conceptual and practical repertoires. Finally, looking at ice from the unusual vantage point of South Asia reveals how imperial consumption, control, and knowledge shaped this powerful but fragile material in ways that matter ever more in the contemporary era of cryospheric crisis. Legacies of empire persist in contemporary visions of ice as geostrategic sites, a crucial water resource for increasingly parched lowland states, and a key instrument for calculating the magnitude of global climate change (Pomeranz, 2009; Pandit, 2017: 213–57; Gagné, 2018; Harris, 2020).

Notes

1 Recent climate-informed histories have enriched our understanding of the effects of the Little Ice Age on global dynamics during the long seventeenth century, but the later period remains underanalysed. See for example Parker (2013); White (2017); Degroot (2018).
2 On ice-destabilising knowledge and practices in another colonial setting, see Cruikshank (2005).
3 Here I draw on Lévi-Strauss's concept of animals becoming totems because they are 'good to think with' rather than 'good to eat': Lévi-Strauss (1963: 89).
4 On 'friction' as a necessary part of, and tool for, critically interrogating global capitalism, see Tsing (2005).
5 On the relevant comparative case of European exploration in South America, see Burnett (2002).
6 On the verticality of ice, see Dodds (2019).
7 'Discovery' here refers to the largely European concept of Europeans garnering and disseminating knowledge accredited by other Europeans. See Kennedy (2013).
8 As they were in Europe: James Forbes stated that he 'often detached *hand specimens* of the ice, which, if they could be preserved in cabinets, would convey the most perfect idea of the structure' (1843: 160).
9 On Forbes's fieldwork and place in discussions of glaciers in nineteenth-century Europe, see Hevly (1996).
10 On a similar entanglement of expanding ice and expanding empires in northwest North America during the late eighteenth century, see Cruikshank (2005: 10).
11 On mobile go-betweens in imperial exploration more generally, see Driver (2018: 455–456).

References

Anon. (1832a) 'Dr. Wise's Ice Manufacture at Húgli', *Journal of the Asiatic Society of Bengal* 1(2): 74.

Anon. (1832b) 'Húgli Ice Manufactory', *Journal of the Asiatic Society of Bengal* 1(5): 204–205.

Anon. (1833) 'The Ice Trade Between America and India', *Journal of the Asiatic Society of Bengal* 2(21): 491–494.

Anon. (1835) 'Ice Trade', *Asiatic Journal* 18: 210.

Anon. (1836) 'The Balloon', *Journal of the Asiatic Society of Bengal* 5(52): 255.

Arnold, D. (2005a) 'Envisioning the Tropics: Joseph Hooker in India and the Himalayas, 1848–1850', in F. Driver and L. Martins, eds, *Tropical Visions in an Age of Empire*. Chicago, IL: University of Chicago Press, 137–155.

Arnold, D. (2005b) *The Tropics and the Traveling Gaze: India, Landscape, and Science 1800–1856*. Delhi: Permanent Black.

Auden, J. B. (1935) 'The Snout of the Biafo Glacier in Baltistan', *Records of the Geological Survey of India* 68: 400–418.

Babur (2006) *Babur Nama: Journal of Emperor Babur*, trans. A. S. Beveridge, ed. D. Hiro. New Delhi: Penguin.

Barker, R. (1775) 'The Process of Making Ice in the East Indies', *Philosophical Transactions* 65: 252–257.

Bennett, J. (2010) *Vibrant Matter: A Political Ecology of Things*. Durham, NC: Duke University Press.

Bergmann, C. (2016) 'Confluent Territories and Overlapping Sovereignties: Britain's Nineteenth-century Indian Empire in the Kumaon Himalaya', *Journal of Historical Geography* 51: 88–98.

Bernier, F. (1891) *Travels in the Mogul Empire AD 1656–1668*, trans. Irving Brock and Archibald Constable. Westminster: Archibald Constable (First published in 1670).

Burnett, D. G. (2002) '"It is Impossible to Make a Step Without the Indians": Nineteenth-century Geographical Exploration and the Amerindians of British Guiana', *Ethnohistory* 49(1): 3–40.

Campbell, J. F. and H. B. Medlicott (1877) 'On Himalayan Glaciation', *Journal of the Asiatic Society of Bengal* 46(2): 1–13.

Chakrabarty, D. (2008) *Provincializing Europe: Postcolonial Thought and Historical Difference*, 2nd ed. Princeton, NJ: Princeton University Press.

Colebrooke, H. T. (1810) 'On the Sources of the Ganges, in the Himadri or Emodus', *Asiatic Researches* 11: 429–445.

Conway, W. M. (1892) 'Mr. Conway's Karakoram Expedition', *Proceedings of the Royal Geographical Society* 14(11): 753–770.

Conway, W. M. (1893) 'Exploration in the Mustagh Mountains', *Geographical Journal* 2(4): 289–299.

Conway, W. M. (1894) *Climbing and Exploration in the Karakoram-Himalayas*. London: T. Fisher Unwin.

Cruikshank, J. (2005) *Do Glaciers Listen? Local Knowledge, Colonial Encounters, and Social Imagination*. Vancouver: University of British Columbia Press.

de Filippi, F. (1911) 'The Expedition of H.R.H. the Duke of Abruzzi to the Karakoram Himalayas', *Geographical Journal* 37(1): 19–30.

Degroot, D. (2018) *The Frigid Golden Age: Climate Change, the Little Ice Age, and the Dutch Republic, 1560–1720*. Cambridge: Cambridge University Press.

Dickason, D. G. (1991) 'The Nineteenth-century Indo-American Ice Trade: An Hyperborean Epic', *Modern Asian Studies* 25(1): 53–89.

Driver, F. (2018) 'Face to Face with Nain Singh: The Schlagintweit Collections and Their Uses', in A. MacGregor, ed., *Naturalists in the Field: Collecting, Recording, and Preserving the Natural World from Fifteenth to the Twenty-First Century*. Leiden: Brill, 441–469.

Eraly, A. (2007) *The Mughal World: Life in India's Last Golden Age*. Delhi: Penguin.

Fleetwood, L. (2019) 'The Vertical Globe: Altitude and Science in the Exploration of the Himalaya, 1800–1850'. PhD dissertation, University of Cambridge.

Forbes, J. D. (1843) *Travels Through the Alps of Savoy and Other Parts of the Pennine Chain with Observations on the Phenomena of Glaciers*. Edinburgh: Adam and Charles Black.

Forbes, J. D. (1856) 'Himalayan Journals', *Edinburgh Review* 103: 55–81.

Forêt, P. (2018) 'Le changement climatique dans les cartes de la Route de la Soie: Les contributions négligées de l'Expedition sino-suédoise de 1927–1935', *Annales de Géographie* 722: 403–428.

Fraser, J. B. (1820) *Journal of a tour through part of the snowy range of the Himala mountains and to the sources of the rivers Jumna and Ganges*. London: Rodwell and Martin.

Freshfield, D. W. (1902) 'The Glaciers of Kangchenjunga', *Geographical Journal* 19(4): 453–472.

G. (1831) 'On the Method of Making Ice at Hooghly', *Gleanings in Science* 3(25): 18–20.

Gagné, K. (2018) *Caring for Glaciers: Land, Animals, and Humanity in the Himalaya*. Seattle, WA: University of Washington Press.

Gardner, K. S. (2021) *The Frontier Complex: Geopolitics and the Making of the India–China Border, 1846–1962*. Cambridge: Cambridge University Press.

Gerard, A. (1841) *Account of Koonawur, in the Himalaya*, ed. George Lloyd. London: James Madden & Co.

Gerard, J. G. (1833) 'Observations on the Spiti Valley and Circumjacent Country Within the Himalaya', *Asiatic Researches* 18: 238–278.

Godwin-Austen, H. H. (1864) 'On the Glaciers of the Mustakh Range', *Journal of the Royal Geographical Society of London* 34: 19–56.

Grant, C. (1862) *Anglo-Indian Domestic Life: A Letter from an Artist in India to his Mother in England*. Calcutta: Thacker, Spink.

Harris, T. (2020) 'Lag: Four-Dimensional Bordering in the Himalayas', in Franck Billé, ed., *Voluminous States: Sovereignty, Materiality, and the Territorial Imagination*. Durham, NC: Duke University Press, 78–90.

Harrison, M. (1999) *Climates and Constitutions: Health, Race, Environment and British Imperialism in India, 1600–1850*. Oxford: Oxford University Press.

Hedin, S. (1910) 'The Kumdan Glaciers in 1902', *Geographical Journal* 36(2): 184–194.

Hevly, B. (1996) 'The Heroic Science of Glacier Motion', *Osiris* 11: 66–86.

Hewitt, K. (2014) *Glaciers of the Karakoram Himalaya: Glacial Environments, Processes, Hazards and Resources*. Heidelberg: Springer.

Hodgson, J. A. (1822) 'Journal of a Survey to the Heads of the Rivers, Ganges and Jumna', *Asiatic Researches* 14: 60–152.

Hooker, J. D. (1854) *Himalayan Journals; or, Notes of a Naturalist in Bengal, the Sikkim and Nepal Himalaya, the Khasia Mountains, &c*. London: John Murray.

Kennedy, D. (2013) *The Last Blank Spaces: Exploring Africa and Australia*. Cambridge, MA: Harvard University Press.

Lévi-Strauss, C. (1963) *Totemism*, trans. Rodney Needham. Boston, MA: Beacon.

Madden, Capt. E. (1847) 'Notes of an Excursion to the Pindree Glacier, in September 1846', *Journal of the Asiatic Society of Bengal* 16(1): 226–266.
Mueggler, E. (2011) *The Paper Road: Archive and Experience in the Botanical Expedition of West China and Tibet*. Berkeley, CA: University of California Press.
Pandit, M. K. (2017) *Life in the Himalaya: An Ecosystem at Risk*. Cambridge, MA: Harvard University Press.
Parker, G. (2013) *Global Crisis: War, Climate Change, and Catastrophe in the Seventeenth Century*. New Haven, CT: Yale University Press.
Parkes, F. (1850) *Wanderings of a Pilgrim in Search of the Picturesque, during four-and-twenty years in the East*. London: Pelham Richardson.
Pomeranz, K. (2009) 'The Great Himalayan Watershed: Water Shortages, Mega-Projects and Environmental Politics in China, India, and Southeast Asia', *Asia Pacific Journal* 7(30), 2: 1–29.
Qasim, S. Z., ed. (1983) *Scientific Report of First Indian Expedition to Antarctica*. New Delhi: Department of Ocean Development.
Raper, Capt. F. V. (1810) 'Narrative of a Survey for the Purpose of Discovering the Sources of the Ganges', *Asiatic Researches* 11: 446–563.
Rockefeller, S. A. (2011) '"Flow"', *Current Anthropology* 52(4): 557–578.
Schlagintweit, A. and R. Schlagintweit (1856) 'Report upon the Progress of the Magnetic Survey of India and of the Researches Connected with it in the Himalaya Mountains, from April to October, 1855', *Journal of the Asiatic Society of Bengal* 25(2): 105–133.
Shaw, R. (1871) *Visits to High Tartary, Yarkand, and Kashghar (formerly Chinese Tartary), and Return Journey over the Karakoram Pass*. London: John Murray.
Simpson, T. (2021) *The Frontier in British India: Space, Science, and Power in the Nineteenth Century*. Cambridge: Cambridge University Press.
Strachey, H. (1853) 'Physical Geography of Western Tibet', *Journal of the Royal Geographical Society of London* 23: 1–69.
Strachey, R. (1847) 'A Description of the Glaciers of the Pindur and Kuphinee Rivers in the Kumaon Himálaya', *Journal of the Asiatic Society of Bengal* 16(2): 794–812.
Thomson, T. (1852) *Western Himalaya and Tibet; A Narrative of a Journey through the Mountains of Northern India, during the years 1847–8*. London: Reeve and Co.
Traill, G. W. (1832) 'Statistical Report on the Bhotia Mehals of Kamaon', *Asiatic Researches* 17: 1–50.
Tsing, A. L. (2005) *Friction: An Ethnography of Global Connection*. Princeton, NJ: Princeton University Press.
von Brescius, M. (2018) *German Science in the Age of Empire: Enterprise, Opportunity and the Schlagintweit Brothers*. Cambridge: Cambridge University Press.
Weller, J. A. (1843) 'Extract from the Journal of Lieut. J.A. Weller', *Journal of the Asiatic Society of Bengal* 12(1): 78–102.
Wells, W. C. (1866) *An Essay on Dew and Several Appearances Connected With It*. London: Longman, Green, Reader, and Dyer (First published 1814).
White, S. (2015) 'Unpuzzling American Climate: New World Experience and the Foundations of a New Science', *Isis* 106(3): 544–566.

White, S. (2017) *A Cold Welcome: The Little Ice Age and Europe's Encounter with North America*. Cambridge, MA: Harvard University Press.

Williams, J. L. (1793) 'Account of the Method of making Ice at Benares', *Philosophical Transactions* 83: 56–58.

Wise, T. A. (1833) 'An Experimental Inquiry into the Means employed by the Natives of Bengal for making Ice', *Journal of the Asiatic Society of Bengal* 2(14): 80–90.

Younghusband, F. E. (1896) *The Heart of a Continent: A Narrative of Travels in Manchuria, across the Gobi Desert, through the Himalayas, the Pamirs, and Chitral*. London: John Murray.

Zilberstein, A. (2016) *A Temperate Empire: Making Climate Change in Early America*. New York: Oxford University Press.

11

Negotiating governable objects: glaciers in Argentina

Jasmin Höglund Hellgren

Climate change has led to a significant decrease of the cryosphere, including a considerable loss of ice sheets and glaciers (IPCC, 2019). Although operating on a planetary scale, climate change has varying local impacts, particularly intense in the high-mountain cryosphere (see also Inkpen this volume). In light of this development, societies have come to revalue glaciers and their related ecosystems, giving rise to new narratives of glaciers as 'an endangered species' and 'natural resources' as well as granting natural features such as rivers 'legal personalities' (Carey, 2007). Today it is estimated that nearly two billion people depend on glaciers for water supplies (Immerzeel *et al.*, 2020), and while glaciers have served as critical water resources in many regions for a long time, the metaphor of glaciers as natural 'water towers' has only recently gained prominence in the public imaginary (French *et al.*, 2015). In addition to their importance as water supplies, glaciers are also ascribed scientific, social, and cultural values. All over the world, receding glaciers have become a powerful symbol for the effects of climate change.

As several scholars have argued, the impacts of climate change on the high-mountain cryosphere are both physical and societal, ranging from loss and damage to livelihoods, natural resources, productivity, culture or even lives (Cruikshank, 2005; Carey, 2010; Huggel *et al.*, 2015, 2019; Inkpen this volume). In addition to unrelenting climate change, glaciers are affected by large-scale economic activities such as mineral and energy extraction at the so-called 'frozen frontier' (Bury, 2015). This trend is only expecting to continue under current and future climate change scenarios as previously inaccessible areas of land and water are uncovered revealing new profitable opportunities for resource extraction, infrastructure, or tourism. Accompanying this development, conflicts around mining projects in high-mountain environments have intensified, often connecting glaciers to issues of water quality and availability, placing them in the centre of mining conflicts.

In Argentina, where an estimated 15 per cent of South America's glaciers are located (IANIGLA, 2010), two mining projects operated by Canadian mining corporation Barrick Gold epitomize this kind of elemental conflict. In the early 2000s, what started as a contested EIA (environmental impact assessment) process regarding mining impacts on nearby glaciers, eventually culminated in the world's first national glacier protection law adopted by Argentina in 2010. In the glacier protection law, glaciers and the periglacial environment are understood as 'strategic reserves of water resources' (Ley Nacional 26.639), and an official national inventory of the country's glaciers was ordered to be carried out by the Argentine Institute for Snow Research, Glaciology and Environmental Sciences (IANIGLA). In May 2018, eight years later, IANIGLA published the national glacier inventory identifying 16,968 glaciers, covering a total area of 8,484 km^2. As a result, this area is off limits for activities such as mining, oil, and gas operations – activities now considered incompatible with the current and future existence of glaciers. Thus, conflicts over two mining projects located in a remote part of the Andes came to have a direct effect on how glaciers are valued, conceptualized, and by extension, governed.

This chapter traces the changing values ascribed to glaciers as they went from being natural objects at the centre of local mining conflicts in the beginning of the 2000s, to becoming institutionalized as governable objects in Argentinean environmental politics by the end of the decade. By focusing on how glaciers were constructed as resources and scaled by different actors during this process, the power relations embedded in glaciers are revealed. Further, the chapter highlights how glaciers assumed the role as a mediator of global climate change and its effect on how Argentina came to value its material environment. The chapter shows how questions of scale were re-actualized and enabled a new multidimensional framing of glaciers in Argentina: glaciers as critical water resources and objects of national governance. As such, the case discussed in this chapter does not only pertain to a particular local context, it is also a reflection of how planetary scale dynamics of climate change and, as a result, a destabilized cryosphere, enter into, alter, and reframe the discussions of what, why, and how environmental objects should and need to be governed.

Protecting glaciers and resourcing ice

Argentina's glacier protection law, its surrounding negotiations and conflicts, has received much attention both among the wider public and by scholars analysing it from various perspectives. Commonly the case has been understood as a resource or environmental conflict, or as an environmental justice issue (see for example Urkidi and Walter, 2011; Taillant, 2013; Khadim, 2016; Christel and Torunczyk, 2017). This chapter adds to such analyses by examining the case as a process of *resource construction* and *rescaling*. What started as a local conflict over two mining projects soon became an animated national debate about what type of resource glaciers are, for whom, and how they should be governed.

So, what are glaciers? And how should they be governed? The answers might seem straightforward. Glaciers are frozen water, H_2O in its solid form, a perennial body of dense ice moving under its own weight, and so on. As for the second question, the answer largely depends on the former – that is, on what glaciers are understood to be. As shown by Mark Carey, the understanding of glaciers has varied through history. During the eighteenth and nineteenth century the dominating narratives described glaciers as dangerous, sublime, symbols of wilderness, scientific laboratories, or as places to be explored and conquered (Carey, 2007: 500). More recently however, in the context of climate change causing global glacier recession, a contemporary narrative of glaciers as an *endangered species* has appeared. In this shifting narrative, glaciers are understood as threatened to disappear and with them the fresh water they store and the climate archives they contain. At the same time, melting glaciers and ice sheets threaten societies and livelihoods both on local and global scales. Rising sea levels are predicted to have devastating global impacts while many glaciers locally provide crucial functions of water storage and regulation for downstream communities and ecosystems. As glaciers decrease or completely disappear, concerns about the source of future water supply has given rise to yet another glacier narrative: glaciers as *natural resources* (Carey, 2007: 516).

In order to understand such narratological change, it is important to recognize resources as social constructs. This perspective works alongside a body of literature emphasizing that the value of natural resources is not inherent or pregiven, but socially constructed and produced at a particular time and place in historical contingent processes (Bridge, 2009; Avango *et al.*, 2013). Drawing on the work of Gavin Bridge (2009), material entities are composed as resources as actors ascribe usefulness and value to them. Accordingly, resources 'are [embedded in] a relational understanding of the non-human world' and change alongside social or technological changes that give, alter, or take away resources' value (Zimmerman in Bridge, 2009: 1220). This relationship is not fixed but a dynamic process that changes over time and space. As such, the same object can be considered a resource by different actors or stakeholders simultaneously as they highlight different values, functions, and uses. Thus, asking how and by whom glaciers are conceptualized as resources within their respective governance structures reveal dominating values and power relations in society (Bridge, 2009:

1220). In a time of global anthropogenic climate change with varying global, regional, and local effects, the context in which resources obtain their valued is rapidly changing, opening for new ways of understanding and valuing the human–glacier relationship.

In addition to the concept of resource construction, the concept of scale – also theorized mainly by scholars in the field of geography – can similarly be understood as socially constructed and produced (Swyngedouw, 1997; Brenner, 2001). Actors use scalar narratives as tools by which their respective interests and agendas with regards to the control and governance of resources can be articulated (McCarthy, 2005). Scale is thus not an independent variable but a strategy for actors to forward their particular agenda (Brown and Purcell, 2005). As such, the scale articulated through the discursive framing of glaciers, or any other resource or environmental problem, implicitly also contains a scalar preference for its level of governance (Lebel *et al.*, 2005). Governance is here understood as the structure of political processes, sets of rules, and decision-making procedures where it, as opposed to government, operate increasingly through the establishment of social norms and institutions (Young, 1997, 2002: 5).

In Argentina, as the mining-glacier conflict emerged in San Juan province, Barrick did not at first acknowledge glaciers close to their projects as *glaciers*, and much less as critical water resources. Provincial authorities in San Juan claimed glaciers were provincial resources and therefore under provincial jurisdiction, and further argued glaciers had to be weighed against mineral resources considering provincial development. Local activists on the other hand claimed glaciers were both critical local water resources, incommensurable to mineral deposits, as well as water resources of critical importance beyond the provincial scale and hence a national matter. In this way, as the conflict progressed, different actors brought forward competing framings of glaciers invoking various and often multiscalar functions and services tied to particular levels of governance. For example, glacial meltwater, depending on how far it travels, became a determinant for the status of glaciers as a local or national water resource, while frozen glaciers were ascribed economic value by the tourist industry or scientific value as a resource for climate scientists all over the world. The construction of glaciers as resources should therefore be understood also as a process of rescaling where scalar politics shaped glaciers as a governable object in the Argentinean Andes.

Glaciers, climate change, and mining in Argentina

The cryosphere in Latin America is represented by thousands of glaciers along the Andes mountain range, from Venezuela in the north to Chile and Argentina in the south. Most of the ice belongs to the Patagonian Ice Sheet located in southern Chile and Argentina. Based on approximate calculations before the Argentinean national glacier inventory, IANIGLA estimated South America's glacier coverage to around 25,500 km^2, of which 75 per cent are located in Chile and 15 per cent, the second largest part, in Argentina (IANIGLA, 2010). At the turn of the twenty-first century, there existed no complete or systematized information on the location, total area, hydrological significance, or history of glaciers or periglacial environments in Argentina (Iza and Rovere, 2006: 17; IANIGLA, 2010: 3). Nevertheless, several studies of individual glaciers show a general pattern of recession over the last decades associated with rising temperatures and decreased precipitation (see for example Leiva, 1999; Villalba *et al.*, 2003). Despite the lack of exact quantifications of glacier runoff contribution to downstream rivers, several reports highlight the importance of seasonal rain, snowfall, and glacier runoff for community livelihoods across the Andean region and in Argentina particularly (Watson *et al.*, 1998: 18; República Argentina, 2007).

Before Argentina's glacier debate took off in 2008 the public imaginary of glacier existence outside of Patagonia was in general very limited despite glaciers' long standing as an important national symbol in the country. One of the biggest tourist attractions, *Parque Nacional Los Glaciares* (The Glaciers National Park) in Santa Cruz province, received almost 668,000 visitors in 2018, of which approximately half were foreign tourists (Administración de Parques Nacionales, 2019). In 1981, UNESCO declared the park a World Heritage Site emphasizing its 'exceptional natural beauty' (UNESCO website, no date). While such descriptions allude to narratives of sublime and recreational glaciers, they also represent significant economic value to the province and the state. At the same time, as political scientist Carlos Escudé reminded his readers, Argentina has for a long time invested in 'patriotic' forms of education that have often emphasized high and cold places in the borderlands in the Andes, offshore islands, and the Antarctic continent as part of the regional territorial imaginary (Escudé, 1988, 1992). In the 1990s, for example, the country was embroiled in a dispute with Chile over ownership of the remote Southern Patagonian Ice Field (*Hielos Continentales*). The Argentine Parliament was resistant to endorsing the proposed borderline between the mountain peaks, watersheds, and glaciers, in part because of specific territorial perceptions of these areas (Allan, 2007 and Figure 11.1).

An additional important context to the increased tensions between mining activities and glaciers in the Andes were the market liberalization reforms implemented across Latin America in the 1980s to promote economic growth and reduce external debt. Within this political strategy mining came to

Figure 11.1 The Grande Glacier, Santa Cruz province, 2008. The Grande Glacier is an example of the typical Argentinean glacier imaginary with its large, uncovered, white surface, majestic beauty, and southern Andean location.

occupy a central place and the sector significantly increased in the region (Bridge, 2004). During the 1990s, Argentina introduced new legal frameworks promoting itself as a 'new mining country'. In 1993, for example, Argentina passed the Investment for Mining Activity law in order to promote foreign mining investments (Prado, 2005). As a result, Argentina ranked ninth in the world of mining investments by value between 1990 and 2001 (Bridge, 2004: 413). A decade later, in 2004, as Argentina experienced one of its worst economic crises, the government launched a National Mining Plan, in which mining had a central role in the country's development plan (Walter and Martinez-Alier, 2010). The resulting boom in mineral exploration and extraction, particularly in the high Andes, has put increased pressure on already scarce water resources and generated conflicts between local communities and mining interests (Bottaro *et al.*, 2014). The mining conflicts in which glaciers have come to occupy a central role should therefore be seen as a continuation of this development where *water* becomes the main point of contestation, in line with an overall trend observed in Latin America (see for example Bebbington and Williams, 2008; Bebbington *et al.*, 2010; Budds and Hinojosa, 2012). However, glaciers proved scalable in a new way compared to water resources. Through the global 'endangered species' narrative, glaciers in the remote Andes came to signify much more than local water resources had previously done in conflicts between local communities and transnational mining interests. Glaciers' planetary dimension and characteristic as a global environmental object made them scalable in new ways, which got picked up by actors and interests wanting to influence national environmental legislation.

Local glaciers: the formation of community concerns (2001–08)

In the early 2000s, the Canadian multinational mining company Barrick Gold was in the middle of developing its binational project Pascua-Lama in the Andes mountains on the border between Chile and Argentina. The project, located at an altitude of 3,800 to 5,200 metres above sea level is estimated to contain 17.8 million ounces of gold and 718 million ounces of silver (White, 2009). For Barrick there was only one problem: the gold and silver deposits were located beneath several glaciers. In their Glacier Management Plan, Barrick estimated that ten hectares of ice, three to five metres thick, 'must be removed' (Barrick, 2001). In parallel, Barrick started to develop a sister project, Veladero, located approximately 10 kilometres to the southeast of Pascua-Lama in the San Juan province of Argentina. Communities downstream from the projects, on both sides of the border, are dominated by agricultural activities destined for export as well as local consumption. As Barrick's glacier removal plan became known to locals, community concerns over future water quality and availability surfaced. Glaciers soon assumed a protagonist role driving social mobilization against the projects. The first part of this analysis examines this early construction of glaciers as resources and the first attempts to rescale the issue and shift to the national scale.

Early local mobilization stemmed from concerns over water resources, following the common trend often observed in mining conflicts across Latin America. In San Juan province, however, glaciers were equated with water as a symbol of life and framed as crucial for local livelihoods (Figure 11.2.). One of the first groups to get organized was *Madres Jachalleras*, who co-organized the *First Congress in Defence of Natural Resources*, in 2004, together with national anti-mining networks in Argentina. Invited experts, mainly biologists and water specialists from universities across the country, but also social scientists, journalists, and lawyers, held presentations stressing the negative effects of large-scale mining on water resources and the impact this could have on nearby agricultural activities. For example, as expressed by one water specialist, 'the mine will mean the loss of sustainable modes of production that have been practiced since forever' (Oro Sucio, 2004). As a result, resistance got more organized and continued with protests across the province and in the provincial capital.

In protesters' discourse, glaciers were framed as water resources integral to sustainability and were pitted against mineral resources through slogans such as 'water is worth more than gold' or 'life yes, gold no.' Glaciers were described as essential for local agricultural production, and by extension local livelihoods, in contrast to mineral resources that mainly held economic value to transnational capital, the national state and, to some extent, the

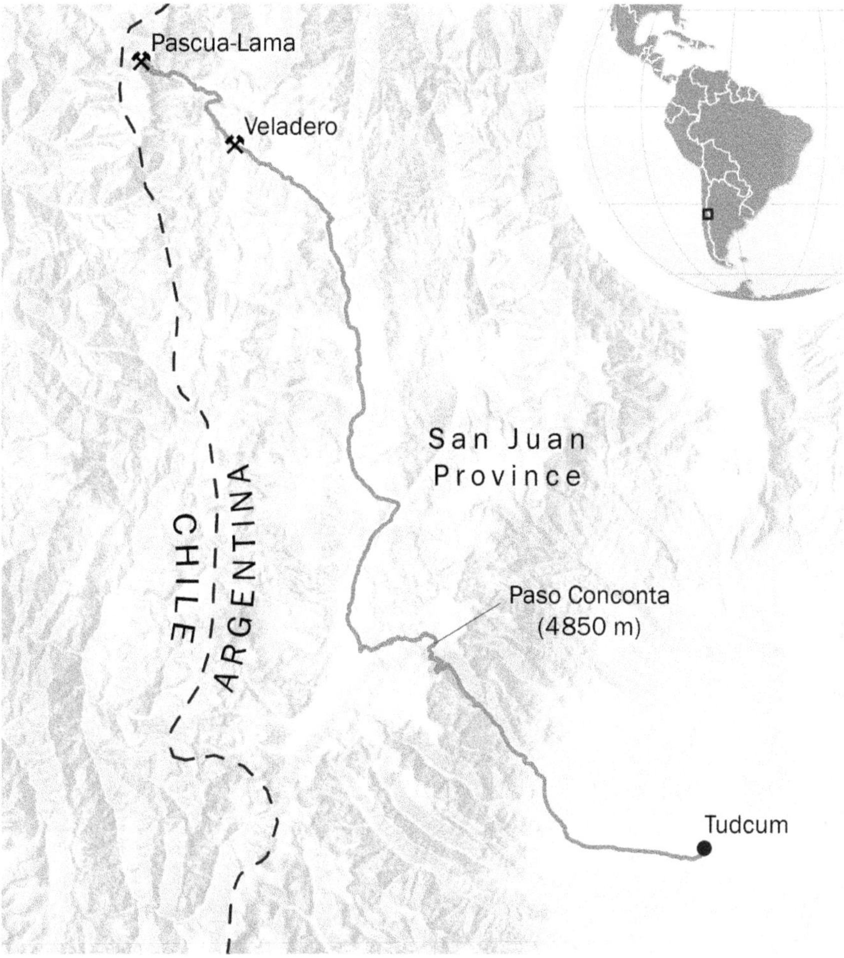

Figure 11.2 Location of the Pascua-Lama and Veladero projects, road construction and nearby communities, San Juan province.

provincial government. The mining company countered this view by arguing their activities would bring economic development to the area. However, for opposing community members the issue was not only one of financial security. Instead, glaciers were linked to community prosperity and survival in a broader sense emphasizing shared cultural values and intergenerational futures. In a letter delivered to the governor of San Juan in 2004, community organizations explained the reason for their opposition by pointing to the devastating social effects (rather than merely economic) the mine would have on their 'people.' The letter invoked arguments of community members

Figure 11.3 The Calingasta Glacier, San Juan province, 2014. This debris glacier might not be recognizable as a glacier to the non-expert but nonetheless contains ice and evidence of movement. This type of glacier is common in the northern dry parts of the Argentinean Andes.

being the legitimate, original, and authentic 'owners of the land' alluding to the land rights Indigenous people have over their ancestral territories (Oro Sucio, 2004). However, these communities have not been formally recognized as having Indigenous status and hence cannot in a judicial setting pursue such argument. Glacier removal and destruction was thus a threat to the local way of life, a culture, and ultimately community survival. This framing also included metaphors of nature as a mother, Pachamama, left wounded and scarred in the form of dirty rivers and torn apart mountains. As such, water – and by extension glaciers – were not only valued as an economic resource, but equally so made into a social, symbolic, and cultural resource.

In 2006, the San Juan-based civil society organization Fundación Ciudadanos Independientes (FuCI) filed a complaint before the provincial ombudsman based on different accounts of glacier existence in two of Barrick's environmental impact statements (EIS) (FuCI, 2006a, 2006b). The different accounts of glacier existence in the same geographical area was interpreted by activists as an intentional 'concealment' of glaciers (FuCI, 2006a). As a result, the provincial Ombudsman recommended the provincial authorities to 'carry out new studies on the existence of glaciers in the area' (FuCI, 2006b). In the legal documents and statements, FuCI defined glaciers as water resources that could be connected both to provincial and national environmental legislation (Bottaro *et al.*, 2014). Emphasis was put on glaciers' hydrological function rather than on aspects such as community rights, the social or cultural value of glaciers. As expressed by a FuCI representative, 'these [glaciers] are not only reserves of San Juan, they are freshwater reserves of Argentina. [...] If this study, which says the water runs through the whole country, is correct our water from the high mountains ends up in the ocean' (Observatorio Latinoamericano de Conflictos Ambientales, 2006). Glaciers were thus positioned as an elemental connector between mountain, river systems, the coastline and ocean implying they were not only local water resources but could be considered a national resource. In this way, a first attempt to scale-jump the issue to the national level was made. Such claims also connected glaciers' function as water reserves and regulators to scientific studies and expertise regarding how glaciers performed and served as water resources across a variety of geographical scales. This differed from previous studies where the focus had been on glaciers existence, location, and whether the mining development and road constructions did impact glaciers negatively or not. This possibility, and scientific uncertainty around glaciers' role in the hydrological system, allowed glaciers to become objects of and for contested knowledge claims, while simultaneously providing an opportunity to shift scale and reach beyond provincial levels of governance.

In response, the provincial government of San Juan turned to science and experts to determine the existence of glaciers, the reach of their ecological function, and eventual impacts of mining on such functions. In the following years, a series of studies were carried out involving scientists with various affiliations under an agreement between the provincial mining authorities and Barrick's subsidiaries in Argentina. One of the major controversies came to revolve around the road construction from the village Tudcum to the mining sites, allegedly running right through a glacier at the Conconta Pass (see Figure 11.1). However, despite the labour of glaciologists and studies brought in to manage uncertainties and settle emerging conflicts, all actors maintained their previous positions. Barrick and provincial authorities asserted that ice removed during road construction did not impact glaciers. However, the same study also indicated that dust from passing trucks might interfere with glacier albedo (Cabrera and Leiva, 2008). Further, Barrick consistently argued that climate change was responsible, while community and civil society groups blamed mining impacts. Despite the unresolved conflict, the Veladero project continued to operate while Chilean authorities on their side of the border suspended all activity around Pascua-Lama. Gradually, local opposition in San Juan weakened. Reports of manipulation of public sector jobs and social transfers in the communities closest to Veladero, as well as large corporate social responsibility programmes carried out by Barrick, are assumed to have led to the decrease in organized resistance (Haslam, 2018: 169).

The early provincial mobilization against mining projects illustrates how glaciers are capable of being conceptualized in a variety of ways and with different scalar perspectives. Simultaneously, glaciers were conceptualized as integral to local resilience, essential to national resource planning, highly symbolic of Argentina's elemental composition, and connected to global climate change.

Political glaciers: negotiating a national resource (2008–10)

The glacier issue reached national level politics in Argentina partly from the community and civil society organizations in San Juan and partly from an Argentinian congress member who, inspired by a glacier protection law proposal in Chile, started to draft an Argentinean version. Various glaciologists, mainly from IANIGLA, got involved and had a significant impact on the early drafting process. In addition to uncovered (white) glaciers they included other landforms acting as water reserves, such as debris glaciers, perennial snow patches, and the periglacial environment (Taillant 2015, 69–70). The glaciologists brought a different perspective on glaciers, their value as resources, and what poses a threat to them. The first glacier law draft mainly valued glaciers according to their role in the hydrological cycle, and glaciologists were particularly interested in how climate change dynamics altered the Argentinean cryosphere. The main threat to glaciers in this perspective was not mining activities at specific localities, even though they might impact individual glaciers to some extent, but climate change and its effects on the entire Andean cryosphere with potentially devastating effects on water availability (locally and across Argentina) was in focus. While the draft law did recognize glaciers as important water resources it did not mention to whom or what activities that should benefit from these (i.e. local livelihoods, agriculture, or industrial activities). Nevertheless, the draft law did list a series of prohibited activities in glacier areas, among them mining and oil exploration and exploitation, while it allowed for interventions related to scientific activities (Maffei *et al.*, 2007).

For IANIGLA glaciologists, the law was ultimately about protecting important water resources in an already dry area expected to become even drier due to continued climate change. In this view, extractive activities were seen as incompatible with glaciers and the periglacial environment because of their already fragile state. However, as an IANIGLA representative emphasized in a debate of the Senate, the law was exclusively about glacier protection and should not be interpreted as an 'anti-mining law' (Cámara de Senadores de la Nación, 2010: 6). As such, the actors behind the first draft law were not exclusively motivated by stopping a particular mining project from materializing; rather, their objective was to protect the Argentinean cryosphere from any type of direct human or industrial interventions that would worsen, in their view, an already serious situation. Such ambition, of course, required national legislation, and a foreseeable problem from the start was the anticipated resistance from provincial governments in mining-heavy areas – the same areas harbouring many of the country's glaciers. Understanding glaciers as objects in a larger global climate system, glaciologists recognized national legislation was not enough to protect them, but nevertheless reasoned that national legislation would lessen the damage.

In late October 2008, the glacier law was passed in both chambers of congress without any major debate or opposition. It has been suggested this was due to a lack of understanding of the vast existence of glaciers outside Patagonia and the encompassing meaning of the periglacial environment among congress representatives (Taillant, 2015: 81–82). However, the law did not survive for long. Before entering into force, the then President Fernández de Kirchner vetoed it arguing the 'absolute prohibition of activities' were problematic for investments and employment opportunities specifically in the mountain provinces (Rosemberg, 2008). The presidential veto was criticized immediately for placing economic interests before those of the environment and future generations (see for example La Nacion, 2008) and for being influenced by foreign business interests (Taillant, 2015: 143). As a way to address the critique, President Fernández de Kirchner established an interdisciplinary work group to discuss how glaciers and the periglacial environment should be protected (Boletín Oficial de la República Argentina No. 31.529, 2008). In this way, even though vetoed by the president, glaciers and their status as resources deserving protection became consolidated as an object of national politics.

The explicit connection that the veto made between glaciers and mining activities opened a window of opportunity for environmental NGOs. Before 2008, glaciers belonged mostly to NGO's advocacy work on climate change.

Figure 11.4 In a Greenpeace campaign on climate change, the organisation compares a photo of the Viedma Glacier, Santa Cruz province, from 1930 to the same place in 2008. The typical image of a white uncovered glacier is used to illustrate the effects of climate change to invoke a sense of urgency and call to action.

Glaciers were often mobilized and used as symbols to pressure the national government to act in global political contexts, in relation to national carbon dioxide emission reduction commitments, or to push for regional commitments in Latin America (Greenpeace Argentina, 2004). Thus far, in an environmental NGO perspective, glacier vulnerability was a scale problem and had to be tackled by the international community. NGO advocacy work against the growing mining industry in Argentina had not yet incorporated glaciers to strengthen their cause (Figure 11.4). With a glacier protection law vetoed due to what could be interpreted as industrial interests, mining activities effectively came to represent an additional threat to glaciers – this one governable at the national level. As such, the multiscalar dimensions of both glacier threats and possible solutions in terms of legislation became apparent (Figure 11.5). As a result, the conflict between glaciers and mining interests articulated by President Fernández de Kirchner would be the subject of intense debate and media coverage for the next two years.

After the presidential veto in 2008 two sides formed, one led by the political opposition advocating a resubmission of the first law, and one led

Figure 11.5 In a Greenpeace manifestation outside the Canadian embassy in Buenos Aires in April 2011, glaciers were mobilized in the organization's campaign against Barrick Gold. The banner reads 'Stop Barrick, save the glaciers'. The manifestation conveys the message of how two resources stand against each other and cannot mutually exist; it is either glaciers or gold.

by the Environment and Sustainable Development Commission of the Senate representing the executive side. The latter consulted with a broader range of stakeholders and provincial representatives aiming to amplify the sectoral interests and perspectives. In somewhat simplified terms, the slightly reworked original version understood glaciers to have an intrinsic value and hence assumed a conservationist approach, while opponents advocated for a more service-oriented approach where glacier resources could be interpreted as a resource to be exploited for the benefit of national or provincial development.

Senators opposed to the resubmission of the original version argued against it on several grounds, mainly because it did not specify what activities were considered legitimate beneficiaries of glacier resources and ecosystem services. In contrast, a version drafted by the Senate explicitly added industrial activities and generation of hydroelectric energy to the law's first article (Filmus, 2009). Opposition was most strongly expressed by representatives from the 'mining provinces' arguing that the law would have devastating effects on provincial economic development and that natural resources (referring to glacier and glacier meltwater), as stated in the national constitution, belong to the provinces (La Nación, 2010). Understood thus, an inclusion of industrial activities would enable glacier resources to be weighed against mining projects in the context of provincial economic development, and hence they would represent 'commensurable' values according to the same 'logic of equivalence' (Li, 2011, 2015). In this way, ice, glaciers, and glacial meltwater were to be treated as any other natural resource out of which other forms of capital could be produced. In contrast, those supporting the original version saw a conflict between the necessity to protect fresh water reserves (in the form of glaciers) and the economic interests invested in mining projects. In their view, these interests represented two incompatible lines of action.

IANIGLA representatives, who had been active in drafting the first version, entered the political debate after the presidential veto, gave media interviews, made public statements, and were invited as experts both to the debates in the lower house and in the Senate. In several public statements, they made a strong case for glaciers as critical water resources and 'the mountains as water providing towers' crucial for the future survival of communities and cities downstream (Rodriguez Pardo, 2008). However, the strong emphasis put on glaciers and the periglacial environment as water resources, valued mainly due to their function and capacity to store and regulate water flow, made the debate focus on a key question: how much water did glaciers and the periglacial environment store and how much did they contribute to downstream water availability? If the question in focus at the local level had been the existence of glaciers, the question at the national level became whether, and to what extent, glaciers and the periglacial environment had a critical role in the hydrological cycle at the national scale. In order to know whether glaciers actually were critical water resources, the opponents argued, this must be studied and answered scientifically rather than just asserted. For instance, Barrick hired a consulting firm to conduct studies during the Pascua-Lama EIA process who, in 2009,[1] published a report recognizing that the project was located in the periglacial environment. However, as the consultants argued in various public seminars touring the country, their findings showed that the hydrological contribution of rock glaciers to the ecosystem was insignificant (Taillant, 2015: 185). They concluded that '[g]lacial melt is only important at the small (<10 km^2) watershed scale but diminishes exponentially with distance downstream and the proportional increase in watershed area' (Arenson *et al.*, 2015: 32). These conclusions stood in sharp contrast to the overall 'glacier at risk' narrative and concerns over downstream water availability put forward by environmental NGOs and other glaciologists.

In the debate, a central point of contestation therefore became the concept of the periglacial environment and whether it significantly contributed to the hydrological cycle or not. For example, for the province of San Juan, all territory above 3,500 metres would belong to this category, including most mining projects in the province. While the original version of the law sought to protect the entire periglacial environment that 'acts as a regulator of water resources' (Maffei *et al.*, 2007), the Senate's version limited protection to rock glaciers within the periglacial environment (Filmus, 2009). From IANIGLA's perspective, the Senate's version had significant advantages as it provided a technical definition easier to implement in the national glacier inventory ordered by the law's third article (Rojas and Wagner, 2019). Further, to calm critical senators, an IANIGLA expert explained that 'technically [but] under different names, both projects are protecting the same area' (Cámara de Senadores de la Nación, 2010: 5). That is, according to IANIGLA representatives, protection of the periglacial environment was in both cases conditioned by the law's first article stating it had to act as 'strategic reserves of water resources' (Ley Nacional 26.639). In practice, IANIGLA asserted, this would limit the protected area to rock glaciers within the periglacial environment in any case.

From the concerns voiced by the executive, Barrick, and pro-mining senators, it is clear how the definitions of glaciers and the periglacial environment were perceived as a threat to corporate and national governmental accumulation strategies. As Steinberg and Kristoffersen (2017) have shown in the Norwegian context, the state of Norway similarly defined the 'ice edge' opportunistically in order to allow continued oil and gas explorations in the Barents Sea (Steinberg and Kristoffersen, 2017). Likewise, Argentinean actors turned to science and experts that fitted their respective political and economic perspectives. As the controversy continued it became clear how definitions, interpretations, and conceptualizations were coloured by political preferences and ideology from both sides. Ice had become politicized and is an illustration of how material objects are constructed through a mix of science, policy, culture, and economic structures.

In October 2010, after what seemed like a standstill between the two law proposals, a negotiated version appeared and subsequently passed through both houses of congress. In the final version – what became the world's first glacier protection law – industrial activities and hydroelectric energy generation were removed from Article 1, while concessions seem to have been made as provinces retained discretion over whether and when to suspend mining activities due to glacier impacts (Taillant, 2015: 196). In a sense, this was a victory for the proponents of the original law, glaciologists, environmental NGOs and, in the beginning, also for the community organizations in San Juan. However, the glacier issue would prove to be far from over. Following the ratification of the law, a series of lawsuits initiated by San Juan province and Barrick were debated in the courts for almost a decade, and the national glacier inventory was delayed several times. In addition, Barrick continues to operate the Veladero project which has suffered no less than three cyanide spills since, for which the inventory has gotten most of the blame. Heavily targeted both by environmental NGOs and community organizations, criticism of the inventory eventually culminated in the former IANIGLA director being charged for not complying with the glacier law and is still awaiting trial (see also Chapter 4 this volume). The aftermath of the glacier law is beyond the scope of this chapter, but most likely we have not seen the last of glacier-related controversy in Argentina.

Conclusion: Governable glaciers and scaling in resource construction

From the early 2000s to the end of 2010, Argentina's glaciers were transformed and got increasingly conceptualized as a multiscale resource. A destabilized cryosphere, linked to glacial disappearance caused by global warming or corporate plans to 'remove ice', made the non-existing regulation of the cryosphere visible. This situation enabled a new kind of glacier requiring a revaluation of the current governance structures. Through contested mining projects, political debates, and negotiations, as well as through the establishment of an official national glacier inventory to be carried out by IANIGLA, glaciers became constructed as critical water resources. Embedded in a global discourse of glaciers as an 'endangered species' (Carey, 2007), the articulation of concerns at the local level were able to rescale and consolidate glaciers as a governable object in Argentinean national politics.

The multiscalar resource dimension of glaciers made them relevant to a wide range of actors, sometimes hundreds or thousands of miles away from their physical location. As President Fernández de Kirchner vetoed the first glacier law, glaciers in Argentina became, almost overnight, a central interest of provincial and national politicians, business actors, environmental NGOs (with ties to transnational advocacy networks), national and international glaciologists, as well as of the general public. Everyone wanted a say about glaciers and positioned themselves as important stakeholders. However, for local actors, engaging in politics of scale is also a risk. In Argentina, even though the issue scaled to the national level, it left local opponents disappointed. While local mobilization eventually led to the sanctioning of a national glacier protection law and official glacier inventory – which has put considerable areas out of reach for further development projects, at least for now – it simultaneously moved glaciers away from the local context. Through this process, local communities lost control over definitions, meanings, and the management of nearby glaciers highlighting how glaciers are embedded in existing power structures and constitute 'sites of power and authority' (see Chapter 4 this volume).

As this short historicizing of Argentina's glacier protection law has argued, glaciers are multidimensional, multiscaled, and highly symbolic of Argentina's elemental composition. As glaciers in the Andes appeared as mediators of global climate change they replaced water as the central issue in mining conflicts enabling the issue to scale in new ways. The particular scaling of glaciers became a crucial constitutive dimension of their construction as resources, and by extension, of their transformation into governable objects. Even though governed at the national level, the glacier law reflects glaciers as a resource also beyond the national scale. These characteristics should not be seen as standing in opposition to each other; rather, together, they make visible how glaciers belong to a broader context, in addition to the political and judicial, and help us better understand how new objects of governance and legislation are formed.

Note

1 The author has not had access to the report itself – BGC Engineering (2009) 'Pascua Lama Permafrost Characterization Study' – but to a chapter in an edited volume written by the same authors, based on the same fieldwork and data, reaching the same conclusions (see Arenson *et al.*, 2015).

References

Administración de Parques Nacionales (2019) *Plan de Gestión del Parque Nacional Los Glaciares*. El Calafate, Santa Cruz: Administración de Parques Nacionales.

Allan, L. (2007) 'Néstor Kirchner, Santa Cruz, and the Hielos Continentales Controversy 1991–1999*', *Journal of Latin American Studies* 39(4): 747–770.

Arenson, L. U., M. Jakob, and P. Wainstein (2015) 'Effects of Dust Deposition on Glacier Ablation and Runoff at the Pascua-Lama Mining Project, Chile and Argentina', in G. Lollino, A. Manconi, J. Clague, S. Wei, and M. Chiarle, eds, *Engineering Geology for Society and Territory, Volume 1*. Switzerland: Springer, 27–32.

Avango, D., A. E. Nilsson, and P. Roberts (2013) 'Assessing Arctic Futures: Voices, Resources and Governance', *Polar Journal* 3(2): 431–446.

Barrick (2001) *Pascua Lama Project – Glacier Management Plan in the Rio El Toro Basin, Pascua Lama Project (EIA) Annex B, February, 2001 (Rev. 0)*. Santiago: Barrick.

Bebbington, A., D. Humphreys Bebbington, and J. Bury (2010) 'Federating and Defending: Water, Territory and Extraction in the Andes', in R. Boelens, D.

Getches, and A. Guevara-Gil, eds, *Out of the Mainstream: Water Rights, Politics and Identity*. London: Earthscan, 307–328.

Bebbington, A. and M. Williams (2008) 'Water and Mining Conflicts in Peru', *Mountain Research and Development* 28(3–4): 190–195.

Boletín Oficial de la República Argentina No. 31.529 (2008) 'Decreto 1837/2008 Obsérvase el Proyecto de Ley registrado bajo el No26.418'. Available at: www.boletinoficial.gov.ar./detalleAviso/primera/9281360/20081111?busqueda=1 (accessed 15 July 2020).

Bottaro, L., A. Latta, and M. Sola (2014) 'La politización del agua en los conflictos por la megaminería: Discursos y resistencias en Chile y Argentina', *European Review of Latin American and Caribbean Studies* 97: 97–115. doi: http://doi.org/10.18352/erlacs.9798.

Brenner, N. (2001) 'The Limits to Scale? Methodological Reflections on Scalar Structuration', *Progress in Human Geography* 25(4): 591–614.

Bridge, G. (2004) 'Mapping the Bonanza: Geographies of Mining Investment in an Era of Neoliberal Reform', *Professional Geographer* 56(3): 406–421.

Bridge, G. (2009) 'Material Worlds: Natural Resources, Resource Geography and the Material Economy', *Geography Compass* 3(3):1217–1244.

Brown, J. C. and M. Purcell (2005) 'There's Nothing Inherent about Scale: Political Ecology, the Local Trap, and the Politics of Development in the Brazilian Amazon', *Geoforum* 36(5): 607–624.

Budds, J. and L. Hinojosa (2012) 'Restructuring and Rescaling Water Governance in Mining Contexts: The Co-production of Waterscapes in Peru', *Water Alternatives* 5(1): 119–139.

Bury, J. (2015) 'The Frozen Frontier: The Extractives Super Cycle in a Time of Glacier Recession', in C. Huggel, M. Carey, J. J. Clague, and A. Kääb, eds, *The High-Mountain Cryosphere: Environmental Changes and Human Risks*. Cambridge: Cambridge University Press, 71–89.

Cabrera, G. A. and J. C. Leiva (2008) *Monitoreo de Glaciares del Paso de Conconta, Iglesia, San Juan, Argentina*. Mendoza, Argentina: CONICET.

Cámara de Senadores de la Nación (2010) *VT-Reunión de la Comisión de Ambiente y Desarollo Sustentable* [transcript from Senate working group]. Congreso de la Nación Argentina – Información Parlamentaria.

Carey, M. (2007) 'The History of Ice: How Glaciers Became an Endangered Species', *Environmental History* 12(3): 497–527.

Carey, M. (2010) *In the Shadow of Melting Glaciers*. New Haven, CT: Oxford University Press.

Christel, L. and D. Torunczyk (2017) 'Sovereignties in Conflict: Socio-environmental Mobilization and the Glaciers Law in Argentina', *European Review of Latin American and Caribbean Studies* 104: 47–67. http://doi.org/10.18352/erlacs.10212.

Cruikshank, J. (2005) *Do Glaciers Listen?* Vancouver: University of British Columbia Press.

Escudé, C. (1988) 'Argentine Territorial Nationalism', *Journal of Latin American Studies* 20(1): 139–165.

Escudé, C. (1992) *Education and Foreign Policy: The Case of Argentina, 1908–1982*. Working Paper Series of the Duke-UNC Program in Latin American Studies. Durham, NC.

Filmus, D. F. *et al.* (2009) Proyecto de Ley: (S-2200/09). Senado de la Nación.

French, A., J. Barandiarán, and C. Rampini (2015) 'Contextualizing Conflict', in C. Huggel, M. Carey, J. J. Clague, and A. Kääb, eds, *The High-Mountain Cryosphere: Environmental Changes and Human Risks*. Cambridge: Cambridge University Press, 315–336.

FuCI (2006a) *Barrick Ocultó la Verdad Acerca de los Glaciares*. Available at: http://fucimineria.blogspot.com/2006/02/barrick-ocult-la-verdad-acerca-de-los.html (accessed 15 July 2020).

FuCI (2006b) *Se Confirma la Denuncia de Ocultamiento de Glaciares*. Available at: http://fucimineria.blogspot.com/2006/03/se-confirma-la-denucia-de-ocultamiento.html (accessed 15 July 2020).

Greenpeace Argentina (2004) *Cambio Climático: Futuro Negro para los Glaciares*. Buenos Aires: Greenpeace Argentina.

Haslam, P. A. (2018) 'The Two Sides of Pascua Lama', *European Review of Latin American and Caribbean Studies* 106: 157–182. http://doi.org/10.32992/erlacs.10398.

Huggel, C., M. Carey, J. J. Clague, and A. Kääb, eds (2015) *The High-Mountain Cryosphere: Environmental Changes and Human Risks*. Cambridge: Cambridge University Press.

Huggel, C., V. Muccione, M. Carey, R. James, C. Jurt, and R. Mechler. (2019) 'Loss and Damage in the Mountain Cryosphere', *Regional Environmental Change* 19(5): 1387–1399.

IANIGLA (2010) *Inventario Nacional de Glaciares y Ambiente Periglacial: Fundamentos y Cronograma de Ejecución*. Mendoza, Argentina: IANIGLA-CONICET.

Immerzeel, W. W., A. F. Lutz, M. Andrade *et al.* (2020) 'Importance and Vulnerability of the World's Water Towers', *Nature* 577(7790): 364–369.

IPCC (2019) 'Summary for Policymakers', in H.-O. Pörtner, D. C. Roberts, V. Masson-Delmotte *et al.*, eds, *IPCC Special Report on the Ocean and Cryosphere in a Changing Climate*. Cambridge, New York: Cambridge University Press, 3–35.

Iza, A. and M. B. Rovere, eds (2006) *Aspectos jurídicos de la conservación de los glaciares*. Gland, Switzerland: Unión Mundial para la Naturaleza (UICN).

Khadim, A. N. (2016) 'Defending Glaciers in Argentina', *Peace Review* 28(1): 65–75.

Lebel, L., P. Garden, and M. Imamura (2005) 'The Politics of Scale, Position, and Place in the Governance of Water Resources in the Mekong Region', *Ecology and Society* 10(2): 18.

Leiva, J. C. (1999) 'Recent Fluctuations of the Argentinian Glaciers', *Global and Planetary Change* 22(1–4): 169–177.

Ley Nacional 26.639 (2010) Régimen de Presupuestos Mínimos para la Preservación de los Glaciares y del Ambiente Periglacial.

Li, F. (2011) 'Engineering Responsibility: Environmental Mitigation and the Limits of Commensuration in a Chilean Mining Project', *Focaal: Journal of Global and Historical Anthropology* 60: 61–73.

Li, F. (2015) *Unearthing Conflict*. Durham, NC: Duke University Press.

Maffei, M. O., C. A. Raimundi, and D. B. Bisutti (2007) Proyecto de Ley: Presupuestos Minimos para la Protección de los Glaciares y del Ambiente Periglacial. Cámara de Diputados de la Nación.

McCarthy, J. (2005) 'Scale, Sovereignty, and Strategy in Environmental Governance', *Antipode* 37(4): 731–753.

La Nacion (2008) 'Sospechoso veto a la ley de glaciares', *La Nacion*, 21 November. Available at: www.lanacion.com.ar/editoriales/sospechoso-veto-a-la-ley-de-glaciares-nid1072316/ (accessed 15 July 2020).

La Nación (2010) 'Debate tenso en el Senado/Minería vs. protección ambiental. Gobernadores rechazan la nueva ley de glaciares', *La Nacion*, 22 September. Available at: www.lanacion.com.ar/politica/gobernadores-rechazan-la-nueva-ley-de-glaciares-nid1307094/ (accessed 15 July 2020).

Observatorio Latinoamericano de Conflictos Ambientales (2006) *Hay más de 40 Glaciares en Veladero y Pascua-Lama*. Available at: www.olca.cl/oca/argentina/mineras23.htm (accessed 15 July 2020).

Oro Sucio (2004) *Jáchal: Primer Congreso en Defensa de los Recursos Naturales*. Available at: https://orosucio.madryn.com/articulos/04_10_14.html (accessed 14 July 2020).

Prado, O. A. (2005) *Situación y perspectivas de la minería metálica en Argentina*. Santiago de Chile: Comisión Económica para America Latina y el Caribe (CEPAL), Naciones Unidas.

República Argentina (2007) *2da Comunicación Nacional de la República Argentina a la Convención Marco de las Naciones Unidas sobre Cambio Climático*. Report to UNFCCC.

Rodriguez Pardo, J. (2008) *Tratamiento del decreto 1837/08 Veto presidencial de la Ley N 26.418 de Protección de Glaciares y del Ambiente Periglaciar*. Available at: www.youtube.com/watch?v=4JC8IidW-Lw&ab_channel=tramalatierra (accessed 15 July 2020).

Rojas, F. and L. Wagner (2019) 'Inventario de glaciares en Argentina: polémicas públicas y disputas de sentido', in M. Miraglia and A. M. Franca, eds, *Paisaje y patrimonio: impresiones de la historia en el ambiente natural*. Buenos Aires: Universidad Nacional de Quilmes – TESEO [in press].

Rosemberg, J. (2008) 'Polémica por el veto de una ley sobre los glaciares', *La Nacion*, 19 November. Available at: www.lanacion.com.ar/politica/polemica-por-el-veto-de-una-ley-sobre-los-glaciares-nid1071556/ (accessed 15 July 2020).

Steinberg, P. and B. Kristoffersen (2017) '"The Ice Edge is Lost … Nature Moved It": Mapping Ice as State Practice in the Canadian and Norwegian North', *Transactions of the Institute of British Geographers* 42(4): 625–641.

Swyngedouw, E. (1997) 'Neither Global nor Local: "Glocalization" and the Politics of Scale', in K. R. Cox, ed., *Spaces of Globalization: Reasserting the Power of the Local*. New York: Guilford Press, 137–166.

Taillant, J. D. (2013) 'The Human Right to Glaciers?', *Journal of Environmental Law and Litigation* 28(1): 59–78.

Taillant, J. D. (2015) *Glaciers: The Politics of Ice*. New York: Oxford University Press.

UNESCO website (n.d.) *Los Glaciares National park*. Available at: https://whc.unesco.org/en/list/145/ (accessed 22 June 2020).
Urkidi, L. and M. Walter (2011) 'Dimensions of Environmental Justice in Anti-gold Mining Movements in Latin America', *Geoforum* 42(6): 683–695.
Villalba, R., A. Lara, J. A. Boninsegna *et al.* (2003) 'Large-scale Temperature Changes Across the Southern Andes: 20th-century Variations in the Context of the Past 400 Years', *Climatic Change* 59(1–2): 177–232.
Walter, M. and J. Martinez-Alier (2010) 'How to be Heard When Nobody Wants to Listen: Community Action Against Mining in Argentina', *Canadian Journal of Development Studies* 30(1–2): 281–301.
Watson, R. T., M. C. Zinyowera, and R. H. Moss, eds (1998) *The Regional Impacts of Climate Change: An Assessment of Vulnerability*. Cambridge: Cambridge University Press.
White, N. J. (2009) *Special Report on Pascua-Lama: A Barrick Gold Report on Responsible Mining*. Barrick Gold Corp. Available at: https://issuu.com/barrickgold/docs/pascua-lama-september-2009 (accessed 15 July 2020).
Young, O. R. (1997) 'Rights, Rules, and Resources in World Affairs', in O. R. Young, ed., *Global Governance: Drawing Insights from the Environmental Experience*. Cambridge, MA: MIT Press, 1–24.
Young, O. R. (2002) *The Institutional Dimensions of Environmental Change: Fit, Interplay, and Scale*. Cambridge, MA: MIT Press.

12

Cryonarratives for warming times: icebergs as planetary travellers

Elizabeth Leane

In the Anthropocene, icebergs have moved from the periphery to the centre of global public consciousness, their ephemerality and mutability ominously signalling the mobile and impermanent nature of the polar regions. With ice sheets and ice shelves increasingly unstable as ocean temperatures rise, humans feel implicated for the first time in the creation of these objects. The calving of a huge tabular berg is now a political event, framed by media headlines worldwide not simply as a visual spectacle but also as a source of communal guilt, fear, anxiety, and anger. Yet such icebergs can endure for decades beyond this sensationalized moment, wandering peripatetically through the oceans and interacting with human and nonhuman actors in diverse and unpredictable ways.

This chapter proposes a new term, 'cryonarrative', as shorthand for the kinds of stories humans are telling about ice in the contemporary period, and looks at the cryonarratives being applied to icebergs. Within media and tourist discourse, icebergs are often subject to reductive narratives that render them as symbols of human doom or aesthetic objects for human consumption. As a remedy to this anthropocentric approach, I argue for the advantages of characterizing and narrating icebergs as travellers on a planetary scale.

Ice humanities and cryonarratives

With environmental ice increasingly recognized as pivotal to the future of humanity and the planet it inhabits, cultural scholarship has shifted in recent years from examining human activities on, in, over, and about ice to understanding ice as a nonhuman force that interacts with humans in unpredictable ways. As Klaus Dodds and Sverker Sörlin observe in their introduction to this volume, a growing number of critics focus on ice's agency, with different emphases and to different degrees: ice is a 'nonhuman actor' (Boorst, 2010), 'a vital force in the universe' (Glasberg, 2011: 232), 'alive, creaturely and desiring' (Duckert, 2013: 71). These approaches sit alongside – and not always comfortably with – calls to historicize ice: to understand icy regions not as a wildernesses, wastes, or forms of emptiness, but as historically complex environments that are both materially and culturally produced (e.g. Antonello, 2017).

This shift towards ice within the humanities has in turn generated new and useful coinages that merge the scientific idea of the cryosphere with human ways of making meaning. These include 'cryo-politics', to take account of how ice functions physically and discursively within geopolitics (Bravo and Rees, 2006: 207–208); 'cryoscape', to describe 'a coming together of human epistemic practices with the physical phenomena that constitute the cryosphere' (Nüsser and Baghel, 2014: 150); and 'cryo-history', to encompass the emerging work within environmental history and anthropology analysing the coupled changes of human societies and ice across time (Sörlin, 2015).

Coming from the perspective of literary studies, I propose a related term, 'cryonarrative', to describe contemporary stories in which ice and humans are closely interconnected. This term could encompass a diverse range of specific ice-centred narratives: fiction, from popular ecothrillers such as such L. A. Larkin's *Thirst* (2012) to literary works such as Ilija Trojanow's *Ice Tau* (2011); travel accounts and memoirs such as Jonathan Franzen's *The End of the End of the Earth* (2016) and Elizabeth Bradfield's poetic memoir *Toward Antarctica* (2019); and screen texts, including documentary films like Werner Herzog's *Encounters at the End of the Earth* (2012), feature films such as Youyin Wu's romance *Till the End of the World* (2018), and television series such as the recent thriller, *The Head* (2020). Depending on how tightly it is defined, the category could potentially comprise a huge array of texts, genres, and subgenres.

Cryonarratives could also refer to broader and less easily pinned down narrative framing strategies that play out repeatedly across mainstream media, informing the stories that individuals place around their own encounters, mediated and material, with icescapes. These include narratives of ruination (Nuttall, 2020: 7), vanishment (Dodds, 2021), and heroic rescue (Glasberg, 2011: 222), all of which have particular social and political implications. By self-consciously paying attention to these cryonarratives, we are better able to work to challenge elements of them where necessary and to identify new ones that better address contemporary challenges.

Although narrative framings are already implicit within the concepts of cryo-politics, cryoscape, and cryo-history, 'cryonarrative' draws our attention to the specifically literary aspects of the stories we tell about ice – particularly the generic patterns that we draw upon. This literary term alerts us to how concepts such as plot, character, and point of view function in stories of ice: who are the protagonists (human and nonhuman), how are they represented, and what familiar or unexpected roles do they play? For example, in a previous analysis of Antarctic-set ecothrillers, I have argued that icescapes are often themselves active characters – or perhaps more accurately, actants in the narratological sense: they are key structural elements in the story, rather than simply 'setting'. In these thrillers, icescapes work in 'ambiguous alliances with both the human heroes and villains', resulting in 'narratives of human/nonhuman alliance that reflect new ways of thinking in the Anthropocene' (Leane, 2019: 88, 97).

One of the potential problems of 'cryo-' terms is an overly homogenous approach, leading to abstract discussions of 'ice' and monolithic representations of complex geographical places. Some of the most influential cryo-analyses, such as Julie Cruikshank's *Do Glaciers Listen?* (2005), not only focus on specific ice features but also demonstrate their rich connections with particular communities and places. Certainly it is possible and productive to identify general characteristics of ice – such as its tendency to be in process, transition, and movement. However, ice occurs on the planet not in the abstract, but in particular places, times, and forms, and narrative accounts suited to one kind of ice in a particular part of the planet (for Indigenous Arctic communities, for example) may not be extrapolatable to 'ice' elsewhere or in general.

Within the short space of this chapter, then, I want to focus on the narrative frames that are and might be attached to one particular kind of ice in one region of the world: Antarctic icebergs. The specific characteristics of icebergs – their ability to be identified as discrete, if mutable, objects; their temporal and spatial extent; and their unpredictable routes – enables them to be apprehended not simply as 'ice' but as individual actants that enact specific planetary trajectories. In the Antarctic, where icebergs are remote from any established human communities, this apprehension is almost always mediated, either remotely through written, visual, and aural texts, or more directly through tourist operators and guides. How, this chapter asks, might these cryonarratives look if we consider icebergs themselves as planetary travellers?

Icebergs and/as travellers in the Anthropocene

The present pandemic period, when human travel has been disrupted, abandoned, or indefinitely postponed, provides a unique moment at which to turn our attention to nonhuman travellers. Travel-writing scholarship has recently started to incorporate the nonhuman as agent rather than backdrop. Much of this work focuses on human travellers' encounters with nonhuman animals, who are occasionally themselves framed as co-travellers rather than simply passive 'travellees' to be seen and experienced by humans (Leane, 2020). Critics have only just begun to consider the role in travel writing of 'not only sentient animals or other biological mechanisms, but also impersonal agents, ranging from water to hurricanes, from mineral to bacteria, from mountains to information networks' (Kantarbaeva, 2018: 8). Work of this kind emphasizes the agency of landscape features, but seldom extends to considering such objects *as* themselves travellers – presumably because such geological features, while far from static, do not tend to move from place to place, at least on human temporal scales.

In this sense, icebergs are unusual. Of course, movement (along with phase transition) is a characteristic of ice in general: glaciers flow, collapse, retreat and advance; sea ice moves on ocean currents, and appears and disappears seasonally. Icebergs, however, are both discrete, autonomous objects and mobile over large distances. While they may ground or remain hemmed in by sea ice for long periods, their lifetimes are usually characterized by journeys. Their slow beginnings as proto-bergs – parts of glaciers and icesheets – can last thousands of years, while after calving, depending on their size and the temperature of the surrounding water and atmosphere, they may last weeks, months, years, or decades. With knowledge of prevailing currents, their journeys can be predicted at a broad although not a detailed level. The journeys of icebergs above a certain size, however, are known precisely, as they are tracked by satellite to avoid posing a hazard to shipping. For centuries, humans have speculated about controlling and diverting this travel, with iceberg-towing schemes to provide water to dry nations having been touted periodically since the nineteenth century (see e.g. Ruiz, 2015; Morgan, 2018). However, apart from small-scale diversion of bergs from oil rigs in the Arctic, such geoengineering schemes have remained fantasies.

The unpredictable mobility of icebergs means that they are readily characterized as travellers. This is a literary rather than a philosophical argument: I do not contend that icebergs can be conceptualized ontologically as travellers in the intentional sense of human or animals. Nor do I suggest that icebergs can be the subject of travel narratives as conventionally conceived, given that these are normally defined as first-person accounts written by author-travellers (Youngs, 2013: 3). Cryonarratives are written by human observers, aided by technologies. However, in the sense that icebergs' journeys are materially encoded in their shape, colour, and composition, as well as scoured out (in the case of large bergs) on the ocean floor, they might be considered in a sense as co-authors of their own stories.

For cryonarratives written in the Anthropocene, there are advantages to rhetorically positioning icebergs as travellers. Unlike narratives of rescue, ruination, vanishment, elegy, or apocalypse, narratives of iceberg-travellers must necessarily be episodic and open-ended. While human impact on calving or melt-rate must be part of any contemporary iceberg's journey, such an approach does not assume that icebergs' only significance lies in their role in an anthropocentric story. Moreover, figuring icebergs as travellers puts them, at a structural level, in a symmetrical relationship with the human travellers who encounter them.

The narrative of idea of iceberg-as-traveller is one that has been utilized effectively before, in Richard Brown's book, *Voyage of the Iceberg: The Story of the Iceberg that Sank the Titanic* (1983). Anticipating the nonhuman turn by a couple of decades, Brown deliberately decentres the human tragedy of the ocean liner in favour of following the iceberg with which it collided. His intention is to challenge demonized stereotypes of the berg by presenting its own story – a story that is thousands of years long (beginning with the berg's likely origins in a Greenland glacier) and in which the *Titanic* collision is one of many human and nonhuman encounters. Inspired by Brown's book, historian Ben Maddison and I published a short cryonarrative that adopted the genre of the biography to tell the story of a large tabular iceberg named B-9B that originally (as part of a larger berg, B-9) calved from the Ross Ice Shelf in 1987 (Leane and Maddison, 2018). We used this genre normally reserved for humans purposefully and provocatively as a way of drawing attention to the berg's individuality, longevity, and role as an actor in various natural-cultural events. However, our biography is simultaneously a travel narrative, as B-9B's long (and ongoing) life was/is inevitably a restless one, with the iceberg driven by deep ocean currents, grounding for years in shallow areas, colliding with and changing other ice formations, occasionally causing havoc to human expeditions, and eventually being drawn into climate change debates.

In the next section, I turn to another giant Antarctic iceberg – one whose 'birth' in 2017 was far more prominent than B-9B's. While this new iceberg was quickly enrolled in an apocalyptic story of catastrophic melting ice, I argue that a more productive and less anthropocentric cryonarrative might be constructed around the idea of this iceberg, and others like it, as planetary travellers whose journeys are entangled with, but not reducible to, human events and activities.

Creating a monster? A-68 as planetary traveller

> People were impressed by the first magnificent photographs of icebergs in the process of creation, but, although no iceberg is quite like any other iceberg, the generic similarity is pronounced. A rather brief period of awe was succeeded

> by the thought that while it was really very clever of science to know all about icebergs and climate and so on, it did not seem to be much good knowing if it could not, resultantly, do something about it.
>
> John Wyndham, *The Kraken Wakes* (231–232).

John Wyndham's cold-war science fiction novel *The Kraken Wakes*, first published in 1953, begins with its protagonists gazing out at a group of grounded icebergs, appearing 'like sudden white mountains' (5). What initially seems an impressive polar icescape is quickly transformed into a far more disturbing and defamiliarizing scene as the reader learns, in the third sentence, that these bergs have floated up the English Channel. The disturbing nature of the view lies not in what the icebergs are but *where* they are. Newspaper reports of the breaking up of polar ice recur throughout the narrative, which comprises a long flashback from this opening scene. In the Antarctic, there is 'news, via New Zealand, of glaciers in Victoria Land shedding huge quantities of bergs into the Ross Sea, and suggestions that the great Ross Ice-Barrier itself might be beginning to break up … The Filchner Barrier … and the Larsen Ice Shelf were both said to be calving bergs in fantastic numbers' (233–234). The spectacular icebergs are greeted by the public initially with an 'attitude of polite patronage of Nature' at the 'clever turns she put on to edify and amuse the human race' (234), but opinions begin to change when the phenomenon is revealed to be the work of an alien civilization living in the deep sea. While Wyndham's novel can clearly be read as a Cold War fable – some believe the icebergs are the work of 'the Russians' before the real alien culprits become apparent (232–233) – it takes on new significance when read in the context of the Anthropocene, with its out-of-place icebergs appearing as a portentous signal of unnatural change on a planetary scale.

Over sixty years after the publication of *The Kraken Wakes*, an iceberg was making headlines very similar to those described in Wyndham's novel. In early 2017, a growing rift in the Larsen C Ice Shelf was reported worldwide in tones of suspense and growing alarm (e.g. Rice, 2017) until, in mid-July, the split produced one of the largest icebergs on record. As in Wyndham's novel, the initial calving of the huge berg generated a reaction something like awe. However, this emotion was tempered by the suspicion of human complicity in the event. Less than two decades earlier, Antarctic travellers could gaze upon icebergs as purely aesthetic objects providing escape from their own lives and problems. Australian writer Helen Garner, in an account of an Antarctic tourist journey taken in the summer of 1997–98, describes herself while watching a calving scene as 'strangely at peace', reflecting that 'None of these gargantuan cataclysms has anything to do with *me*. Nothing is *my fault*' (2001: 22; original italics). In 2017, such detachment was no longer possible. As Bruno Latour writes, in the contemporary period 'there are no more spectators, because there is no shore that has not been mobilized in the drama of geohistory. Because there are no more tourists, the feeling of the sublime has disappeared along with the safety of the onlookers' (2017: 34). When the *Titanic* and the Arctic iceberg collided in 1912, this could be interpreted as a hubristic tragedy of humanity against nature and the iceberg itself vilified as an 'icy demon' (Brown, 1983: 7). By contrast, while A-68 was dubbed a 'monster' by the press (Amos, 2018), the potential implication of human action in its creation made it seem more like Frankenstein's monster, with humanity's hubris found not in challenging nature but in tangling with it.

Over the subsequent weeks, then, as announcements of A-68's arrival turned to interpretations of its significance, reports of the berg took on an increasingly concerned and politicized tone. Vice News gave readers '3 reasons to worry about that huge iceberg that broke off Antarctica' (the instability of the shelf, sea-level rise and the event being 'a sign of things to come') (Lubben, 2017). In an article in *WIRED* headlined 'Antarctic icebergs and crushing existential dread', Adam Rogers mused on whether humanity could recognize the calving event 'not just as megasized glaciological action but as yet another piece of the global weirdness ... that tells us Earth is getting hotter, the seas are rising and we are all in trouble', and asked whether the iceberg might 'galvanize action to fight climate' (Rogers, 2017). The climate activist organization 350.org strove to make this vision a reality, petitioning the US National Ice Center to name the berg '#Exxonknew', in reference to their Twitter campaign to draw attention to the oil and gas company ExxonMobil's contribution to climate change (Packard, 2017). A-68 was 'the first iceberg to become a star on social media' (Amos, 2021).

Throughout the calving event, scientists were at best divided over whether to attribute the iceberg's existence to climate change (Mosher, 2017). Collapses strongly linked to planetary warming, such as those at the neighbouring Larsen A and B, had taken a different form, with pooling meltwater on the ice surface causing sudden disintegration rather than the calving of a single giant berg from a rift. Even while acknowledging the instability of the Antarctic icescape, glaciologist Helen Fricker, writing in the *Guardian*, pointed to the breaking-off of large bergs as a component of the 'normal processes of a healthy icesheet ... part of the background against which we must look for real change' (2017). Thus while in an affective sense A-68 acted as a symbol of the enormity of climate change, scientifically the casual connection was less straightforward and was tied to the iceberg's earlier history of travel, prior to its autonomous existence, when it was part of the Larsen C Ice Shelf.

The calving of this giant iceberg was certainly, in Sörlin's phrase (2015: 327), a 'cryo-historical moment', but its existence extended well beyond this point. Glaciologists had been following its gestation – the growing rift in the ice shelf – for several years (Luckman, 2017). It was this broader narrative context that enabled them to understand the berg in the framework of the ice shelf as a whole, rather than a sudden object of fear in a human-centric drama. And while the iceberg's fifteen minutes of fame passed, it nonetheless retained media salience as its journey continued.

During the following years reports on the progress of the iceberg – by now named, more conventionally, A-68, according to its location and timing of origin – appeared periodically in the news as it floated off in a peripatetic but broadly predictable northeasterly path. As A-68 was very thin compared to its area, parts of the berg sheared away to form separate icebergs – A-68B and A-68C – with the larger remaining part now technically A-68A. By 2020, even as the impact of the COVID-19 pandemic began to be felt across the globe, the berg's activities were still making world news as it launched 'a run for it' (Amos, 2020b), heading past the sea-ice limit and into 'Iceberg Alley', and threatening to further disintegrate (see also e.g. Hambling, 2020; Carbone, 2020). Media reports of A-68's wanderings were often framed in the context of the warming and destabilizing ice of the Peninsula, and other icebergs that were far smaller but more immediately linked to climate change – such as B-49, which calved from thinning and retreating Pine Island Glacier (Amos, 2020a) – generated attention, benefitting from A-68's ongoing media halo. By late 2020, the iceberg remnant was approaching the island of South Georgia, and media reports began to focus on the potential impact of it grounding near the island and blocking the foraging routes of penguins and seals (Hansen, 2020) – effectively spoiling the travel plans of other nonhumans. On 10 April, however, it began disintegrating into hundreds of smaller bergs – too small to be systematically tracked. At this point, according to an Australian Bureau of Meteorology bulletin, 'Iceberg A-68A ha[d] completed its journey' – a journey which covered more than 4,000 km and continued for nearly four years (Lieser 2022: 15).

While the narrator of *The Kraken Wakes* might consider all icebergs as generically similar, A-68 – which, at over a trillion tonnes at its calving, must have been the largest single mobile object on Earth – had a size, singularity, and longevity that marked it as a distinct planetary actant with an individual history and future trajectory. As a traveller whose journey was long-term and episodic rather than just a sensational media event, A-68 was an ongoing reminder of impact on Antarctic ice, but was also more than only a threatening 'monster' or an avatar of human doom.

While A-68's path lay east of typical cruise ship itineraries in the Antarctic Peninsula, its peregrinations were not irrelevant to leisured human travellers. In late December 2019 – not long before COVID-19 put a pause on almost all Antarctic tourism – the *National Geographic Orion*, a vessel operated by Lindblad Expeditions, ventured into the Weddell Sea and sailed along the edge of the massive berg (Bluestein, 2019). In the next section, I turn to the iceberg tourism industry, asking how human visitors to the Antarctic might be encouraged to consider A-68 and all icebergs as fellow travellers with their own individual stories, rather than simply visual commodities to be aesthetically admired and photographically captured.

The end of travel? Iceberg tourism

> We'd rather have the iceberg than the ship,
> although it meant the end of travel.
> Elizabeth Bishop, 'The Imaginary Iceberg' (1983: 4)

These famous opening lines of Elizabeth Bishop's poem 'The Imaginary Iceberg', first published in 1935, have a strange prescience when one reads them today. Contemporary poet and Antarctic tour guide Elizabeth Bradfield notes in her prose-poem memoir *Toward Antarctica* (2019: 67) that she recalls these lines whenever she sees an iceberg, and elsewhere admits to reciting the poem, 'transgressive and true', aloud to the guests she guides in a Zodiac around polar ice (2016: n.p.). The travel blog Adventure365 quotes the entirety of Bishop's poem in celebration of World Poetry Day, adding that 'it's got us all inspired to venture to Antarctica and join a sailing expedition aboard a tall ship!', and providing the hyperlinks to direct customers to just such an adventure (Another World Adventures, 2020).

When Bishop's poem first appeared, only two decades after the *Titanic* disaster, icebergs would have been present in the public mind particularly as a hazard to shipping. They were dangerous and unpredictable objects that could put an end to travel very literally. This sense of threat is latent in the poem, in which the iceberg appears slightly sinister as well as beautiful, its self-adornments 'like jewelry from a grave'. Yet Bishop's poem anticipates a time in which icebergs would be less a danger than an attraction for ships – the tourist ships that ply the polar regions. In the poem's second stanza, the iceberg (along with the sailor who watches it) is figured as a theatre actor whose curtain rises 'on finest rope that airy twists of snow provide', one who 'dares […] upon a shifting stage'. The iceberg performs for its human audience, who, despite their stated communal preference, inevitably choose the ship, as it departs for warmer climes. Bishop's iceberg, however, is not simply a passive object of the human gaze. The iceberg itself returns the gaze – it 'stands and stares' at the observer – and, although 'stock-still' in repose, it threatens to 'wake' and 'pasture' upon the surrounding snowy ice, its 'field'. Figured as a nonhuman animal, the iceberg has an incipient liveliness and agency.

In the twenty-first century, the choice posed at the beginning of Bishop's poem between the iceberg and the ship takes on a disturbingly literal meaning. With their construction of a choice between the means and object of travel, her lines seem to inadvertently point up the ironies of this industry. Antarctica received almost 75,000 tourists in the summer of 2019–20, the vast majority of them travelling by cruise ship (IAATO, 2020a). This was a steep increase on the previous year's figures, continuing an exponential climb that has characterized the industry – with a dip during the Global Financial Crisis – since it began in earnest in the early 1990s. Travelling to Antarctica requires, for almost all tourists, first travelling by plane or ship to the bottom of South America, and then taking a cruise journey to the northern end of the Antarctic Peninsula. In doing so, they create a carbon footprint that has been estimated at 5.44t of CO_2 per passenger (Farreny *et al.*, 2011) – higher than the global average of emissions per person per year (Friedlingstein *et al.*, 2019: 1810). Cruise tourists thus contribute to the demise of the Peninsula's icescape even while they enjoy its aesthetic pleasures. Read in this context, 'The Imaginary Iceberg' becomes an unexpected poem of last-chance tourism.

In early 2020, the choice between iceberg and ship became suddenly redundant when the majority of global travel was put to an abrupt end by the COVID-19 pandemic. With cruise ship tourism – including in the Antarctic – a significant vector in the pandemic, the future of the industry is currently uncertain. Whether the so-called 'Anthropause' (Rutz *et al.*, 2020) will produce, like the global financial crisis, only a hiccup in an otherwise steady rise of the sector, or will prompt an entire rethinking of vessels, destinations, itineraries, and experiences, remains to be seen (Frame and Hemmings, 2020). We are thus in a forced period of inaction and reflection on the cultural significance and physical implications of the ice continent as a tourist destination.

Iceberg tourism has been the subject of very little critical attention to date. While there is a robust body of scholarship focused on the polar tourism industry, few studies look in any detail at the particular attractions and impact of encountering ice itself as part of this experience. One exception is Anita Lam and Matthew Tegelberg's 'Dark Tourism in Iceberg Alley', which focuses on sightseeing off the Newfoundland coast (along with the commercial harvesting of Arctic icebergs for bottled water and other drinks).[1] Lam and Tegelberg use the literary framework of noir crime narratives to approach their topic, making a connection between the 'mean streets' of hardboiled fiction and the 'alley' along the coast of Canada where Greenlandic icebergs collect, to consider 'how the human conquest of nature appears as a form of slow violence in the Anthropocene noir' (146). Within this literary framework, Lam and Tegelberg read iceberg sightseeing as a form of 'dark tourism': 'melting icebergs are transformed into nonhuman death spectacles for sightseeing tourists … Couched in a narrative of purity, [iceberg consumption] has been sold in ways that conceal the dark, ecological costs associated with commodifying iceberg deaths' (2020: 148).

While Lam and Tegelberg's analysis is provocative and stimulating, it also provides an instructive example of the construction of a cryonarrative around a pre-existing generic framework, the noir crime narrative – one which imposes particular roles on the material environment, in this case figuring icebergs primarily as passive victims of human consumerism. In keeping with the materialist turn, Lam and Tegelberg contend that ice is agential with its own 'vitality and liveliness' (2020: 154), but they do so in order to interpret the melting of icebergs as equivalent to human death. In this conceptualization icebergs, which are melting from (or before) the moment they calve, are therefore always dying.

Iceberg tourism in Antarctica necessarily takes a different form to that described by Lam and Tegelberg. While tourists in Newfoundland can 'hunt' icebergs on their own itineraries along coastal routes, guided by marketing material and an online app (Lam and Tegelberg, 2020: 160), in Antarctica icebergs are almost entirely inaccessible to the independent tourist, aside from a small number of private yachts. Rather, Antarctic iceberg encounters are inevitably curated, with passengers viewing icebergs from a ship and, for more close-up encounters, in Zodiacs driven by guides. And while Lam and Tegelberg point to the ways that iceberg consumption is nationally fostered through government economic policies, Antarctica's international status and governance provides different opportunities and risks. Operators must abide by the environmental specifications of the Antarctic Treaty System, but at present the industry itself – organized by the International Association of Antarctica Tour Operators (IAATO) – is largely responsible for setting standards around environmental impact.

This curation of the Antarctic experience opens the possibility that iceberg tourism could be deliberately framed in a way that both acknowledges the 'darkness' of the experience while remaining open-ended enough to turn this knowledge into positive action. Is it feasible that, as Adam Rogers hoped for A-68, icebergs might be used to galvanize tourists' environmental consciousness through purposeful narrative framing? There are admittedly reasons to be sceptical of such a vision. Despite the long-standing idea, vigorously promoted by IAATO and individual operators, that Antarctic tourism creates 'ambassadors' for the continent, the evidence of this claim is mixed at best (e.g. Vila *et al.*, 2016). And although tour operators already run onboard lectures raising awareness of anthropogenic warming, the timing, nature, extent, and efficacy of such endeavours is not yet clear.

Nonetheless, the curated and self-organized nature of Antarctic tourism does open the possibility for purposeful storytelling, and existing citizen science activities suggest the potential for systematically engaging tourists in environmental endeavours (Farmer, 2019). IAATO's advice to its members goes beyond the practicalities of visitor numbers per site and approach distances to animals to include less tangible aspects of the experience, including how the operator mediates and markets it (IAATO, 2020b and 2020c). The discourse of 'last chance' tourism, for example, is explicitly discouraged by IAATO in favour of an emphasis on protection. IAATO's guidelines do not yet extend as far as advice to tour guides and onboard lecturers on framing particular encounters. It is possible, however, to imagine a tour experience that, through mediation, not only enables tourists to admire the aesthetics of icebergs as photogenic objects but also encourages them to contextualize their features – location, size, shape, colours, debris – within a travel narrative. Where might an iceberg have travelled from and where is it headed to? How might human actions – particularly anthropogenic warming – have contributed to its source, its route, and its fate? 'Celebrity' icebergs such as A-68 could be tied into this narrative in order to incorporate a broader context of thinning and destabilizing iceshelfs against a background of natural calving.

Moreover, polar tourists' encounters with icebergs are not only impacted by the way in which these objects are marketed and mediated, but also by the 'pre-imaginaries' they bring to their journeys – that is, 'the narratives used to make sense of the touristic experiences' (Bislev and Smed, 2018: 39, 36). One study found that icebergs were the second biggest element in tourists' images of the Antarctic prior to their travel (after penguins) (Bauer, 2001: 156); people bring to Antarctica their own 'imaginary icebergs'. The cryonarratives that circulate in culture – via literature, film, non-fiction accounts, and the media – provide a framework within which human–iceberg encounters take place. One important area of future research into the form and impact of cryonarratives is qualitative analysis of the stories that tourists from different cultural backgrounds bring with them, and how these combine with both their own direct experiences of polar ice and the mediation offered by the operator. Here, I suggest that cryonarratives in which icebergs are characterized as continually travelling, rather than continually dying, provide tourists with a richer and more open-ended framework against which to consider their own environmental behaviour.

This chapter has put forward some reasons for characterizing and narrating icebergs as dynamic co-travellers with their own histories and futures. The interest in the journey of the giant iceberg A-68 points to an appetite for such narratives, which could in turn be drawn on to contextualize and frame tourist encounters with polar ice. Through such open-ended cryonarratives of travel, icebergs can be transformed from symbols of human destruction and doom, or static objects of the tourist gaze, to mobile, active, and unpredictable planetary actors whose journeys are interconnected with our own.

Note

1 There are 'iceberg alleys' in both the North and South Atlantic, with currents sweeping icebergs from the Arctic and Antarctic respectively down the coast of Labrador towards Newfoundland, and northeast from the top of the Antarctic Peninsula.

References

Amos, J. (2018) 'The "Monster" Iceberg: What Happened Next?', BBC News, 9 July. Available at: www.bbc.com/news/science-environment-44745734 (accessed 19 October 2020).

Amos, J. (2020a) 'Antarctica's Big New Iceberg: Up Close with B49', BBC News, 12 February. Available at: www.bbc.com/news/science-environment-51479329 (accessed 19 October 2020).

Amos, J. (2020b) 'World's Biggest Iceberg Makes a Run for it', BBC News, 5 February. Available at: www.bbc.com/news/science-environment-51389690 (accessed 19 October 2020).

Amos, J. (2021) 'A68: Iceberg that Became a Social Media Star Melts Away', BBC News, 18 April. Available at: www.bbc.com/news/science-environment-56779687 (accessed 24 May 2020).

Another World Adventures (2020) 'The Imaginary Iceberg', Another World Adventures website, 20 March. Available at: www.anotherworldadventures.com/the-imaginary-iceberg1/ (accessed 19 October 2020).

Antonello, A. (2017) 'Engaging and Narrating the Antarctic Ice Sheet: The History of an Earthly Body', *Environmental History* 22(1): 77–100.

Bauer, T. (2001) *Tourism in the Antarctic: Opportunities, Constraints, and Future Prospects*. New York: Haworth Hospitality Press.

Bishop, E. (1983) 'The Imaginary Iceberg', in *The Complete Poems: 1927–1979*. New York: Farrar Strauss Giroux, 4.

Bislev, A. and K. Smed (2018) 'When will the Iceberg Melt? Narrating the Arctic Among Chinese and Danish Tourists Aboard a Cruise Ship in Greenland', *Arctic Yearbook* 2018: 29–42.

Bluestein, S. (2019) 'Paulet Island and Tabular Iceberg A-68-A'. Lindblad Expeditions – National Geographic website, 14 December. Available at: https://au.expeditions.com/daily-expedition-reports/191475/ (accessed 19 October 2020).
Boorst, L. R. (2010) 'The Tip of the Iceberg: Ice as a Non-human Actor in the Climate Change Debate'. *Études/Inuit/Studies et le changement climatique* 34(1): 131–150.
Bradfield, E. (2016) 'Fluid States: Ocean as Place and Poetic', *West Branch Wired* 83: n.p.
Bradfield, E. (2019) *Toward Antarctica: An Exploration*. Pasadena, CA: Boreal Books, Red Hen Press.
Bravo, M. and G. Rees (2006) 'Cryo-politics: Environmental Security and the Future of Arctic Navigation', *Brown Journal of World Affairs* 8(1): 205–215.
Brown, R. (1983) *Voyage of the Iceberg: The Story of the Iceberg that Sank the Titanic*. London: Bodley Head.
Carbone, C. (2020) 'World's Largest Iceberg the Size of Delaware is Streaming Toward Open Water and Could Endanger Shipping Vessels'. Fox News, 7 Feb. Available at: www.foxnews.com/science/worlds-largest-iceberg-streaming-toward-open-water-endanger-shipping-vessels (accessed 19 October 2020).
Cruikshank, J. (2005) *Do Glaciers Listen? Local Knowledge, Colonial Encounters and Social Imagination*. Vancouver and Toronto: UBC; Seattle, WA: University of Washington Press.
Dodds, K. (2021) 'Geopolitics and the Ice Humanities: Elemental, Metaphorical and Volumetric Reverberations', *Geopolitics* 26(4): 1121–1149.
Duckert, L. (2013) 'Glacier', *postmedieval: a journal of medieval cultural studies* 4(1): 68–79.
Farmer, L. (2019) 'Empowering Citizens at the Poles', *Narrative Inquiry in Bioethics* 9(1): 11–13.
Farreny, R., J. Oliver-Solà, M. Lamers *et al.* (2011) 'Carbon Dioxide Emissions of Antarctic Tourism', *Antarctic Science* 23(6): 556–566.
Frame, B. and A. D. Hemmings (2020) 'Coronavirus at the End of the World: Antarctica Matters', *Social Sciences & Humanities Open* 2: 1–5.
Fricker, H. (2017) 'Melting and Cracking – is Antarctica Falling Apart?', *Guardian*, 23 June. Available at: www.theguardian.com/science/2017/jun/23/melting-and-cracking-is-antarctica-falling-apart-climate-change (accessed 18 October 2020).
Friedlingstein, P., M. W. Jones, M. O'Sullivan *et al.* (2019) 'Global Carbon Budget 2019', *Earth System Science Data* 11(4): 1783–1838.
Garner, H. (2001) 'Regions of Thick-ribbed Ice', in *The Feel of Steel*. Sydney: Picador-Pan Macmillan, 13–34.
Glasberg, E. (2011) '"Living ice": Rediscovery of the Poles in an Era of Climate Change', *WSQ: Women's Studies Quarterly* 39(3–4): 221–246.
Hambling, D. (2020) 'Weatherwatch: Giant Iceberg on the Move and Picking Up Speed', *Guardian*, 18 July. Available at: www.theguardian.com/news/2020/jul/17/weatherwatch-giant-iceberg-on-the-move-and-picking-up-speed (accessed 18 October 2020).
Hansen, K. (2020) 'Iceberg A-68A nears South Georgia', National Space and Aeronautical Administration Earth Observatory website, 16 November. Available

at: https://earthobservatory.nasa.gov/images/147535/iceberg-a-68a-nears-south-georgia?linkId=104616459 (accessed 19 October 2020).

IAATO (International Association of Antarctica Tourism Operators) (2020a) 'Data and Statistics', *IAATO*, n.d. Available at: https://iaato.org/information-resources/data-statistics/ (accessed 19 October 2020).

IAATO (International Association of Antarctica Tourism Operators) (2020b) 'IAATO Annual Marketing and PR Self-audit', *IAATO*, n.d. Available at: https://iaato.org/wp-content/uploads/2020/04/Agenda-Item-11b.-IAATO-Marketing-PR-self-Audit.pdf (accessed 19 October 2020).

IAATO (International Association of Antarctica Tourism Operators) (2020c) 'Marketing Antarctica', *IAATO*, n.d. Available at: https://iaato.org/wp-content/uploads/2020/04/IAATO_Internal-Marketing-Antarctica.compressed.pdf (accessed 19 October 2020).

Kantarbaeva-Bill, I. (2018) 'Introduction', *Caliban: The French Journal of English Studies* 59: 7–17.

Lam, A. and M. Tegelberg (2020) 'Dark Tourism in Iceberg Alley: The Hidden Costs of Consuming Iceberg Deaths', in A. Lam and M. Tegelberg, *Criminal Anthroposcenes: Media and Crime in the Vanishing Arctic*. Cham, Switzerland: Palgrave Macmillan, 145–187.

Latour, B. (2017) *Facing Gaia: Eight Lectures on the New Climatic Regime*, trans. Catherine Porter. Cambridge and Medford, MA: Polity.

Leane, E. (2019) 'Ice and the Ecothriller: Popular Representations of Antarctica in the Anthropocene', in E. Leane and J. McGee, eds, *Anthropocene Antarctica: Perspectives from the Humanities, Law and Social Sciences*. London: Routledge, 87–100.

Leane, E. (2020) 'Animals', in A. Pettinger and T. Youngs, eds, *The Routledge Research Companion to Travel Writing*. London and New York: Routledge, 305–317.

Leane, E. and B. Maddison (2018) 'A Biography of Iceberg B09B', *Australian Humanities Review* 63: 99–115.

Lieser, J. L. (2022) 'Ice Bulletin: Iceberg A-68A', in *Ice Bulletins for the Antarctic Shipping Season 2021–2022*. Hobart, Tasmania: Bureau of Meteorology.

Lubben, A. (2017) '3 Reasons to Worry about that Huge Iceberg that Broke Off Antarctica', Vice News, 12 July. Available at: www.vice.com/en/article/9kd8da/3-reasons-to-worry-about-that-huge-iceberg-that-broke-off-antarctica (accessed 19 October 2020).

Luckman, A. (2017) 'I've Studied Antarctica's Giant Iceberg for Years and it's not a Simple Story of Climate Change', The Conversation, 13 July. Available at: https://theconversation.com/ive-studied-larsen-c-and-its-giant-iceberg-for-years-its-not-a-simple-story-of-climate-change-80529 (accessed 19 October 2020).

Morgan, R. A. (2018) 'Dry Continent Dreaming: Australian Visions of using Antarctic Icebergs for Water Supplies', *International Review of Environmental History* 4(1): 145–166.

Mosher, D. (2017) 'Why some Scientists won't Link Antarctica's Giant Iceberg to Global Warming', *Business Insider Australia*, 19 July. Available at: www.businessinsider.com.au/iceberg-a68-did-climate-change-cause-crack-2017-7 (accessed 19 October 2020).

Nüsser, M. and R. Baghel (2014) 'The Emergence of the Cryoscape: Contested Narratives of Himalayan Glacier Dynamics and Climate Change', in B. Schuler, ed., *Environmental and Climate Change in South and Southeast Asia: How are Local Cultures Coping?* Leiden and Boston: Brill, 138–156.

Nuttall, M. (2020) 'Water, Ice and Climate Change in Northwest Greenland', *WIREs Water* 7(3): 1–10 (e1433).

Packard, A. (2017) '10,000 People Join Call for the Iceberg that Just Broke Off Antarctica to be Called #Exxonknew', The Climate Centre, 14 July. Available at: https://theclimatecenter.org/10000-people-join-call-iceberg-just-broke-off-antarctica-called-exxonknew/ (accessed 19 October 2020).

Rice, D. (2017) 'The Massive Crack in the Antarctic Ice Shelf is Hanging on by a 12-mile "Thread"'. *USA Today*, 3 April. Available at: www.usatoday.com/story/tech/sciencefair/2017/04/03/antarctic-ice-crack-iceberg-larsen-c/99984132/ (accessed 19 October 2020).

Rogers, A. (2017) 'Giant Antarctic Icebergs and Crushing Existential Dread', WIRED, 13 July. Available at: www.wired.com/story/giant-antarctic-icebergs-and-crushing-existential-dread/ (accessed 19 October 2020).

Ruiz, R. (2015) 'Media Environments: Icebergs/Screens/History', *Journal of Northern Studies* 9(1): 33–50.

Rutz, C., M.-C. Loretto, A. E. Bates *et al.* (2020) 'COVID-19 Lockdown Allows Researchers to Quantify the Effects of Human Activity on Wildlife', *Nature Ecology and Evolution* 4(9): 1156–1159.

Sörlin, S. (2015) 'Cryo-history: Narratives of Ice and the Emerging Arctic Humanities', in B. Evengård, J. Nyman Larsen, and O. Paasche, eds, *The New Arctic*. Cham, Switzerland: Springer, 327–339.

Vila, M., G. Costa, C. Angulo-Preckler, R. Sarda, and C. Avila (2016) 'Contrasting Views on Antarctic Tourism: 'Last Chance Tourism' or 'Ambassadorship' in the Last of the Wild', *Journal of Cleaner Production* 111(B): 451–460.

Wyndham, J. (1953) *The Kraken Wakes*. London: Michael Joseph.

Youngs, T. (2013) *The Cambridge Introduction to Travel Writing*. Cambridge: Cambridge University Press.

13

Frozen archives on the go: ice cores and the temporalization of Earth system science

Erik Isberg

In 1975, ice made an appearance in a seemingly strange place. Wallace Broecker, a geophysicist at Lamont-Doherty Geological Observatory, had barely worked with ice or glaciological research before, but in his article 'Climatic Change: Are We on the Brink of a Pronounced Global Warming?' ice served as the empirical foundation for his argument (Broecker, 1975). The argument did not have much to do with ice either: it concerned the temperature of the planet and how it could be rising due to increased amounts of carbon dioxide in the atmosphere. Yet, Broecker relied on the work of Danish ice core scientist Willi Dansgaard and his ice core data from Greenland in order to prognosticate the future of the world's climate. In the years following Broecker's article, ice cores began to appear in the company of other paleoarchives – deep sea cores, lake sediments, tree rings, corals – in order to temporalize the dynamics of the entire Earth system (Oeschger, 1985). These archives differed in many ways – they had different durations, qualities, materialities, and geographical boundaries – but were placed together as a part of the effort to produce planetary-scale environmental knowledge, rather than represent a particular geography.

Ice appears, in this setting, as a proxy, an *Earth system object* rather than a scientific object bound to particular geographies or practices, defined by its function within a planetary-scale environmental system. In the form of the ice core, ice has become a crucial element in the work of Earth system scientists in their effort to configure the planet as an environmental object, which can be quantified, measured, and governed. According to Dipesh Chakrabarty, Earth system science (ESS) has enabled the emergence of a 'new regime of historicity' which brings together 'the history of the planet, the history of life on the planet, and the history of the globe made by the logics of empires, capital, and technology' (Chakrabarty, 2019: 2) It is, however, less evident how this new regime of historicity came about in practice. Under which conditions did the history of the planet encounter the history of the globe? The integration of multiple human and nonhuman histories through ESS is in itself a historical process unfolding in the postwar era.

In this chapter, I trace the temporalities of ice core science in relation to the production of planetary-scale environmental knowledge and the institutionalization of ESS. During the 1970s and 1980s ice, in the form of ice core data, was spatially and conceptually decoupled from the cryosphere and entered a new scientific inquiry into the entire planet as one coherent system. Samples from ice sheets in the cryosphere appeared in the company of other, equally despatialized, paleoarchives, in fields that had just a few decades before been seen completely detached from glaciological issues and methodologies. In order to conceptualize the Earth system as an object of knowledge, the temporal properties of the system had to be defined (Koselleck, 1997; Isberg, 2020). Ice cores played a key role in this process of temporalization by providing ways to measure rates of change in planetary dynamics. During this time, ice appeared in multiple mediations – as ice core, timescale, climate model, dataset – in the production of environmental knowledge. The history of postwar ice core drilling as a history of *environing media* is investigated, highlighting how multiple mediations of ice have been part of conceptualizing environmental epistemologies on a planetary scale (Sörlin and Wormbs, 2018; Wickberg and Gärdebo, 2020).[1] The rise of Earth system science made a new kind of ice possible: one that mediated deep pasts and futures and moved from the cryosphere to the entire planet.

Ice cores have, during the last decade, gathered significant attention from humanist scholars because of their work as 'climate change messengers' (Elzinga, 2017). Calls for critical studies of ice core temporalities have appeared in geography and environmental history, because of the role they play in producing not only planetary pasts, but increasingly also human futures (Yusoff, 2008; Antonello and Carey, 2017). The history of ice core drilling has been studied in both an Arctic and Antarctic context (O'Reilly, 2016; Achermann, 2020) and the work of Danish glacial physicist Willi Dansgaard has also been addressed by several historians as well as in his own autobiography (Lolck, 2004; Dansgaard, 2005; Martin-Nielsen, 2013a; Dry, 2019). However, the interaction between ice cores and the broader community of scientists working on planetary-scale environmental knowledge is still not widely studied.

Drilling in ice: a brief historical overview

When Willi Dansgaard, widely regarded as the founder of modern ice core science, began drilling into Greenland icebergs in the 1950s, he entered a longer history of drilling into ice and interest in the interiority of glaciers.

Drilling had been an Indigenous technology in the Arctic for centuries in order to access fishing waters, but in the nineteenth century it also began gathering scientific interest. Louis Agassiz, who was one of the founders of glacial theory in the mid 1800s, tried to drill through the Unteraargletscher Glacier in the Alps, but his rudimentary drill never managed to get all the way through to bedrock (Talalay, 2016).

In early twentieth century glacier research, the inner life of glaciers remained largely unknown. Rather, most of the scientific work on glaciers and cryosphere environments concerned mapping and surveying. Glaciers were studied for their geomorphological impact, volume, surface extension, and location. The gaze was mostly horizontal and the glaciers were studied in their entirety as coherent scientific objects (Sörlin, 2011). During the interwar period, some glaciologists began to revisit what Agassiz had tried but failed to do in the nineteenth century due to insufficient drilling technology: scientifically investigate the interior of a glacier. Ernst Sorge, a German glaciologist and part of Alfred Wegener's third and last expedition to Greenland's *Eismitte*, began to take an interest in the way the ice sheet was structured vertically (Martin-Nielsen, 2013b). He dug out large pits in the snow, which allowed him to quantitatively study the near-surface snow strata, and notice how older snow was visible further down, while snow that fell more recently was closer to the surface. Standing in the pit, immersed by snow all around him, Sorge could perhaps be understood to stand at the threshold of a vertical understanding of glaciers. He was still concerned with the glacier in its entirety, maintaining his horizontal outlook while being lowered into the snow, but he was beginning to uncover a vertical dimension of stratigraphically ordered layers within the glacier itself. Sorge had – symbolically and literally – merely scratched the surface of the ice sheet's vertical properties. In order for the ice core to emerge, and the deep interiors of the ice sheet to be made visible, an entirely different research programme was necessary.

It was in the postwar era, when the Arctic region appeared as a site of great geopolitical interest, that the ice core itself could materialize. The International Geophysical Year in 1957–58 marked the beginning of institutionalized large-scale ice core drilling, and mobilized manpower, scientific interest, and technology in order for the expensive venture into the interior of the ice sheet to become possible (Collis and Dodds, 2008). Ice, as a scientific object, was in this format different from the vast spaces that early twentieth century explorers had encountered. It was a vertical object rather than a horizontal landscape and the International Geophysical Year provided enough institutional and technological infrastructure for the ideas about verticality in glaciology to materialize and previous pitfalls of broken ice core samples to be avoided (Langway, 1958).

Less than a decade after the International Geophysical Year, in 1966, ice core drilling took a leap forward with the successful drilling to bedrock at the military base Camp Century on Greenland (Petersen, 2011; Nielsen *et al.*, 2014; Doel *et al.*, 2016). The ice core that emerged from the drilling was 1,387 metres long and of high quality, which enabled the construction of long, unbroken timescales and provided proof that the ice sheet was frozen all the way down to bedrock. For Willi Dansgaard, the Camp Century ice core provided him with the opportunity to utilize his isotope-dating methodology, which he had been developing since the 1950s (Dansgaard, 1953). By studying the ratio of two different oxygen isotopes – $\delta^{18}O$ and ^{16}O – in the layers of the ice core, he could determine the temperature at the time the snow fell onto the ice sheet. Lower temperatures would lead to a lower ratio of the heavier ^{18}O compared to the lighter ^{16}O (Dansgaard, 2005). Adding temperature as a category for what could be studied through ice made ice core drilling relevant for the growing fields of paleoclimatology and theoretical geochemistry (Summerhayes, 2020: 189–190). It also served as a way to connect ice core drilling with climate modelling, a practice which was gathering interest by utilizing computer technology to predict future climatic changes. In 1966, the massive core itself and the relevance it gained through isotope dating intersected and led to a large interest from several actors in Europe and the US (Dansgaard, 2005).

After the ice core had been recovered from the ice sheet, a complex infrastructural project began: Dansgaard wanted access to his ice cores, but some of them were stored in Buffalo, NY, at the Cold Regions Research and Engineering Laboratory (CRREL) where Dansgaard's collaborator at Camp Century, C. C. Langway was in charge of the ice core samples. In 1968, Dansgaard was able to get some smaller samples sent to him and his Copenhagen colleague Jorgen Møller travelled to CRREL shortly after and brought back some 1,000 samples (Achermann, 2020). The vertical form of the ice core samples not only enabled new temporal conceptualizations of ice, they also, through their portability, produced *new epistemic geographies of the cryosphere* (Mahony and Hulme, 2018).[2] Ice core storage facilities provided access to remote geographical areas through the ice cores, which could serve as proxies for entire ice sheets and cryospheric environments.

In the years following the Camp Century ice core, Dansgaard and his colleagues published their results extensively in glaciological journals as well as in broad-spectrum, high-impact scientific journals such as *Nature* and *Science*. Wallace Broecker, who was at the time working on ocean circulation and deep sea core drilling, noted how the ice core data provided higher reliability in its data on past climatic conditions than the information he could access through deep sea cores (Dry, 2019). The role of the ice core as proxy had, in Broecker's work, expanded even more: it was now functioning as a proxy for climatic changes taking place on a planetary scale, therefore of relevance to researchers who were not at all concerned with ice as a scientific object or the cryosphere as a geographical area. If, in the 1950s and 1960s, ice cores had travelled geographically from the ice sheets to ice core storage facilities, the 1970s marked the beginning of a period of disciplinary travel from glaciology into climatology, oceanography, geophysics, and the growing interdisciplinary community that would, in the 1980s, form Earth system science.

Calendars, archives and time machines: ice as a temporal imaginary in glaciology

Time has been an intrinsic part of glaciological inquiry since its foundation as a scientific field. However, the temporal qualities of ice have varied over time, rendering different conceptualizations of glaciological temporalities visible (Yusoff, 2005; Antonello, 2017). A significant shift in temporal imagination was, in Dania Achermann's words, the turn from a geography oriented two-dimensional approach to a geophysics oriented three-dimensional approach, where the latter came with the emergence of ice core drilling (Achermann, 2020). Reading time through ice became a different activity in the vertical ice core compared to the horizontal landscape. Time, as W. J. T. Mitchell has pointed out, only becomes legible through spatial mediation; time can be 'long' or 'short', linear or circular, become visible in clocks, timelines. and stratigraphy (Mitchell, 1980). In glaciology, time has also appeared as mediated through a particular form – ice – but depending on how this ice has been researched and conceptualized, the temporal imaginaries have differed over time.

For early polar explorers, the temporality of the vast ice sheets in the Arctic and Antarctica was defined by their absence of time: distant, white geographies, remnants from the last ice age, frozen in time (Antonello, 2017; Heuer, 2019). In the interwar period, glaciologists such as Hans Ahlmann, professor at Stockholms Högskola, began tracking glacier movements in real time. His annual measurements was comprised into glaciological 'calendars', as he put it in 1935, measuring annual glacier movements over decadal length through annual measurements (Ahlmann, 1935). The modus operandi of the kind of work Ahlmann was conducting was waiting: the temporality of ice was performed in real time, glaciers shrunk or grew annually and were measured in their entirety by Ahlmann and his colleagues as well as local inhabitants (Sörlin, 2018). The multisite, comparative method that Ahlmann championed enabled glacier ice to acquire a temporal dimension, but, at the same time, the horizontal perspective did not allow for connections with other geochronologies or with non-gradualist understandings of environmental change (Sörlin and Isberg, 2021).[3] The time he was tracking in the ice was spatially bound to the glaciers themselves and he did not transpose the slow extensive times of glacier mobility that he established to other geographies or scientific temporalities in neighbouring disciplines.

When ice core drilling appeared as a new, promising scientific endeavour in the postwar years, a new temporal concept began to circulate in glaciology: ice as archive (e.g. Sullivan, 1958). The vertical dimension enabled a stratigraphic outlook on glacier ice, which in turn made ice possible to connect to other kinds of geological archives. Compared to Ahlmann's calendars, the ice-as-archive imaginary opened the door to other disciplinary connections and other sciences of the archive (Daston, 2017). The possibility to utilize ice as an archive unsettled the boundaries of glaciology: what could really be scientifically studied through ice? In a 1973 entry in the Danish scientific periodical *Meddelelser om Grønland* (Communications on Greenland), Dansgaard, along with colleagues from the University of Copenhagen, asserted that: 'The development of ice core drilling technique has led to a broad variety of studies reaching far beyond glaciology itself' (Dansgaard *et al.*, 1973). In the article, they listed a wide range of scientific disciplines to engage with, ranging from meteorology and climatology to solar physics and atmospheric chemistry. Lorraine Daston identifies the work of making data commensurable as one key role of a scientific archive, as 'methods, instruments, records and observers must be calibrated across polities, epochs and genres' (Daston, 2017). This process of calibration is visible in the practices surrounding ice core science too.

In one of the first major publications on the Camp Century ice core, in *Journal of Glaciology* (1969), Willi Dansgaard and S. J. Johnsen presented a 'time scale' that the ice core had made possible (Dansgaard and Johnsen, 1969). This was the first time that 'time scale' was in a title in *Journal of Glaciology* and the visualization of the ice core timescale entered the visual language of geological stratigraphy as well as the history of timelines itself (Rosenberg and Grafton, 2010; Simonetti, 2015). According to Daston, almost all scientific archives follow a similar four step evolution: acquisition, retrieval, reconfiguration, and transcription (2017: 9). The work of Dansgaard and his colleagues, from the first attempts of using isotope dating on glacier ice in the late 1950s to the publication of ice core data in a standardized 'time scale' in 1969, can be seen yet another example of this process. In turning glaciological practices towards the archival, Dansgaard was also, at the same time, unsettling the disciplinary boundaries of the field itself.

Following Daston's notion of the scientific archive as a site which makes historical data commensurable, the role of the ice core can also be understood to destabilize the demarcations of glaciological research and connect ice – in a quantified and abstracted form – with other chronologies and geographies. In the process of making planetary-scale environmental knowledge, the commensurability of ice core data would be key to connecting ice with geophysical planetary processes and the bourgeoning work of calculating and projecting the Earth's climate.

By opening up the temporal domain of the deep past, ice cores were, in the early 1970s, also increasingly seen as reference points against which current or future climatic changes could be measured (Rosol, 2017). Shortly after the ice cores had expanded the temporal boundaries into the deep past, Dansgaard and other ice core scientists were increasingly looking in another temporal direction: the future. In a 1972 article in *Quaternary Research*, Dansgaard, along with S. J. Johnsen, C. C. Langway, and H. B. Clausen, used their results from Camp Century to 'speculate' about future glaciations (Dansgaard *et al.*, 1972). The article raised concern regarding the risk for a new, rapidly developing glaciation and asked whether human impact could be a force comparable with geophysical phenomena such as ice surges or intense volcanic activity. The data from the ice they recovered in 1966 could, in this article, be used to project planetary futures, rather than merely archive Arctic pasts, connecting ice with the emergent scientific study of the future (Andersson, 2018). This remarkably quick epistemological expansion, unfolding in less than a decade, enabled Dansgaard and his colleagues to appear as experts in areas that would previously have been unthinkable for glaciologists to claim. Dansgaard's isotope dating methodology, which he had been developing since the 1950s, was key for this expansion to occur, as it brought in global temperature as a phenomenon which could be studied through the ice. The temporal scope, which by the early 1970s had grown into the deep past as well as the future, also came with political implications: were humans interfering with the planetary dynamics visible in the ice cores?

An indicative example of this temporal, epistemological, and political expansion of ice core research is a 1975 article in *Nature* entitled 'Climatic Changes, Norsemen and Modern Man'. Dansgaard was the lead author of the article, which was co-written with colleagues from the Geophysical Isotope Laboratory in Copenhagen. By using ice core data from Greenland, the authors argued that they could track the climatic changes that caused the demise of the Norse settlement on Greenland in the fifteenth century. Additionally, they went on to connect the climatic changes that they argue caused the settlement to collapse with present-day environmental problems (Dansgaard *et al.*, 1975). Climate change appears as an exterior force, which can hit societies in different points in time.[4] This way of conceptualizing drastic climatic shifts, as Alessandro Antonello and Mark Carey have shown, is common in ice core futurities, and tends to obscure the role of politics and societal organization while giving primacy to climate as an exterior force. But the article also highlights the increasingly complicated epistemologies and temporalities of ice core research at the time. By connecting ice cores to human pasts – the Greenland settlement – and human futures – climate modelling – Dansgaard and his co-writers were simultaneously expanding the political, temporal, and epistemological boundaries of their research.

The article made a particular appeal to climate modellers: paleoarchives such as ice cores could be used in 'diagnosing the processes that cause climatic changes, and for checking the validity of the models on all timescales between 10 and 10^6 yr' (Dansgaard *et al.*, 1975). The early years of the 1970s is described by Matthias Heymann and Amy Dahan Dalmedico as a period of 'deep cultural shifts' in climate modelling, as climate change had risen on the scientific agenda and an increased political and scientific interest in prognosticating future climates rather than just acquiring a better physical understanding of the atmosphere (Edwards, 2010; Heymann and Dahan Dalmedico, 2019). Ice cores were increasingly picked up in these new attempts to model future climates, as they were providing an empirical basis for computerized modelling and a proxy record of climatic changes over vast periods of time.

If the ice as archive-imaginary had taken hold in the 1960s, a new kind of temporal imaginary began appearing towards the end of the 1970s: ice as time machine. The perceived temporal one-directedness of the archive was increasingly reconceptualized as the temporal framework expanded into the future, and ice cores could serve as proxies for both past and future climates. In ice core scientist Richard B. Alley's 2000 monography on his work in Antarctica in the 1990s, *The Two-Mile Time Machine: Ice Cores, Abrupt Climate Change and Our Future*, this view is particularly evident, but as the 1975 article by Dansgaard *et al.* shows, this idea of ice cores as mediators of future temporalities had, by 2000, already existed for decades (Dansgard *et al.*, 1975; Alley, 2000).

Paleoarchives in the Earth system

As ice cores began entering the scientific framework of climate modelling, the chronologies made visible through ice core drilling could appear in a new context. Other paleoarchives had, during the same period, caught the interest of the climate modelling community: ice was put side by side with tree rings, lake sediments, deep sea cores, and corals in a kind of elemental diorama. Similar interests in using sedimentation to track past climatic changes had been visible in other disciplines, such as quaternary geology and palynology, since the early twentieth century (De Geer, 1912), but the emerging interest in planetary-scale environmental knowledge created a demand for these alternate temporalities to merge into a singular temporality of the Earth system itself.[5] Ice, along with the other materialities which could serve as proxies for past planetary dynamics, became part of the work of configuring the planet as an environmental object.

Even though ice cores are most often studied as objects of glaciology, the expansion of ice cores into climate modelling led other scientists to engage with ice outside of glaciology. The ice they dealt with was far from the vast, field-based observed element of earlier glaciology. Rather, ice appeared here as a dataset from a specific ice core, which could then be utilized for studies in a wide array of fields, much like Dansgaard had predicted in 1973. In the 1970s, data from the Camp Century ice core is visible in articles in geo- and solar physics, meteorology, climatology, oceanography, and geology.[6] But it was particularly in matters concerning planetary-scale environmental knowledge that ice cores began to appear more frequently during the second half of the 1970s. In addition to the previously mentioned article on global warming by Wallace Broecker, Dansgaard's ice cores appeared in a 1975 report by the National Academy of Sciences entitled 'Understanding Climatic Change: A Program for Action'. The report was written by a committee from the Global Atmospheric Research Program (GARP) and the members came from different geophysical, climatological, mathematical and oceanographic institutions. Among its members were Broecker, but also the mathematician Francis Bretherton, who would later become central to the formation of Earth system science, oceanographer Henry Stommel, and atmospheric scientist Joseph Smagorinsky (United States Committee for the Global Atmospheric Research Program, 1975).

Dansgaard's Camp Century ice core was in this context one of many proxy archives listed by the committee as possible ways of tracking past climatic changes. The authors stated that 'Proxy data come from a wide variety of sources; potentially, any biological, chemical, or physical characteristic that responds to climate may provide proxy data useful in the reconstruction of past climates' (ibid.). Ice, as a scientific object, is defined by its utility in relation to other paleoarchives and its contribution to planetary-scale environmental knowledge. The cryosphere in its entirety is perceived as a mechanism in the planetary system rather than an environment in itself. Dansgaard's work appears in 'Understanding Climatic Change' in a setting that would have been impossible for earlier field-based research on glaciers and polar environments. He is referenced not along with other glaciologists, but with meteorologists such as William Kellogg, palynologist Margaret Bryan Davis, and oceanographer John Imbrie (ibid.).

This way of gathering data from a wide array of geographies, materialities, and temporalities to produce a coherent perception of Earth system dynamics would become commonplace with the institutionalization of Earth system science in the 1980s, but could also, I argue, be seen as a continuation of an emergent 'vertical glaciology' (Achermann, 2020). The verticality of the ice core enabled it to travel beyond glaciology and to translate layers of ice and snow into data sets recording changes over time. Similar enterprises in other fields, such as oceanography, dendrochronology, and palynology, conducted in a similar vertical manner, rendered translations between materialities and geographies possible. In this sense, the work of the committee writing 'Understanding Climatic Change' was to synchronize multiple paleoarchives and their temporalities into an emergent understanding of a quantifiable Earth system (Jordheim, 2014; Jordheim, 2017).

By the beginning of the 1980s, ice core scientists themselves were part of this work of synchronization and actively engaged with a network of scientists interested in planetary dynamics and the Earth system. In 1982, a group of scientists – geophysicists, oceanographers, glaciologists, and geochemists – gathered at the Lamont-Doherty Geological Observatory outside New York for the fourth biennial Maurice Ewing Symposium. The theme of the symposium was 'Climate Processes and Climate Sensitivity' and the objective was to bring 'together the work of modellers and others involved in the analysis of observational data. The approach considered climate variations and climate processes (focusing on climate feedback processes) on a broad range of time scales' (Hansen and Takashi, 1984). Even though the volume, which comprises the papers presented at the symposium, is divided into different areas, such as atmosphere, cryosphere, and ocean, the editors James Hansen and Taro Takashi note that these areas are bound together through feedback processes and thereby 'overlap extensively' (ibid.).

The volume includes entries by Willi Dansgaard and his fellow ice core colleague Hans Oeschger, as well as Wallace Broecker, and points to the increased interconnection between disciplines that had previously been separated by different objects of study. The vertical outlook and planetary scope made new connections possible. Temporality was, as the climate variations and processes were studied on 'a broad range of time scales', a central issue. How could different temporalities align in a way that made the system legible and predictable over time? For the ice core scientists themselves, this led to a necessity of situating their work in relation to the larger scientific effort of configuring an Earth system and other paleoarchives from other disciplines. In a 1985 entry in the volume 'Greenland Ice Core: Geophysics, Geochemistry, and The Environment', which was published as a Maurice Ewing Volume by the American Geophysical Union, Hans Oeschger discussed ice core timelines in relation to the 'environmental system' as well as other paleoarchives. The Dye 3 ice core – recovered in Greenland in 1973 – was put next to other chronologies from adjacent disciplines to produce planetary temporalities (Oeschger, 1985: Figure 13.1).

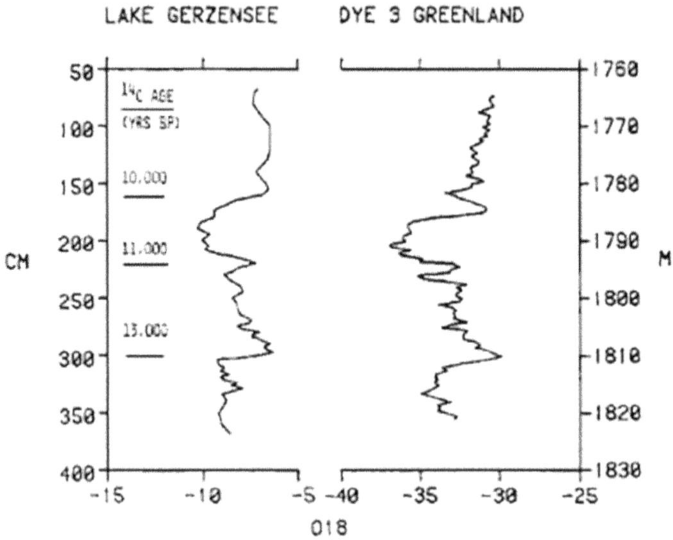

Figure 13.1 Comparing environmental records from Lake Gerzensee Switzerland and Dye 3 Greenland.

In this figure, Oeschger put the Dye 3 ice core chronology next to a lake carbonate record from Gerzensee in his native Switzerland. Despite their different materiality and geography of origin, the two timelines could be aligned through a similar vertical format as well as the utilization of the same isotope dating technology. The goal with this practice, Oeschger explained in the article, was to produce coherent and systematic knowledge about 'the environmental system'. This system encompassed 'the entirety of physical, chemical and biological processes action in the earth's surface and in the atmosphere' (ibid.). Further emphasizing the practical qualities of the concept, Oeschger asserted that the environmental system should be seen as a 'research approach' (ibid.). Working with environmental systems, in the way Oeschger defined them, was in other words as much about defining an object of study as it was a way to conduct research, a scientific practice.

Returning to the concept *environing media*, the environmental systems-approach laid out by Oeschger appears as a particular kind of environing activity, which aims to align timescales from different disciplines in order to create a coherent planetary-scale environmental system. Ice cores, as an environing medium, can here be seen to produce environmental knowledge which emphasize the unity of the planetary system and the possibilities to translate environmental data across spatial and temporal scales. Ice cores, in Oeschger's work, appear as the intermediary between human and planet, enabling a particular epistemology of the planetary environment to take form.

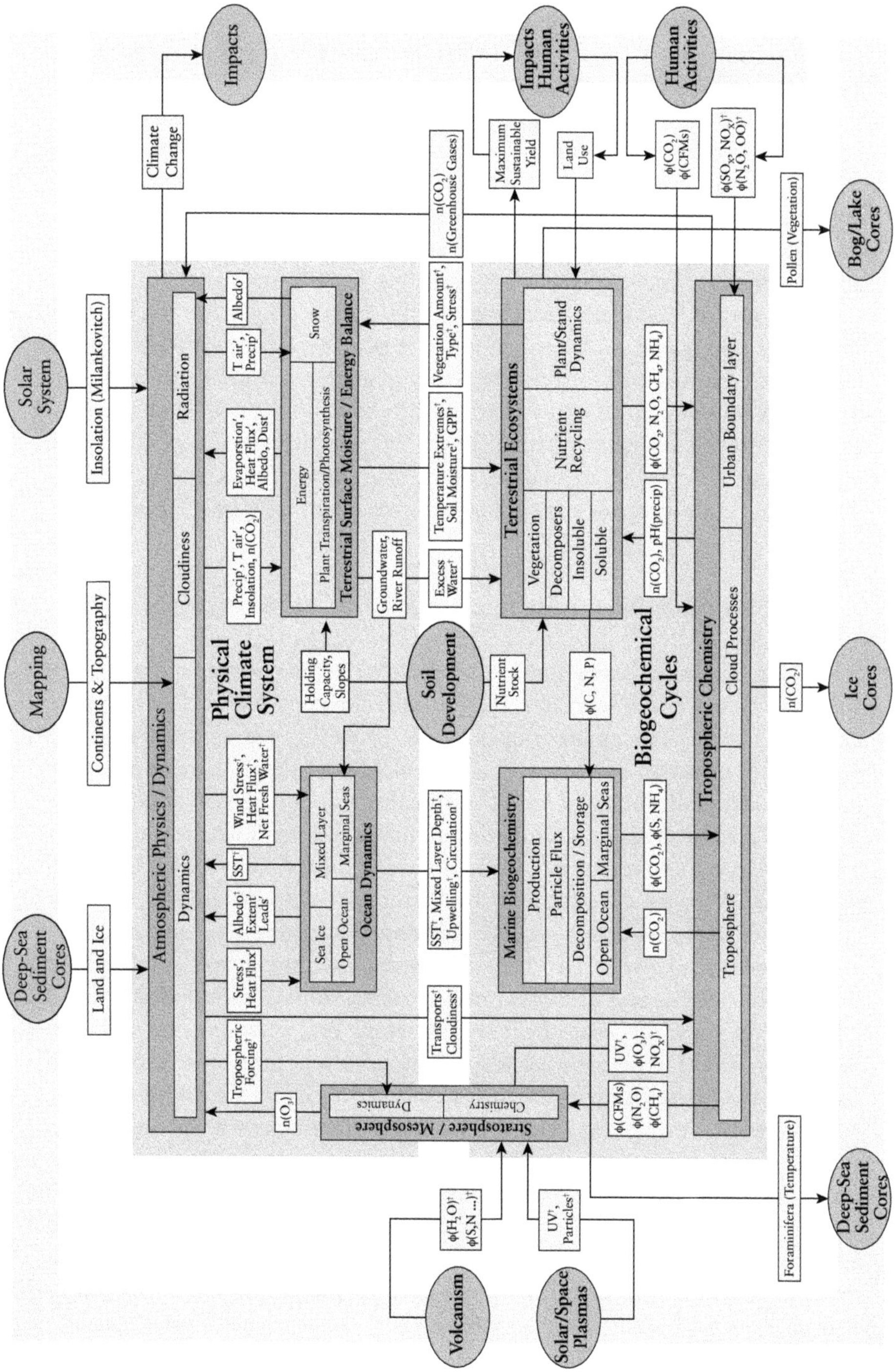

′ = on timescale of hours to days † = on timescale of months to seasons φ = flux n = concentration

Figure 13.2 The Bretherton Diagram.

In the 1986 Bretherton diagram, published as a part of NASA's Earth system science efforts in the late 1980s, the entire Earth system appears as an integrated, calculable system.

Ice cores are part of the diagram as one out of many inputs that are used to determine the system's stability on 'a timescale from decades to centuries' (NASA, 1986). Here, the hard labour of drilling through ice sheets in the cryosphere has been mediated into a humble box with 'ice cores' written on it, appearing next to an array of similar looking boxes in a vast and abstract model. Vertical glaciology did, this heavily mediated form of ice shows, not only open new temporal horizons, but also made the ice itself calculable and translatable, which made it possible to export ice core data far beyond the disciplinary boundaries of glaciology. It furthermore points to the ways in which ice cores have, at different moments in time, realized particular forms of environmental knowledge, which have depended both on the materiality and mediation of the ice cores themselves as well as disciplinary and institutional frameworks in which they have appeared.

Conclusion: New epistemic geographies of the cryosphere

Following ice cores from the ice sheets to the Bretherton diagram, the complex history of integrating ice into the concept of the Earth system becomes visible. The environing qualities of ice have drastically changed during the twentieth century and a crucial part of these changes has been the temporal malleability of ice in various forms (Christensen *et al.*, 2013). Ahlmann's calendar-approach, and the gradualist, locally situated temporalities it enabled, produced a different form of environmental knowledge compared to later work in ice core drilling. By turning ice into an archive, another kind of environing became possible: the verticality allowed for quantification and translation between ice cores and other disciplines and geographies. During the 1970s and 1980s, transformative shifts in the human–Earth relationship took place as the environmental understanding increasingly shifted towards the planetary scale, through both political institutions and scientific enterprises such as Earth system science (Edwards, 2010; Borowy, 2014; Warde *et al.*, 2018). The shift towards the planetary was not only a spatial upscaling of the environment, it was also a temporal negotiation that aimed to synchronize various environmental objects and geographies in order to produce a planetary scale environmental system. Ice has been a consistent part of this process of synchronization. If the rise of Earth system science has enabled the globe to encounter the planet, as Dipesh Chakrabarty argues, historical studies of ice core drilling can teach us more regarding the conditions under which this encounter could occur in the first place (Chakrabarty, 2019).

The integration of ice into the Earth system speaks to larger historiographical questions regarding the entangled temporalities of the Anthropocene (Swanson, 2016). Since its appearance in the early twenty-first century, the Anthropocene concept has enabled a profound temporal integration of geological and human timescales and been the subject of debate in a vast number of disciplines. The last decade has also seen an abundance of debates regarding potentially reductionist tendencies of the Anthropocene concept, as critics have pointed out that the Anthropocene might obscure political and geographical differences (e.g. Haraway, 2016; Moore, 2016; Yusoff 2019). Yet, despite the central role time and temporality play in these discussions, the origin of Anthropocene temporalities themselves are rarely taken into account and remain taken for granted. By following ice core drilling, from its early years in the 1950s to its implementation in Earth system science in the 1980s, the production and synchronization of human and planetary timescales can appear as a historical process in itself.

When Wallace Broecker, and other paleoclimatologists, engaged with ice core research in the 1970s, it marked the appearance of a new kind of ice. These scientists were not dependent on their own field-based observations; they did not have to set foot in the cryosphere; instead they approached the ice in an already mediated form, recovered and quantified in a way which rendered it possible for non-glaciologists to engage with it. This new kind of ice did not self-evidently belong in its origin geography in the cryosphere, but rather among other paleoarchives, made from mud, rock, pollen, and wood. When ice appeared as a spatially ambiguous and temporally standardized archive in the Bretherton diagram, it had become something else compared to earlier scientific conceptions. As an environing medium, ice cores gradually came to mediate planetary pasts and futures and functioned within a larger process of configuring epistemologies for planetary-scale environmental knowledge. These temporal and spatial transformations altered both previous understandings of what ice could be, as well as who could study it scientifically.

Notes

1 The concept 'environing media' points to the ways in which environments are actively constructed – materially and conceptually – through human intervention and how environmental knowledge is dependent on mediation (Sörlin and Wormbs, 2018; Wickberg and Gärdebo, 2020).

2 Martin Mahony and Mike Hulme (2018) use the term 'epistemic geographies' to show how global knowledge about climate change is grounded in particular geographies that are made to speak for the entire globe. I use the same concept for the cryosphere in order to show how it, too, became spatially ambiguous as it became an interest for those who studied the planetary environment. It was, in the postwar era, not self-evident that the cryosphere could be ideally studied *in situ*, but that it could be researched in mediated forms in other places as well.
3 Ahlmann made some attempts to map the vertical properties of glaciers, but his outlook remained primarily horizontal (Sörlin and Isberg, 2021).
4 It should be noted that Dansgaard was not the first to ascribe climatic fluctuations as the cause of the settlement's demine. However, he was the first to utilize ice core data to make this case. See for example Utterström (1955).
5 Deep Sea Drilling Project 1968–83 and the Decade of Ocean Exploration in the 1970s are examples of similar activities in oceanography. These programmes consisted, to a significant degree, of attempts to scale up oceanographic knowledge to a planetary level, both spatially and temporally (National Research Council, 1969; McNutt, 2000).
6 A brief survey of *Nature* and *Science* between 1969 and 1975 shows a wide array of articles utilizing Camp Century ice core data for non-glaciological inquiries (Kleinman and Volchok, 1969; Bray, 1970; Jay Gould, 1970; Wilson and C. H. Hendy, 1971; Schneider and Mass, 1975; Roosen, *et al.*, 1976; Mercer, 1978).

References

Achermann, D. (2020) 'Vertical Glaciology: The Second Dimension of the Third Dimension in Climate Research', *Centaurus* 62(4): 720–743.

Ahlmann, H. W. (1935) 'Contribution to the Physics of Glaciers', *Geographical Journal* 86(2): 97–107.

Alley, R. B. (2000) *The Two-Mile Time Machine: Ice Cores, Abrupt Climate Change and Our Future*. Princeton, NJ: Princeton University Press.

Andersson, J. (2018) *The Future of the World: Futurology, Futurists, and the Struggle for the Post-Cold War Imagination*. Oxford: Oxford University Press.

Antonello, A. (2017) 'Engaging and Narrating the Antarctic Ice Sheet: The History of an Earthly Body', *Environmental History* 22(1): 77–100.

Antonello, A. and M. Carey (2017) 'Ice Cores and the Temporalities of the Global Environment', *Environmental Humanities* 9(2): 181–203.

Borowy, I. (2014) *Defining Sustainable Development for our Common Future: A History of the World Commission on Environment and Development (Brundtland Commission)*. London: Routledge.

Bray, J. R. (1970) 'Solar Activity Index: Validity Supported by Isotope Dating', *Science* 168(3931): 571–572.

Broecker, W. (1975) 'Climatic Change: Are We on the Brink of a Pronounced Global Warming?', *Science* 189(4201): 460–463.

Chakrabarty, D. (2019) 'The Planet: An Emergent Humanist Category', *Critical Inquiry* 41(1): 1–31.

Christensen, M., A. E. Nilsson, and N. Wormbs, eds (2013) *Media and the Politics of Arctic Climate Change: When the Ice Breaks*. New York: Palgrave Macmillan.

Collis, C. and K. Dodds (2008) 'Assault on the Unknown: The Historical and Political Geographies of the International Geophysical Year', *Journal of Historical Geography* 34(4): 555–567.

Dansgaard, W. (1953) 'The Abundance of O^{18} in Atmospheric Water and Water Vapour', *Tellus* 5(4): 461–469.

Dansgaard, W. (2005) *Frozen Annals: Greenland Ice Cap Research*. Copenhagen: Niels Bohr Institute.

Dansgaard, W. and S. J. Johnsen (1969) 'Flow Model and A Time Scale for the Ice Core from Camp Century, Greenland', *Journal of Glaciology* 8(59): 215–223.

Dansgaard, W., S. J. Johnsen, H. B. Clausen, and N. Gundestrup (1973) 'Stable Isotope Glaciology', *Meddelelser om Grønland* 197(2): 1–53.

Dansgaard, W., S. J. Johnsen, H. B. Clausen, and C. C. Langway (1972) 'Speculations about the Next Glaciation', *Quaternary Research* 2(3): 396–398.

Dansgaard, W., S. J. Johnsen, N. Reeh, N. Gundestrup, H. B. Clausen, and C. U. Hammer (1975) 'Climatic Changes, Norsemen, and Modern Man', *Nature* 255(5503): 24–28.

Daston, L. (2017) 'Introduction: Third Nature', in L. Daston, ed., *Science in the Archive: Pasts, Presents, Futures*. Chicago, IL: Chicago University Press, 1–17.

De Geer, G. (1912) 'A Geochronology of the Last 12,000 Years', in *Compte Rendu 11 Congrées Geologique International*, 11th International Geological Congress (1910). Stockholm: Norstedts, 241–253.

Doel, R. E., K. C. Harper, and M. Heymann, eds (2016) *Exploring Greenland: Cold War Science and Technology on Ice*. New York: Palgrave Macmillan.

Dry, S. (2019) *Waters of the World*. Chicago, IL: Chicago University Press.

Edwards, P. (2010) *A Vast Machine: Computer Models, Climate Data, and the Politics of Global Warming*. Cambridge, MA: MIT Press.

Elzinga, A. (2017) 'Polar Ice Cores. Climate Change Messengers', in B. Bensuade Vincent, S. Loeve, A. Nordmann, and A. Schwartz, eds, *Research Objects in their Technological Setting*. London: Routledge, 215–232.

Hansen, J. and T. Takashi (1984) 'Preface', in J. Hansen and T. Takashi, eds, *Climate Processes and Climate Sensitivity*. Washington, DC: American Geophysical Union, vii–l.

Haraway, D. (2016) *Staying with the Trouble: Making Kin in the Chthulucene*. Durham, NC: Duke University Press.

Heuer, C. P. (2019) *Into the White: The Renaissance Arctic and the End of the Image*. New York: Zone Books.

Heymann, M. and A. Dahan Dalmedico (2019) 'Epistemology and Politics in Earth System Modeling: Historical Perspectives', *Journal of Advances in Modeling Earth Systems* 11(5): 1139–1152.

Isberg, E. (2020) 'Multiple Temporalities in a New Geological Age: Revisiting Koselleck's Zeitschichten', *Geschichte und Gesellschaft* 46(4): 729–735.

Jay Gould, S. (1970) 'Coincidence of Climatic and Faunal Fluctuations in Pleistocene Bermuda', *Science* 168(3931): 572–573.

Jordheim, H. (2014) 'Introduction: Multiple Times and the Work of Synchronization', *History and Theory* 53(4): 498–518.

Jordheim, H. (2017) 'Synchronizing the World: Synchronism as Historiographical Practice, Then and Now', *History of the Present* 7(1): 59–95.

Kleinman, M. T. and H. L. Volchok (1969) 'Radionuclide Concentration in Surface Air: Direct Relationship to Global Fallout', *Science* 166(3903): 376–377.

Koselleck, R. (1997) 'The Temporalisation of Concepts', *Finnish Yearbook of Political Thought* 1(1): 16–24.

Langway, J. R. and C. Chester (1958) 'A 400 Meter Deep Ice Core in Greenland: A Preliminary Report', *Journal of Glaciology* 3(23): 217–218.

Lolck, M. (2004) *Klima, Kold Krig og Iskerner.* Aarhus: Steno Instituttet, Aarhus Universitet.

Mahony, M. and M. Hulme (2018) 'Epistemic Geographies of Climate Change: Science, Space and Politics', *Progress in Human Geography* 42(3): 395–424.

Martin-Nielsen, J. (2013a) '"The Deepest and Most Rewarding Hole Ever Drilled": Ice Cores and the Cold War in Greenland', *Annals of Science* 70(1): 47–70.

Martin-Nielsen, J. (2013b) *Eismitte in the Scientific Imagination: Knowledge and Politics at the Center of Greenland.* London: Palgrave Macmillan.

McNutt, M. K. (2000) 'Achievements in Marine Geology and Geophysics', in National Research Council Ocean Studies Board, *50 Years of Ocean Discovery: National Research Council 1950–2000.* Washington, DC: National Academies Press, 51–64.

Mercer, J. H. (1978) 'West Antarctica Ice Sheet and CO_2 Greenhouse Effect: A Threat of Disaster', *Nature* 271(5643): 321–325.

Mitchell, W. J. T. (1980) 'Spatial Form in Literature: Towards a General Theory', *Critical Inquiry* 6(3): 539–567.

Moore, J. W. (2016) *Anthropocene or Capitalocene? Nature, History and the Crisis of Capitalism.* Oakland, CA: PM Press.

NASA Earth System Sciences Committee (1986) *Earth System Science Overview: A Program for Global Change.* Washington, DC: National Academies Press.

National Research Council (1969) *An Oceanic Quest: The International Decade of Ocean Exploration.* Washington, DC: National Academies Press.

Nielsen, K. H., H. Nielsen, and J. Martin-Nielsen (2014) 'City Under the Ice: The Closed World of Camp Century in Cold War Science', *Science as Culture* 23(4): 443–464.

Oeschger, H. (1985) 'The Contribution of Ice Core Studies to the Understanding of Environmental Process', in J. R. Langway, C. Chester, H. Oeschger, and W. Dansgaard, eds, *Greenland Ice Cores: Geophysics, Geochemistry and the Environment.* Washington, DC: American Geophysical Union, 9–19.

O'Reilly, J. (2016) 'Sensing the Ice: Field Science, Models, and Expert Intimacy with Knowledge', *Journal of the Royal Anthropological Institute* 22(1): 27–45.

Petersen, N. (2011) 'SAC at Thule: Greenland in the US Polar Strategy', *Journal of Cold War Studies* 13(2): 90–115.

Roosen, R. G., R. S. Harrington, J. Giles, and I. Browning (1976) 'Earth Tides, Volcanoes and Climate Change', *Nature* 261(5562): 680–682.
Rosenberg, D. and A. Grafton (2010) *Cartographies of Time: A History of the Timeline*. Princeton, NJ: Princeton University Press.
Rosol, C. (2017) 'Data, Models and Earth History in Deep Convolution: Paleoclimate Simulations and their Epistemological Unrest', *Berichte zur Wissenschaftsgeschichte* 40(2): 120–139.
Schneider, S. H. and C. Mass (1975) 'Volcanic Dust, Sunspots, and Temperature Trends', *Science* 190(4216): 741–746.
Simonetti, C. (2015) 'The Stratification of Time', *Time & Society* 24(2): 139–162.
Sörlin, S. (2011) 'The Anxieties of a Science Diplomat: Field Co-production of Climate Knowledge and the Rise and Fall of Hans Ahlmann's "Polar Warming"', *OSIRIS* 26(1): 66–88.
Sörlin, S. (2018) 'A Microgeography of Authority: Glaciology and Climate Change at the Tarfala Station 1945–1980', in H. Ekerholm, K. Grandin, C. Nordlund, and P. Schell, eds, *Understanding Field Science Institutions*. Sagamore Beach, MA: Science History Publications, 255–285.
Sörlin, S. and E. Isberg (2021) 'Synchronizing Earthly Timescales: Ice, Pollen, and the Making of Proto-Anthropocene Knowledge in the North-Atlantic Region', *Annals of the American Association of Geographers* 111(3): 717–728.
Sörlin, S. and N. Wormbs (2018) 'Environing Technologies: A Theory of Making Environment', *History and Technology* 34(2): 101–125.
Sullivan, W. (1968) 'History in an Ice Core', *New York Times*, February 11, 7.
Summerhayes, C. P. (2020) *Paleoclimatology: From Snowball Earth to the Anthropocene*. Hoboken, NJ: Wiley Blackwell.
Swanson, H. A. (2016) 'Anthropocene as Political Geology: Current Debates of How to Tell Time', *Science as Culture* 25(1): 157–163.
Talalay, P. G. (2016) *Mechanical Ice Drilling Technology*. Singapore: Springer Geophysics.
United States Committee for the Global Atmospheric Research Program (1975) *Understanding Climate Change: A Program for Action*. Washington, DC: National Academy of Sciences.
Utterström, G. (1955) 'Climatic Fluctuations and Population Problems in Early Modern History', *Scandinavian Economic History Review* 1(3): 3–47.
Warde, P., L. Robin, and S. Sörlin (2018) *The Environment: A History of the Idea*. Baltimore, MD: Johns Hopkins University Press.
Wickberg, A. and J. Gärdebo (2020) 'Where Humans and the Planetary Conflate: An Introduction to Environing Media', *Humanities* 9(3): 1–13.
Wilson, A. T. and C. H. Hendy (1971) 'Past Wind Strength from Isotope Studies', *Nature* 234(5328): 344–345.
Yusoff, K. (2005) 'Visualizing Antarctica as a Place in Time: From Geological Sublime to "Real Time"', *Space and Culture* 8(4): 381–398.
Yusoff, K., ed. (2008) *BiPolar*. London: Arts Catalyst.
Yusoff, K. (2019) *A Billion Black Anthropocenes or None*. Minneapolis, MN: University of Minnesota Press.

Index

EU authorised representative for GPSR:
Easy Access System Europe, Mustamäe tee 50,
10621 Tallinn, Estonia
gpsr.requests@easproject.com

www.ingramcontent.com/pod-product-compliance
Ingram Content Group UK Ltd.
Pitfield, Milton Keynes, MK11 3LW, UK
UKHW021840100926
8556IPUK00004B/462
9781526157775